Numerical Methods in Engineering with Python 3

This book is an introduction to numerical methods for students in engineering. It covers the usual topics found in an engineering course: solution of equations, interpolation and data fitting, solution of differential equations, eigenvalue problems, and optimization. The algorithms are implemented in Python 3, a high-level programming language that rivals MATLAB® in readability and ease of use. All methods include programs showing how the computer code is utilized in the solution of problems.

The book is based on *Numerical Methods in Engineering with Python*, which used Python 2. Apart from the migration from Python 2 to Python 3, the major change in this new text is the introduction of the Python plotting package *Matplotlib*.

Jaan Kiusalaas is a Professor Emeritus in the Department of Engineering Science and Mechanics at Pennsylvania State University. He has taught computer methods, including finite element and boundary element methods, for more than 30 years. He is also the co-author or author of four books – *Engineering Mechanics: Statics*; *Engineering Mechanics: Dynamics*; *Mechanics of Materials*; *Numerical Methods in Engineering with MATLAB* (2nd edition); and two previous editions of *Numerical Methods in Engineering with Python*.

NUMERICAL METHODS IN ENGINEERING WITH PYTHON 3

Jaan Kiusalaas

The Pennsylvania State University

CAMBRIDGE
UNIVERSITY PRESS

CAMBRIDGE
UNIVERSITY PRESS

University Printing House, Cambridge CB2 8BS, United Kingdom

One Liberty Plaza, 20th Floor, New York, NY 10006, USA

477 Williamstown Road, Port Melbourne, VIC 3207, Australia

4843/24, 2nd Floor, Ansari Road, Daryaganj, Delhi - 110002, India

79 Anson Road, #06-04/06, Singapore 079906

Cambridge University Press is part of the University of Cambridge.

It furthers the University's mission by disseminating knowledge in the pursuit of education, learning and research at the highest international levels of excellence.

www.cambridge.org
Information on this title: www.cambridge.org/9781107033856

First published 2013

A catalogue record for this publication is available from the British Library

Library of Congress Cataloging in Publication data
Kiusalaas, Jaan.
Numerical methods in engineering with Python 3 / Jaan Kiusalaas.
  pages cm
Includes bibliographical references and index.
ISBN 978-1-107-03385-6
1. Engineering mathematics – Data processing.   2. Python (Computer program language)   I. Title.
TA345.K58   2013
620.00285'5133–dc23   2012036775

ISBN 978-1-107-03385-6 Hardback

Additional resources for this publication at www.cambridge.org/kiusalaaspython.

Contents

Preface

This book is targeted toward engineers and engineering students of advanced standing (juniors, seniors, and graduate students). Familiarity with a computer language is required; knowledge of engineering mechanics (statics, dynamics, and mechanics of materials) is useful, but not essential.

The primary purpose of the text is to teach numerical methods. It is not a primer on Python programming. We introduce just enough Python to implement the numerical algorithms. That leaves the vast majority of the language unexplored.

Most engineers are not programmers, but problem solvers. They want to know what methods can be applied to a given problem, what their strengths and pitfalls are, and how to implement them. Engineers are not expected to write computer code for basic tasks from scratch; they are more likely to use functions and subroutines that have been already written and tested. Thus, programming by engineers is largely confined to assembling existing bits of code into a coherent package that solves the problem at hand.

The "bit" of code is usually a function that implements a specific task. For the user the details of the code are unimportant. What matters are the interface (what goes in and what comes out) and an understanding of the method on which the algorithm is based. Because no numerical algorithm is infallible, the importance of understanding the underlying method cannot be overemphasized; it is, in fact, the rationale behind learning numerical methods.

This book attempts to conform to the views outlined earlier. Each numerical method is explained in detail and its shortcomings are pointed out. The examples that follow individual topics fall into two categories: hand computations that illustrate the inner workings of the method, and small programs that show how the computer code is utilized in solving a problem. Problems that require programming are marked with ■.

The material consists of the usual topics covered in an engineering course on numerical methods: solution of equations, interpolation and data fitting, numerical differentiation and integration, solution of ordinary differential equations, and eigenvalue problems. The choice of methods within each topic is tilted toward relevance to engineering problems. For example, there is an extensive discussion of symmetric, sparsely populated coefficient matrices in the solution of simultaneous equations.

In the same vein, the solution of eigenvalue problems concentrates on methods that efficiently extract specific eigenvalues from banded matrices.

An important criterion used in the selection of methods was clarity. Algorithms requiring overly complex bookkeeping were rejected regardless of their efficiency and robustness. This decision, which was taken with great reluctance, is in keeping with the intent to avoid emphasis on programming.

The selection of algorithms was also influenced by current practice. This disqualified several well-known historical methods that have been overtaken by more recent developments. For example, the secant method for finding roots of equations was omitted as having no advantages over Ridder's method. For the same reason, the multistep methods used to solve differential equations (e.g., Milne and Adams methods) were left out in favor of the adaptive Runge-Kutta and Bulirsch-Stoer methods.

Notably absent is a chapter on partial differential equations. It was felt that this topic is best treated by finite element or boundary element methods, which are outside the scope of this book. The finite difference model, which is commonly introduced in numerical methods texts, is just too impractical in handling multidimensional boundary value problems.

As usual, the book contains more material than can be covered in a three-credit course. The topics that can be skipped without loss of continuity are tagged with an asterisk (*).

What Is New in This Edition

This book succeeds *Numerical Methods in Engineering with Python*, which was based on Python 2. As the title implies, the new edition migrates to Python 3. Because the two versions are not entirely compatible, almost all computer routines required some code changes.

We also took the opportunity to make a few changes in the material covered:

- An introduction to the Python plotting package `matplotlib.pyplot` was added to Chapter 1. This package is used in numerous example problems, making the book more graphics oriented than before.
- The function `plotPoly`, which plots data points and the corresponding polynomial interpolant, was added to Chapter 3. This program provides a convenient means of evaluating the fit of the interpolant.
- At the suggestion of reviewers, the Taylor series method of solving initial value problems in Chapter 7 was dropped. It was replaced by Euler's method.
- The Jacobi method for solving eigenvalue problems in Chapter 9 now uses the threshold method in choosing the matrix elements marked for elimination. This change increases the speed of the algorithm.
- The adaptive Runge-Kutta method in Chapter 7 was recoded, and the Cash-Karp coefficients replaced with the Dormand-Prince coefficients. The result is a more efficient algorithm with tighter error control.

- Twenty-one new problems were introduced, most of them replacing old problems.
- Some example problems in Chapters 4 and 7 were rearranged or replaced with new problems. The result of these changes is better coordination of examples with the text.

The programs listed in the book were tested with Python 3.2 under Windows 7. The source codes are available at *www.cambridge.org/kiusalaaspython*.

1   Introduction to Python

1.1 General Information

Quick Overview

This chapter is not a comprehensive manual of Python. Its sole aim is to provide sufficient information to give you a good start if you are unfamiliar with Python. If you know another computer language, and we assume that you do, it is not difficult to pick up the rest as you go.

Python is an object-oriented language that was developed in the late 1980s as a scripting language (the name is derived from the British television series, *Monty Python's Flying Circus*). Although Python is not as well known in engineering circles as are some other languages, it has a considerable following in the programming community. Python may be viewed as an emerging language, because it is still being developed and refined. In its current state, it is an excellent language for developing engineering applications.

Python programs are not compiled into machine code, but are run by an *interpreter*.[1] The great advantage of an interpreted language is that programs can be tested and debugged quickly, allowing the user to concentrate more on the principles behind the program and less on the programming itself. Because there is no need to compile, link, and execute after each correction, Python programs can be developed in much shorter time than equivalent Fortran or C programs. On the negative side, interpreted programs do not produce stand-alone applications. Thus a Python program can be run only on computers that have the Python interpreter installed.

Python has other advantages over mainstream languages that are important in a learning environment:

- Python is an open-source software, which means that it is *free*; it is included in most Linux distributions.
- Python is available for all major operating systems (Linux, Unix, Windows, Mac OS, and so on). A program written on one system runs without modification on all systems.

---

[1] The Python interpreter also compiles *byte code*, which helps speed up execution somewhat.

- Python is easier to learn and produces more readable code than most languages.
- Python and its extensions are easy to install.

Development of Python has been clearly influenced by Java and C++, but there is also a remarkable similarity to MATLAB[R] (another interpreted language, very popular in scientific computing). Python implements the usual concepts of object-oriented languages such as classes, methods, inheritance etc. We do not use object-oriented programming in this text. The only object that we need is the N-dimensional *array* available in the module numpy (this module is discussed later in this chapter).

To get an idea of the similarities and differences between MATLAB and Python, let us look at the codes written in the two languages for solution of simultaneous equations $\mathbf{Ax} = \mathbf{b}$ by Gauss elimination. Do not worry about the algorithm itself (it is explained later in the text), but concentrate on the semantics. Here is the function written in MATLAB:

```
function x = gaussElimin(a,b)
n = length(b);
for k = 1:n-1
    for i= k+1:n
        if a(i,k) ~= 0
            lam = a(i,k)/a(k,k);
            a(i,k+1:n) = a(i,k+1:n) - lam*a(k,k+1:n);
            b(i)= b(i) - lam*b(k);
        end
    end
end
for k = n:-1:1
    b(k) = (b(k) - a(k,k+1:n)*b(k+1:n))/a(k,k);
end
x = b;
```

The equivalent Python function is

```
from numpy import dot
def gaussElimin(a,b):
    n = len(b)
    for k in range(0,n-1):
        for i in range(k+1,n):
            if a[i,k] != 0.0:
                lam = a [i,k]/a[k,k]
                a[i,k+1:n] = a[i,k+1:n] - lam*a[k,k+1:n]
                b[i] = b[i] - lam*b[k]
    for k in range(n-1,-1,-1):
        b[k] = (b[k] - dot(a[k,k+1:n],b[k+1:n]))/a[k,k]
    return b
```

The command `from numpy import dot` instructs the interpreter to load the function `dot` (which computes the dot product of two vectors) from the module numpy. The colon (`:`) operator, known as the *slicing operator* in Python, works the same way as it does in MATLAB and Fortran90—it defines a slice of an array.

The statement `for k = 1:n-1` in MATLAB creates a loop that is executed with $k = 1, 2, \ldots, n - 1$. The same loop appears in Python as `for k in range(n-1)`. Here the function `range(n-1)` creates the sequence $[0, 1, \ldots, n - 2]$; k then loops over the elements of the sequence. The differences in the ranges of k reflect the native offsets used for arrays. In Python all sequences have *zero offset*, meaning that the index of the first element of the sequence is always 0. In contrast, the native offset in MATLAB is 1.

Also note that Python has no `end` statements to terminate blocks of code (loops, subroutines, and so on). The body of a block is defined by its *indentation*; hence indentation is an integral part of Python syntax.

Like MATLAB, Python is *case sensitive*. Thus the names n and N would represent different objects.

Obtaining Python

The *Python interpreter* can be downloaded from

$$\text{http} : //\text{www.python.org/getit}$$

It normally comes with a nice code editor called *Idle* that allows you to run programs directly from the editor. If you use Linux, it is very likely that Python is already installed on your machine. The download includes two extension modules that we use in our programs: the numpy module that contains various tools for array operations, and the matplotlib graphics module utilized in plotting.

The Python language is well documented in numerous publications. A commendable teaching guide is *Python* by Chris Fehly (Peachpit Press, CA, 2nd ed.). As a reference, *Python Essential Reference* by David M. Beazley (Addison-Wesley, 4th ed.) is highly recommended. Printed documentation of the extension modules is scant. However, tutorials and examples can be found on various websites. Our favorite reference for numpy is

$$\text{http://www.scipy.org/Numpy\_Example\_List}$$

For matplotlib we rely on

$$\text{http://matplotlib.sourceforge.net/contents.html}$$

If you intend to become a serious Python programmer, you may want to acquire *A Primer on Scientific Programming with Python* by Hans P. Langtangen (Springer-Verlag, 2009).

1.2   Core Python

Variables

In most computer languages the name of a variable represents a value of a given type stored in a fixed memory location. The value may be changed, but not the type. This is not so in Python, where variables are *typed dynamically*. The following interactive session with the Python interpreter illustrates this feature (>>> is the Python prompt):

```
>>> b = 2          # b is integer type
>>> print(b)
2
>>> b = b*2.0  # Now b is float type
>>> print(b)
4.0
```

The assignment b = 2 creates an association between the name b and the *integer* value 2. The next statement evaluates the expression b*2.0 and associates the result with b; the original association with the integer 2 is destroyed. Now b refers to the *floating* point value 4.0.

The pound sign (#) denotes the beginning of a *comment*—all characters between # and the end of the line are ignored by the interpreter.

Strings

A string is a sequence of characters enclosed in single or double quotes. Strings are *concatenated* with the plus (+) operator, whereas *slicing* (:) is used to extract a portion of the string. Here is an example:

```
>>> string1 = 'Press return to exit'
>>> string2 = 'the program'
>>> print(string1 + ' ' + string2)  # Concatenation
Press return to exit the program
>>> print(string1[0:12])            # Slicing
Press return
```

A string can be split into its component parts using the split command. The components appear as elements in a list. For example,

```
>>> s = '3 9 81'
>>> print(s.split())    # Delimiter is white space
['3', '9', '81']
```

A string is an *immutable* object—its individual characters cannot be modified with an assignment statement, and it has a fixed length. An attempt to violate immutability will result in TypeError, as follows:

```
>>> s = 'Press return to exit'
>>> s[0] = 'p'
Traceback (most recent call last):
  File ''<pyshell#1>'', line 1, in ?
    s[0] = 'p'
TypeError: object doesn't support item assignment
```

Tuples

A *tuple* is a sequence of *arbitrary objects* separated by commas and enclosed in parentheses. If the tuple contains a single object, a final comma is required; for example, x = (2,). Tuples support the same operations as strings; they are also immutable. Here is an example where the tuple rec contains another tuple (6,23,68):

```
>>> rec = ('Smith','John',(6,23,68))    # This is a tuple
>>> lastName,firstName,birthdate = rec  # Unpacking the tuple
>>> print(firstName)
John
>>> birthYear = birthdate[2]
>>> print(birthYear)
68
>>> name = rec[1] + ' ' + rec[0]
>>> print(name)
John Smith
>>> print(rec[0:2])
('Smith', 'John')
```

Lists

A list is similar to a tuple, but it is *mutable*, so that its elements and length can be changed. A list is identified by enclosing it in brackets. Here is a sampling of operations that can be performed on lists:

```
>>> a = [1.0, 2.0, 3.0]      # Create a list
>>> a.append(4.0)            # Append 4.0 to list
>>> print(a)
[1.0, 2.0, 3.0, 4.0]
>>> a.insert(0,0.0)          # Insert 0.0 in position 0
>>> print(a)
[0.0, 1.0, 2.0, 3.0, 4.0]
>>> print(len(a))            # Determine length of list
5
>>> a[2:4] = [1.0, 1.0, 1.0] # Modify selected elements
>>> print(a)
[0.0, 1.0, 1.0, 1.0, 1.0, 4.0]
```

If *a* is a mutable object, such as a list, the assignment statement b = a does not result in a new object *b*, but simply creates a new reference to *a*. Thus any changes made to *b* will be reflected in *a*. To create an independent copy of a list *a*, use the statement c = a[:], as shown in the following example:

```
>>> a = [1.0, 2.0, 3.0]
>>> b = a                 # 'b' is an alias of 'a'
>>> b[0] = 5.0            # Change 'b'
>>> print(a)
[5.0, 2.0, 3.0]          # The change is reflected in 'a'
>>> c = a[:]             # 'c' is an independent copy of 'a'
>>> c[0] = 1.0           # Change 'c'
>>> print(a)
[5.0, 2.0, 3.0]          # 'a' is not affected by the change
```

Matrices can be represented as nested lists, with each row being an element of the list. Here is a 3×3 matrix *a* in the form of a list:

```
>>> a = [[1, 2, 3], \
         [4, 5, 6], \
         [7, 8, 9]]
>>> print(a[1])          # Print second row (element 1)
[4, 5, 6]
>>> print(a[1][2])       # Print third element of second row
6
```

The backslash (\) is Python's *continuation character*. Recall that Python sequences have zero offset, so that a[0] represents the first row, a[1] the second row, etc. With very few exceptions we do not use lists for numerical arrays. It is much more convenient to employ *array objects* provided by the numpy module. Array objects are discussed later.

Arithmetic Operators

Python supports the usual arithmetic operators:

+	Addition
−	Subtraction
∗	Multiplication
/	Division
∗∗	Exponentiation
%	Modular division

Some of these operators are also defined for strings and sequences as follows:

```
>>> s = 'Hello '
>>> t = 'to you'
```

```
>>> a = [1, 2, 3]
>>> print(3*s)                  # Repetition
Hello Hello Hello
>>> print(3*a)                  # Repetition
[1, 2, 3, 1, 2, 3, 1, 2, 3]
>>> print(a + [4, 5])           # Append elements
[1, 2, 3, 4, 5]
>>> print(s + t)                # Concatenation
Hello to you
>>> print(3 + s)                # This addition makes no sense
Traceback (most recent call last):
  File "<pyshell#13>", line 1, in <module>
    print(3 + s)
TypeError: unsupported operand type(s) for +: 'int' and 'str'
```

Python also has *augmented assignment operators*, such as $a += b$, that are familiar to the users of C. The augmented operators and the equivalent arithmetic expressions are shown in following table.

a += b	a = a + b
a -= b	a = a - b
a *= b	a = a*b
a /= b	a = a/b
a **= b	a = a**b
a %= b	a = a%b

Comparison Operators

The comparison (relational) operators return `True` or `False`. These operators are

<	Less than
>	Greater than
<=	Less than or equal to
>=	Greater than or equal to
==	Equal to
!=	Not equal to

Numbers of different type (integer, floating point, and so on) are converted to a common type before the comparison is made. Otherwise, objects of different type are considered to be unequal. Here are a few examples:

```
>>> a = 2               # Integer
>>> b = 1.99            # Floating point
>>> c = '2'             # String
>>> print(a > b)
True
```

```
>>> print(a == c)
False
>>> print((a > b) and (a != c))
True
>>> print((a > b) or (a == b))
True
```

Conditionals

The `if` construct

```
if condition:
    block
```

executes a block of statements (which must be indented) if the condition returns `True`. If the condition returns `False`, the block is skipped. The `if` conditional can be followed by any number of `elif` (short for "else if") constructs

```
elif condition:
    block
```

that work in the same manner. The `else` clause

```
else:
    block
```

can be used to define the block of statements that are to be executed if none of the `if`-`elif` clauses are true. The function `sign_of_a` illustrates the use of the conditionals.

```
def sign_of_a(a):
    if a < 0.0:
        sign = 'negative'
    elif a > 0.0:
        sign = 'positive'
    else:
        sign = 'zero'
    return sign

a = 1.5
print('a is ' + sign_of_a(a))
```

Running the program results in the output

```
a is positive
```

Loops

The while construct

```
while condition:
    block
```

executes a block of (indented) statements if the condition is True. After execution of the block, the condition is evaluated again. If it is still True, the block is executed again. This process is continued until the condition becomes False. The else clause

```
else:
    block
```

can be used to define the block of statements that are to be executed if the condition is false. Here is an example that creates the list [1, 1/2, 1/3, ...]:

```
nMax = 5
n = 1
a = []                    # Create empty list
while n < nMax:
    a.append(1.0/n)   # Append element to list
    n = n + 1
print(a)
```

The output of the program is

```
[1.0, 0.5, 0.33333333333333331, 0.25]
```

We met the for statement in Section 1.1. This statement requires a target and a sequence over which the target loops. The form of the construct is

```
for target in sequence:
    block
```

You may add an else clause that is executed after the for loop has finished.

The previous program could be written with the for construct as

```
nMax = 5
a = []
for n in range(1,nMax):
    a.append(1.0/n)
print(a)
```

Here n is the target, and the *range object* [1, 2, ..., $nMax - 1$] (created by calling the range function) is the sequence.

Any loop can be terminated by the

```
break
```

statement. If there is an `else` cause associated with the loop, it is not executed. The following program, which searches for a name in a list, illustrates the use of `break` and `else` in conjunction with a `for` loop:

```
list = ['Jack', 'Jill', 'Tim', 'Dave']
name = eval(input('Type a name: '))  # Python input prompt
for i in range(len(list)):
    if list[i] == name:
        print(name,'is number',i + 1,'on the list')
        break
else:
    print(name,'is not on the list')
```

Here are the results of two searches:

```
Type a name: 'Tim'
Tim is number 3 on the list

Type a name: 'June'
June is not on the list
```

The

```
continue
```

statement allows us to skip a portion of an iterative loop. If the interpreter encounters the `continue` statement, it immediately returns to the beginning of the loop without executing the statements that follow `continue`. The following example compiles a list of all numbers between 1 and 99 that are divisible by 7.

```
x = []                       # Create an empty list
for i in range(1,100):
  if i%7 != 0: continue   # If not divisible by 7, skip rest of loop
  x.append(i)                # Append i to the list
print(x)
```

The printout from the program is

```
[7, 14, 21, 28, 35, 42, 49, 56, 63, 70, 77, 84, 91, 98]
```

Type Conversion

If an arithmetic operation involves numbers of mixed types, the numbers are automatically converted to a common type before the operation is carried out.

Type conversions can also achieved by the following functions:

`int(a)`	Converts a to integer
`float(a)`	Converts a to floating point
`complex(a)`	Converts to complex $a + 0j$
`complex(a,b)`	Converts to complex $a + bj$

These functions also work for converting strings to numbers as long as the literal in the string represents a valid number. Conversion from a float to an integer is carried out by truncation, not by rounding off. Here are a few examples:

```
>>> a = 5
>>> b = -3.6
>>> d = '4.0'
>>> print(a + b)
1.4
>>> print(int(b))
-3
>>> print(complex(a,b))
(5-3.6j)
>>> print(float(d))
4.0
>>> print(int(d))   # This fails: d is a string
Traceback (most recent call last):
  File "<pyshell#30>", line 1, in <module>
    print(int(d))
ValueError: invalid literal for int() with base 10: '4.0'
```

Mathematical Functions

Core Python supports only the following mathematical functions:

`abs(a)`	Absolute value of a
`max(`*sequence*`)`	Largest element of *sequence*
`min(`*sequence*`)`	Smallest element of *sequence*
`round(a,n)`	Round a to n decimal places
`cmp(a,b)`	Returns $\begin{cases} -1 \text{ if } a < b \\ 0 \text{ if } a = b \\ 1 \text{ if } a > b \end{cases}$

The majority of mathematical functions are available in the `math` module.

Reading Input

The intrinsic function for accepting user input is

$$\boxed{\texttt{input}(\textit{prompt})}$$

It displays the prompt and then reads a line of input that is converted to a *string*. To convert the string into a numerical value use the function

$$\boxed{\texttt{eval}(\textit{string})}$$

The following program illustrates the use of these functions:

```
a = input('Input a: ')
print(a, type(a))          # Print a and its type
b = eval(a)
print(b,type(b))           # Print b and its type
```

The function `type(a)` returns the type of the object a; it is a very useful tool in debugging. The program was run twice with the following results:

```
Input a: 10.0
10.0 <class 'str'>
10.0 <class 'float'>


Input a: 11**2
11**2 <class 'str'>
121 <class 'int'>
```

A convenient way to input a number and assign it to the variable a is

$$\boxed{\texttt{a = eval(input}(\textit{prompt}))}$$

Printing Output

Output can be displayed with the *print function*

$$\boxed{\texttt{print}(\textit{object1, object2, ...})}$$

that converts *object1*, *object2*, and so on, to strings and prints them on the same line, separated by spaces. The *newline* character ' \n ' can be used to force a new line. For example,

```
>>> a = 1234.56789
>>> b = [2, 4, 6, 8]
>>> print(a,b)
1234.56789 [2, 4, 6, 8]
>>> print('a =',a, '\nb =',b)
a = 1234.56789
b = [2, 4, 6, 8]
```

The `print` function always appends the newline character to the end of a line. We can replace this character with something else by using the keyword argument end. For example,

$$\boxed{\texttt{print}(\textit{object1, object2, ...},\texttt{end=' ')}}$$

replaces \n with a space.

Output can be formatted with the *format method*. The simplest form of the conversion statement is

$$'\{:fmt1\}\{:fmt2\}\ldots'\,.\texttt{format}(arg1,arg2,\ldots)$$

where *fmt1*, *fmt2*,... are the format specifications for *arg1*, *arg2*,..., respectively. Typically used format specifications are

wd	Integer
$w.d$f	Floating point notation
$w.d$e	Exponential notation

where w is the width of the field and d is the number of digits after the decimal point. The output is right justified in the specified field and padded with blank spaces (there are provisions for changing the justification and padding). Here are several examples:

```
>>> a = 1234.56789
>>> n = 9876
>>> print('{:7.2f}'.format(a))
1234.57
>>> print('n = {:6d}'.format(n))    # Pad with spaces
n =   9876
>>> print('n = {:06d}'.format(n))   # Pad with zeros
n =009876
>>> print('{:12.4e} {:6d}'.format(a,n))
  1.2346e+03   9876
```

Opening and Closing a File

Before a data file on a storage device (e.g., a disk) can be accessed, you must create a *file object* with the command

$$\textit{file\_object}\ =\ \texttt{open}(\textit{filename, action})$$

where *filename* is a string that specifies the file to be opened (including its path if necessary) and *action* is one of the following strings:

`'r'`	Read from an existing file.
`'w'`	Write to a file. If *filename* does not exist, it is created.
`'a'`	Append to the end of the file.
`'r+'`	Read to and write from an existing file.
`'w+'`	Same as `'r+'`, but *filename* is created if it does not exist.
`'a+'`	Same as `'w+'`, but data is appended to the end of the file.

It is good programming practice to close a file when access to it is no longer required. This can be done with the method

$$\textit{file\_object}\,.\texttt{close}()$$

Reading Data from a File

There are three methods for reading data from a file. The method

$$\boxed{\textit{file\_object}.\texttt{read}(n)}$$

reads n characters and returns them as a string. If n is omitted, all the characters in the file are read.

If only the current line is to be read, use

$$\boxed{\textit{file\_object}.\texttt{readline}(n)}$$

which reads n characters from the line. The characters are returned in a string that terminates in the newline character \n. Omission of n causes the entire line to be read.

All the lines in a file can be read using

$$\boxed{\textit{file\_object}.\texttt{readlines}()}$$

This returns a list of strings, each string being a line from the file ending with the newline character.

A convenient method of extracting all the lines one by one is to use the loop

```
for line in file_object:
    do something with line
```

As an example, let us assume that we have a file named `sunspots.txt` in the working directory. This file contains daily data of sunspot intensity, each line having the format (year/month/date/intensity), as follows:

```
1896 05 26 40.94
1896 05 27 40.58
1896 05 28 40.20
        etc.
```

Our task is to read the file and create a list x that contains only the intensity. Since each line in the file is a string, we first split the line into its pieces using the `split` command. This produces a list of strings, such as `['1896','05','26','40.94']`. Then we extract the intensity (element `[3]` of the list), evaluate it, and append the result to x. Here is the algorithm:

```
x = []
data = open('sunspots.txt','r')
for line in data:
    x.append(eval(line.split()[3]))
data.close()
```

Writing Data to a File

The method

$$\textit{file\_object}.\texttt{write}(\textit{string})$$

writes a string to a file, whereas

$$\textit{file\_object}.\texttt{writelines}(\textit{list\_of\_strings})$$

is used to write a list of strings. Neither method appends a newline character to the end of a line.

As an example, let us write a formatted table of k and k^2 from $k = 101$ to 110 to the file `testfile`. Here is the program that does the writing:

```
f = open('testfile','w')
for k in range(101,111):
    f.write('{:4d} {:6d}'.format(k,k**2))
    f.write('\n')
f.close()
```

The contents of `testfile` are

```
101   10201
102   10404
103   10609
104   10816
105   11025
106   11236
107   11449
108   11664
109   11881
110   12100
```

The `print` function can also be used to write to a file by redirecting the output to a file object:

$$\texttt{print}(\textit{object1}, \textit{object2}, \ldots, \texttt{file} = \textit{file\_object})$$

Apart from the redirection, this works just like the regular `print` function.

Error Control

When an error occurs during execution of a program an exception is raised and the program stops. Exceptions can be caught with `try` and `except` statements:

```
try:
    do something
except error:
    do something else
```

where *error* is the name of a built-in Python exception. If the exception *error* is not raised, the `try` block is executed; otherwise the execution passes to the `except` block. All exceptions can be caught by omitting *error* from the `except` statement.

The following statement raises the exception `ZeroDivisionError`:

```
>>> c = 12.0/0.0
Traceback (most recent call last):
  File "<pyshell#0>", line 1, in <module>
    c=12.0/0.0
ZeroDivisionError: float division by zero
```

This error can be caught by

```
try:
    c = 12.0/0.0
except ZeroDivisionError:
    print('Division by zero')
```

1.3 Functions and Modules

Functions

The structure of a Python function is

```
def func_name( param1, param2,...) :
    statements
    return return_values
```

where *param1, param2,...* are the parameters. A parameter can be any Python object, including a function. Parameters may be given default values, in which case the parameter in the function call is optional. If the `return` statement or *return_values* are omitted, the function returns the null object.

The following function computes the first two derivatives of $f(x)$ by finite differences:

```
def derivatives(f,x,h=0.0001):   # h has a default value
    df =(f(x+h) - f(x-h))/(2.0*h)
    ddf =(f(x+h) - 2.0*f(x) + f(x-h))/h**2
    return df,ddf
```

Let us now use this function to determine the two derivatives of $\arctan(x)$ at $x = 0.5$:

```
from math import atan
df,ddf = derivatives(atan,0.5)     # Uses default value of h
print('First derivative  =',df)
print('Second derivative =',ddf)
```

Note that `atan` is passed to `derivatives` as a parameter. The output from the program is

```
First derivative   =  0.799999999573
Second derivative  = -0.639999991892
```

The number of input parameters in a function definition may be left arbitrary. For example, in the following function definition

$$\text{def func(x1,x2,*x3)}$$

`x1` and `x2` are the usual parameters, also called *positional parameters*, whereas `x3` is a tuple of arbitrary length containing the *excess parameters*. Calling this function with

$$\text{func(a,b,c,d,e)}$$

results in the following correspondence between the parameters:

$$\text{a} \longleftrightarrow \text{x1}, \quad \text{b} \longleftrightarrow \text{x2}, \quad \text{(c,d,e)} \longleftrightarrow \text{x3}$$

The positional parameters must always be listed before the excess parameters.

If a mutable object, such as a list, is passed to a function where it is modified, the changes will also appear in the calling program. An example follows:

```
def squares(a):
    for i in range(len(a)):
        a[i] = a[i]**2


a = [1, 2, 3, 4]
squares(a)
print(a)  # 'a' now contains 'a**2'
```

The output is

```
[1, 4, 9, 16]
```

Lambda Statement

If the function has the form of an expression, it can be defined with the lambda statement

> *func_name* = `lambda` *param1, param2, . . .* : *expression*

Multiple statements are not allowed.

Here is an example:

```
>>> c = lambda x,y : x**2 + y**2
>>> print(c(3,4))
25
```

Modules

It is sound practice to store useful functions in modules. A module is simply a file where the functions reside; the name of the module is the name of the file. A module can be loaded into a program by the statement

> from *module_name* import *

Python comes with a large number of modules containing functions and methods for various tasks. Some of the modules are described briefly in the next two sections. Additional modules, including graphics packages, are available for downloading on the Web.

1.4 Mathematics Modules

`math` Module

Most mathematical functions are not built into core Python, but are available by loading the `math` module. There are three ways of accessing the functions in a module. The statement

> from math import *

loads *all* the function definitions in the `math` module into the current function or module. The use of this method is discouraged because it is not only wasteful but can also lead to conflicts with definitions loaded from other modules. For example, there are three different definitions of the *sine* function in the Python modules `math`, `cmath`, and `numpy`. If you have loaded two or more of these modules, it is unclear which definition will be used in the function call `sin(x)` (it is the definition in the module that was loaded last).

A safer but by no means foolproof method is to load selected definitions with the statement

> from math import *func1, func2,...*

as illustrated as follows:

```
>>> from math import log,sin
>>> print(log(sin(0.5)))
-0.735166686385
```

Conflicts can be avoided altogether by first making the module accessible with the statement

> import math

and then accessing the definitions in the module by using the module name as a prefix. Here is an example:

```
>>> import math
>>> print(math.log(math.sin(0.5)))
-0.735166686385
```

A module can also be made accessible under an *alias*. For example, the `math` module can be made available under the alias `m` with the command

```
import math as m
```

Now the prefix to be used is `m` rather than `math`:

```
>>> import math as m
>>> print(m.log(m.sin(0.5)))
-0.735166686385
```

The contents of a module can be printed by calling `dir(`*module*`)`. Here is how to obtain a list of the functions in the `math` module:

```
>>> import math
>>> dir(math)
['__doc__', '__name__', 'acos', 'asin', 'atan',
 'atan2', 'ceil', 'cos', 'cosh', 'e', 'exp', 'fabs',
 'floor', 'fmod', 'frexp', 'hypot', 'ldexp', 'log',
 'log10', 'modf', 'pi', 'pow', sign, sin', 'sinh',
 'sqrt', 'tan', 'tanh']
```

Most of these functions are familiar to programmers. Note that the module includes two constants: π and e.

`cmath` Module

The `cmath` module provides many of the functions found in the `math` module, but these functions accept complex numbers. The functions in the module are

```
['__doc__', '__name__', 'acos', 'acosh', 'asin', 'asinh',
 'atan', 'atanh', 'cos', 'cosh', 'e', 'exp', 'log',
 'log10', 'pi', 'sin', 'sinh', 'sqrt', 'tan', 'tanh']
```

Here are examples of complex arithmetic:

```
>>> from cmath import sin
>>> x = 3.0 -4.5j
>>> y = 1.2 + 0.8j
>>> z = 0.8
>>> print(x/y)
(-2.56205313375e-016-3.75j)
>>> print(sin(x))
(6.35239299817+44.5526433649j)
>>> print(sin(z))
(0.7173560909+0j)
```

1.5   numpy **Module**

General Information

The numpy module[2] is not a part of the standard Python release. As pointed out earlier, it must be installed separately (the installation is very easy). The module introduces *array objects* that are similar to lists, but can be manipulated by numerous functions contained in the module. The size of an array is immutable, and no empty elements are allowed.

The complete set of functions in numpy is far too long to be printed in its entirety. The following list is limited to the most commonly used functions.

```
['complex', 'float', 'abs', 'append', arccos',
'arccosh', 'arcsin', 'arcsinh', 'arctan', 'arctan2',
'arctanh', 'argmax', 'argmin', 'cos', 'cosh', 'diag',
'diagonal', 'dot', 'e', 'exp', 'floor', 'identity',
'inner, 'inv', 'log', 'log10', 'max', 'min',
'ones', 'outer', 'pi', 'prod' 'sin', 'sinh', 'size',
'solve', 'sqrt', 'sum', 'tan', 'tanh', 'trace',
'transpose', 'vectorize','zeros']
```

Creating an Array

Arrays can be created in several ways. One of them is to use the array function to turn a list into an array:

$$\boxed{\text{array}(\textit{list}, \textit{type})}$$

Following are two examples of creating a 2×2 array with floating-point elements:

```
>>> from numpy import array
>>> a = array([[2.0, -1.0],[-1.0, 3.0]])
>>> print(a)
[[ 2. -1.]
 [-1.  3.]]
>>> b = array([[2, -1],[-1, 3]],float)
>>> print(b)
[[ 2. -1.]
 [-1.  3.]]
```

Other available functions are

$$\boxed{\text{zeros}((\textit{dim1}, \textit{dim2}), \textit{type})}$$

which creates a *dim1* × *dim2* array and fills it with zeroes, and

$$\boxed{\text{ones}((\textit{dim1}, \textit{dim2}), \textit{type})}$$

which fills the array with ones. The default type in both cases is float.

---

[2] *NumPy* is the successor of older Python modules called *Numeric* and *NumArray*. Their interfaces and capabilities are very similar. Although *Numeric* and *NumArray* are still available, they are no longer supported.

Finally, there is the function

$$\boxed{\texttt{arange}(\mathit{from}, \mathit{to}, \mathit{increment})}$$

which works just like the `range` function, but returns an array rather than a sequence. Here are examples of creating arrays:

```
>>> from numpy import *
>>> print(arange(2,10,2))
[2 4 6 8]
>>> print(arange(2.0,10.0,2.0))
[ 2.  4.  6.  8.]
>>> print(zeros(3))
[ 0.  0.  0.]
>>> print(zeros((3),int))
[0 0 0]
>>> print(ones((2,2)))
[[ 1.  1.]
 [ 1.  1.]]
```

Accessing and Changing Array Elements

If a is a rank-2 array, then `a[i,j]` accesses the element in row i and column j, whereas `a[i]` refers to row i. The elements of an array can be changed by assignment as follows:

```
>>> from numpy import *
>>> a = zeros((3,3),int)
>>> print(a)
[[0 0 0]
 [0 0 0]
 [0 0 0]]
>>> a[0] = [2,3,2]        # Change a row
>>> a[1,1] = 5            # Change an element
>>> a[2,0:2] = [8,-3]     # Change part of a row
>>> print(a)
[[ 2  3  2]
 [ 0  5  0]
 [ 8 -3  0]]
```

Operations on Arrays

Arithmetic operators work differently on arrays than they do on tuples and lists—the operation is *broadcast* to all the elements of the array; that is, the operation is applied to each element in the array. Here are examples:

```
>>> from numpy import array
>>> a = array([0.0, 4.0, 9.0, 16.0])
```

```
>>> print(a/16.0)
[ 0.      0.25    0.5625  1.    ]
>>> print(a - 4.0)
[ -4.   0.   5.  12.]
```

The mathematical functions available in numpy are also broadcast, as follows:

```
>>> from numpy import array,sqrt,sin
>>> a = array([1.0, 4.0, 9.0, 16.0])
>>> print(sqrt(a))
[ 1.  2.  3.  4.]
>>> print(sin(a))
[ 0.84147098 -0.7568025   0.41211849 -0.28790332]
```

Functions imported from the math module will work on the individual elements, of course, but not on the array itself. An example follows:

```
>>> from numpy import array
>>> from math import sqrt
>>> a = array([1.0, 4.0, 9.0, 16.0])
>>> print(sqrt(a[1]))
2.0
>>> print(sqrt(a))
Traceback (most recent call last):

    ⋮

TypeError: only length-1 arrays can be converted to Python scalars
```

Array Functions

There are numerous functions in numpy that perform array operations and other useful tasks. Here are a few examples:

```
>>> from numpy import *
>>> A = array([[4,-2,1],[-2,4,-2],[1,-2,3]],float)
>>> b = array([1,4,3],float)
>>> print(diagonal(A))        # Principal diagonal
[ 4.  4.  3.]
>>> print(diagonal(A,1))      # First subdiagonal
[-2. -2.]
>>> print(trace(A))           # Sum of diagonal elements
11.0
>>> print(argmax(b))          # Index of largest element
1
>>> print(argmin(A,axis=0))   # Indices of smallest col. elements
[1 0 1]
```

```
>>> print(identity(3))          # Identity matrix
[[ 1.   0.   0.]
 [ 0.   1.   0.]
 [ 0.   0.   1.]]
```

There are three functions in numpy that compute array products. They are illustrated by the following program. For more details, see Appendix A2.

```
from numpy import *
x = array([7,3])
y = array([2,1])
A = array([[1,2],[3,2]])
B = array([[1,1],[2,2]])

# Dot product
print("dot(x,y) =\n",dot(x,y))          # {x}.{y}
print("dot(A,x) =\n",dot(A,x))          # [A]{x}
print("dot(A,B) =\n",dot(A,B))          # [A][B]

# Inner product
print("inner(x,y) =\n",inner(x,y))      # {x}.{y}
print("inner(A,x) =\n",inner(A,x))      # [A]{x}
print("inner(A,B) =\n",inner(A,B))      # [A][B_transpose]

# Outer product
print("outer(x,y) =\n",outer(x,y))
print("outer(A,x) =\n",outer(A,x))
print("outer(A,B) =\n",outer(A,B))
```

The output of the program is

```
dot(x,y) =
17
dot(A,x) =
[13 27]
dot(A,B) =
[[5 5]
 [7 7]]
inner(x,y) =
17
inner(A,x) =
[13 27]
inner(A,B) =
[[ 3   6]
 [ 5 10]]
outer(x,y) =
```

```
[[14   7]
 [ 6   3]]
outer(A,x) =
[[ 7   3]
 [14   6]
 [21   9]
 [14   6]]
Outer(A,B) =
[[1 1 2 2]
 [2 2 4 4]
 [3 3 6 6]
 [2 2 4 4]]
```

Linear Algebra Module

The numpy module comes with a linear algebra module called linalg that contains routine tasks such as matrix inversion and solution of simultaneous equations. For example,

```
>>> from numpy import array
>>> from numpy.linalg import inv,solve
>>> A = array([[ 4.0,  -2.0,   1.0], \
               [-2.0,   4.0,  -2.0], \
               [ 1.0,  -2.0,   3.0]])
>>> b = array([1.0, 4.0, 2.0])
>>> print(inv(A))                          # Matrix inverse
[[ 0.33333333  0.16666667  0.         ]
 [ 0.16666667  0.45833333  0.25       ]
 [ 0.          0.25        0.5        ]]
>>> print(solve(A,b))                      # Solve [A]{x} = {b}
[ 1. ,  2.5,   2. ]
```

Copying Arrays

We explained earlier that if *a* is a mutable object, such as a list, the assignment statement b = a does not result in a new object *b*, but simply creates a new reference to *a*, called a *deep copy*. This also applies to arrays. To make an independent copy of an array *a*, use the copy method in the numpy module:

$$b = a.copy()$$

Vectorizing Algorithms

Sometimes the broadcasting properties of the mathematical functions in the numpy module can be used to replace loops in the code. This procedure is known as vectorization. Consider, for example, the expression

$$s = \sum_{i=0}^{100} \sqrt{\frac{i\pi}{100}} \sin \frac{i\pi}{100}$$

The direct approach is to evaluate the sum in a loop, resulting in the following "scalar" code:

```
from math import sqrt,sin,pi
x = 0.0; s = 0.0
for i in range(101):
    s = s + sqrt(x)*sin(x)
    x = x + 0.01*pi
print(s)
```

The vectorized version of the algorithm is

```
from numpy import sqrt,sin,arange
from math import pi
x = arange(0.0, 1.001*pi, 0.01*pi)
print(sum(sqrt(x)*sin(x)))
```

Note that the first algorithm uses the scalar versions of `sqrt` and `sin` functions in the `math` module, whereas the second algorithm imports these functions from numpy. The vectorized algorithm executes much faster, but uses more memory.

1.6   **Plotting with** `matplotlib.pyplot`

The module `matplotlib.pyplot` is a collection of 2D plotting functions that provide Python with MATLAB-style functionality. Not being a part of core Python, it requires separate installation. The following program, which plots sine and cosine functions, illustrates the application of the module to simple *xy* plots.

```
import matplotlib.pyplot as plt
from numpy import arange,sin,cos
x = arange(0.0,6.2,0.2)

plt.plot(x,sin(x),'o-',x,cos(x),'^-')    # Plot with specified
                                         # line and marker style
plt.xlabel('x')                          # Add label to x-axis
plt.legend(('sine','cosine'),loc = 0)    # Add legend in loc. 3
plt.grid(True)                           # Add coordinate grid
plt.savefig('testplot.png',format='png') # Save plot in png
                                         # format for future use
plt.show()                               # Show plot on screen
input("\nPress return to exit")
```

The line and marker styles are specified by the string characters shown in the following table (only some of the available characters are shown).

`'-'`	Solid line
`'--'`	Dashed line
`'-.'`	Dash-dot line
`':'`	Dotted line
`'o'`	Circle marker
`'^'`	Triangle marker
`'s'`	Square marker
`'h'`	Hexagon marker
`'x'`	x marker

Some of the location (`loc`) codes for placement of the legend are

0	"Best" location
1	Upper right
2	Upper left
3	Lower left
4	Lower right

Running the program produces the following screen:

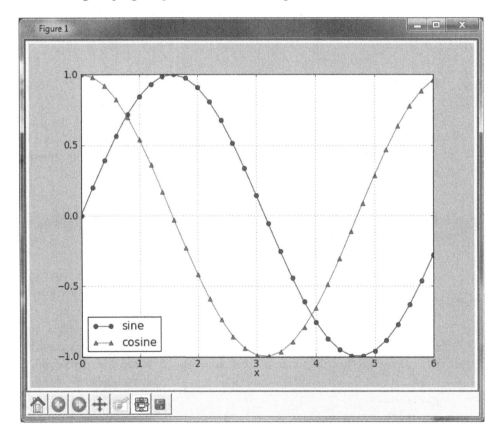

It is possible to have more than one plot in a figure, as demonstrated by the following code:

```
import matplotlib.pyplot as plt
from numpy import arange,sin,cos
x = arange(0.0,6.2,0.2)

plt.subplot(2,1,1)
plt.plot(x,sin(x),'o-')
plt.xlabel('x');plt.ylabel('sin(x)')
plt.grid(True)
plt.subplot(2,1,2)
plt.plot(x,cos(x),'^-')
plt.xlabel('x');plt.ylabel('cos(x)')
plt.grid(True)
plt.show()
input("\nPress return to exit")
```

The command `subplot(`*rows,cols,plot_number*`)` establishes a subplot window within the current figure. The parameters *row* and *col* divide the figure into *row* × *col* grid of subplots (in this case, two rows and one column). The commas between the parameters may be omitted. The output from this above program is

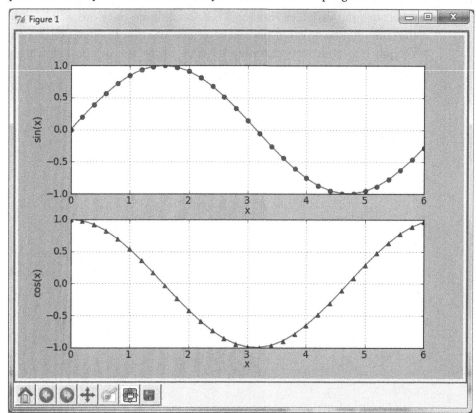

1.7 Scoping of Variables

Namespace is a dictionary that contains the names of the variables and their values. Namespaces are automatically created and updated as a program runs. There are three levels of namespaces in Python:

1. Local namespace is created when a function is called. It contains the variables passed to the function as arguments and the variables created within the function. The namespace is deleted when the function terminates. If a variable is created inside a function, its scope is the function's local namespace. It is not visible outside the function.
2. A global namespace is created when a module is loaded. Each module has its own namespace. Variables assigned in a global namespace are visible to any function within the module.
3. A built-in namespace is created when the interpreter starts. It contains the functions that come with the Python interpreter. These functions can be accessed by any program unit.

When a name is encountered during execution of a function, the interpreter tries to resolve it by searching the following in the order shown: (1) local namespace, (2) global namespace, and (3) built-in namespace. If the name cannot be resolved, Python raises a `NameError` exception.

Because the variables residing in a global namespace are visible to functions within the module, it is not necessary to pass them to the functions as arguments (although it is good programming practice to do so), as the following program illustrates:

```
def divide():
    c = a/b
    print('a/b =',c)
a = 100.0
b = 5.0
divide()
```

```
a/b = 20.0
```

Note that the variable `c` is created inside the function `divide` and is thus not accessible to statements outside the function. Hence an attempt to move the print statement out of the function fails:

```
def divide():
    c = a/b
a = 100.0
b = 5.0
divide()
print('a/b =',c)
```

```
Traceback (most recent call last):
  File "C:\Python32\test.py", line 6, in <module>
    print('a/b =',c)
NameError: name 'c' is not defined
```

1.8   Writing and Running Programs

When the Python editor *Idle* is opened, the user is faced with the prompt >>>, indicating that the editor is in interactive mode. Any statement typed into the editor is immediately processed on pressing the enter key. The interactive mode is a good way both to learn the language by experimentation and to try out new programming ideas.

Opening a new window places Idle in the batch mode, which allows typing and saving of programs. One can also use a text editor to enter program lines, but Idle has Python-specific features, such as color coding of keywords and automatic indentation, which make work easier. Before a program can be run, it must be saved as a Python file with the .py extension (e.g., myprog.py). The program can then be executed by typing python myprog.py; in Windows; double-clicking on the program icon will also work. But beware: The program window closes immediately after execution, before you get a chance to read the output. To prevent this from happening, conclude the program with the line

input('press return')

Double-clicking the program icon also works in Unix and Linux if the first line of the program specifies the path to the Python interpreter (or a shell script that provides a link to Python). The path name must be preceded by the symbols #!. On my computer the path is /usr/bin/python, so that all my programs start with the line #!/usr/bin/python. On multi-user systems the path is usually /usr/local/bin/python.

When a module is loaded into a program for the first time with the import statement, it is compiled into bytecode and written in a file with the extension .pyc. The next time the program is run, the interpreter loads the bytecode rather than the original Python file. If in the meantime changes have been made to the module, the module is automatically recompiled. A program can also be run from Idle using *Run/Run Module* menu.

It is a good idea to document your modules by adding a *docstring* at the beginning of each module. The docstring, which is enclosed in triple quotes, should explain what the module does. Here is an example that documents the module error (we use this module in several of our programs):

```
## module error
''' err(string).
    Prints 'string' and terminates program.
```

```
'''
import sys
def err(string):
    print(string)
    input('Press return to exit')
    sys.exit()
```

The docstring of a module can be printed with the statement

$$\boxed{\texttt{print}(\textit{module\_name}.\texttt{\_\_doc\_\_})}$$

For example, the docstring of `error` is displayed by

```
>>> import error
>>> print(error.__doc__)
 err(string).
    Prints 'string' and terminates program.
```

Avoid backslashes in the docstring because they confuse the Python 3 interpreter.

2 Systems of Linear Algebraic Equations

$$\boxed{\text{Solve the simultaneous equations } \mathbf{Ax} = \mathbf{b}.}$$

2.1 Introduction

In this chapter we look at the solution of n linear, algebraic equations in n unknowns. It is by far the longest and arguably the most important topic in the book. There is a good reason for its importance—it is almost impossible to carry out numerical analysis of any sort without encountering simultaneous equations. Moreover, equation sets arising from physical problems are often very large, consuming a lot of computational resources. It usually possible to reduce the storage requirements and the run time by exploiting special properties of the coefficient matrix, such as sparseness (most elements of a sparse matrix are zero). Hence there are many algorithms dedicated to the solution of large sets of equations, each one being tailored to a particular form of the coefficient matrix (symmetric, banded, sparse, and so on). A well-known collection of these routines is LAPACK—Linear Algebra PACKage, originally written in Fortran77.[1]

We cannot possibly discuss all the special algorithms in the limited space available. The best we can do is to present the basic methods of solution, supplemented by a few useful algorithms for banded coefficient matrices.

Notation

A system of algebraic equations has the form

$$A_{11}x_1 + A_{12}x_2 + \cdots + A_{1n}x_n = b_1$$
$$A_{21}x_1 + A_{22}x_2 + \cdots + A_{2n}x_n = b_2 \tag{2.1}$$
$$\vdots$$
$$A_{n1}x_1 + A_{n2}x_2 + \cdots + A_{nn}x_n = b_n$$

---

[1] LAPACK is the successor of LINPACK, a 1970s and 80s collection of Fortran subroutines.

where the coefficients A_{ij} and the constants b_j are known, and x_i represents the unknowns. In matrix notation the equations are written as

$$\begin{bmatrix} A_{11} & A_{12} & \cdots & A_{1n} \\ A_{21} & A_{22} & \cdots & A_{2n} \\ \vdots & \vdots & \ddots & \vdots \\ A_{n1} & A_{n2} & \cdots & A_{nn} \end{bmatrix} \begin{bmatrix} x_1 \\ x_2 \\ \vdots \\ x_n \end{bmatrix} = \begin{bmatrix} b_1 \\ b_2 \\ \vdots \\ b_n \end{bmatrix} \tag{2.2}$$

or simply

$$\mathbf{Ax} = \mathbf{b}. \tag{2.3}$$

A particularly useful representation of the equations for computational purposes is the *augmented coefficient matrix* obtained by adjoining the constant vector $\mathbf{b}$ to the coefficient matrix $\mathbf{A}$ in the following fashion:

$$\begin{bmatrix} \mathbf{A} \, | \, \mathbf{b} \end{bmatrix} = \left[\begin{array}{cccc|c} A_{11} & A_{12} & \cdots & A_{1n} & b_1 \\ A_{21} & A_{22} & \cdots & A_{2n} & b_2 \\ \vdots & \vdots & \ddots & \vdots & \vdots \\ A_{n1} & A_{n2} & \cdots & A_{n3} & b_n \end{array} \right] \tag{2.4}$$

Uniqueness of Solution

A system of n linear equations in n unknowns has a unique solution, provided that the determinant of the coefficient matrix is *nonsingular*; that is, $|\mathbf{A}| \neq 0$. The rows and columns of a nonsingular matrix are *linearly independent* in the sense that no row (or column) is a linear combination of other rows (or columns).

If the coefficient matrix is *singular*, the equations may have an infinite number of solutions or no solutions at all, depending on the constant vector. As an illustration, take the equations

$$2x + y = 3 \qquad 4x + 2y = 6$$

Since the second equation can be obtained by multiplying the first equation by two, any combination of x and y that satisfies the first equation is also a solution of the second equation. The number of such combinations is infinite. In contrast, the equations

$$2x + y = 3 \qquad 4x + 2y = 0$$

have no solution because the second equation, being equivalent to $2x + y = 0$, contradicts the first one. Therefore, any solution that satisfies one equation cannot satisfy the other one.

Ill Conditioning

The obvious question is, What happens when the coefficient matrix is almost singular (i.e., if $|\mathbf{A}|$ is very small). To determine whether the determinant of the coefficient matrix is "small," we need a reference against which the determinant can be measured. This reference is called the *norm* of the matrix and is denoted by $\|\mathbf{A}\|$. We can then say that the determinant is small if

$$|\mathbf{A}| << \|\mathbf{A}\|$$

Several norms of a matrix have been defined in existing literature, such as the *Euclidan norm*

$$\|\mathbf{A}\|_e = \sqrt{\sum_{i=1}^{n}\sum_{j=1}^{n} A_{ij}^2} \qquad (2.5a)$$

and the *row-sum norm*, also called the *infinity norm*

$$\|\mathbf{A}\|_\infty = \max_{1 \le i \le n} \sum_{j=1}^{n} |A_{ij}| \qquad (2.5b)$$

A formal measure of conditioning is the *matrix condition number*, defined as

$$\text{cond}(\mathbf{A}) = \|\mathbf{A}\|\,\|\mathbf{A}^{-1}\| \qquad (2.5c)$$

If this number is close to unity, the matrix is well conditioned. The condition number increases with the degree of ill conditioning, reaching infinity for a singular matrix. Note that the condition number is not unique, but depends on the choice of the matrix norm. Unfortunately, the condition number is expensive to compute for large matrices. In most cases it is sufficient to gauge conditioning by comparing the determinant with the magnitudes of the elements in the matrix.

If the equations are ill conditioned, small changes in the coefficient matrix result in large changes in the solution. As an illustration, take the equations

$$2x + y = 3 \qquad 2x + 1.001y = 0$$

that have the solution $x = 1501.5$, $y = -3000$. Since $|\mathbf{A}| = 2(1.001) - 2(1) = 0.002$ is much smaller than the coefficients, the equations are ill conditioned. The effect of ill conditioning can verified by changing the second equation to $2x + 1.002y = 0$ and re-solving the equations. The result is $x = 751.5$, $y = -1500$. Note that a 0.1% change in the coefficient of y produced a 100% change in the solution!

Numerical solutions of ill-conditioned equations are not to be trusted. The reason is that the inevitable roundoff errors during the solution process are equivalent to introducing small changes into the coefficient matrix. This in turn introduces large errors into the solution, the magnitude of which depends on the severity of ill conditioning. In suspect cases the determinant of the coefficient matrix should be computed so that the degree of ill conditioning can be estimated. This can be done during or after the solution with only a small computational effort.

Linear Systems

Linear, algebraic equations occur in almost all branches of numerical analysis. But their most visible application in engineering is in the analysis of linear systems (any system whose response is proportional to the input is deemed to be linear). Linear systems include structures, elastic solids, heat flow, seepage of fluids, electromagnetic fields, and electric circuits (i.e., most topics taught in an engineering curriculum).

If the system is discrete, such as a truss or an electric circuit, then its analysis leads directly to linear algebraic equations. In the case of a statically determinate truss, for example, the equations arise when the equilibrium conditions of the joints are written down. The unknowns $x_1, x_2, \ldots, x_n$ represent the forces in the members and the support reactions, and the constants $b_1, b_2, \ldots, b_n$ are the prescribed external loads.

The behavior of continuous systems is described by differential equations, rather than algebraic equations. However, because numerical analysis can deal only with discrete variables, it is first necessary to approximate a differential equation with a system of algebraic equations. The well-known finite difference, finite element, and boundary element methods of analysis work in this manner. They use different approximations to achieve the "discretization," but in each case the final task is the same: solve a system (often a very large system) of linear, algebraic equations.

In summary, the modeling of linear systems invariably gives rise to equations of the form $\mathbf{Ax} = \mathbf{b}$, where $\mathbf{b}$ is the input and $\mathbf{x}$ represents the response of the system. The coefficient matrix $\mathbf{A}$, which reflects the characteristics of the system, is independent of the input. In other words, if the input is changed, the equations have to be solved again with a different $\mathbf{b}$, but the same $\mathbf{A}$. Therefore, it is desirable to have an equation-solving algorithm that can handle any number of constant vectors with minimal computational effort.

Methods of Solution

There are two classes of methods for solving systems of linear, algebraic equations: direct and iterative methods. The common characteristic of *direct methods* is that they transform the original equations into *equivalent equations* (equations that have the same solution) that can be solved more easily. The transformation is carried out by applying the following three operations. These so-called *elementary operations* do not change the solution, but they may affect the determinant of the coefficient matrix as indicated in parenthesis.

1. Exchanging two equations (changes sign of $|\mathbf{A}|$)
2. Multiplying an equation by a nonzero constant (multiplies $|\mathbf{A}|$ by the same constant)
3. Multiplying an equation by a nonzero constant and then subtracting it from another equation (leaves $|\mathbf{A}|$ unchanged)

Iterative, or *indirect methods,* start with a guess of the solution **x** and then repeatedly refine the solution until a certain convergence criterion is reached. Iterative methods are generally less efficient than their direct counterparts because of the large number of iterations required. Yet they do have significant computational advantages if the coefficient matrix is very large and sparsely populated (most coefficients are zero).

Overview of Direct Methods

Table 2.1 lists three popular direct methods, each of which uses elementary operations to produce its own final form of easy-to-solve equations.

Method	Initial form	Final form
Gauss elimination	$\mathbf{Ax} = \mathbf{b}$	$\mathbf{Ux} = \mathbf{c}$
LU decomposition	$\mathbf{Ax} = \mathbf{b}$	$\mathbf{LUx} = \mathbf{b}$
Gauss-Jordan elimination	$\mathbf{Ax} = \mathbf{b}$	$\mathbf{Ix} = \mathbf{c}$

Table 2.1. Three Popular Direct Methods

In Table 2.1 **U** represents an upper triangular matrix, **L** is a lower triangular matrix, and **I** denotes the identity matrix. A square matrix is called *triangular* if it contains only zero elements on one side of the leading diagonal. Thus a 3×3 upper triangular matrix has the form

$$\mathbf{U} = \begin{bmatrix} U_{11} & U_{12} & U_{13} \\ 0 & U_{22} & U_{23} \\ 0 & 0 & U_{33} \end{bmatrix}$$

and a 3×3 lower triangular matrix appears as

$$\mathbf{L} = \begin{bmatrix} L_{11} & 0 & 0 \\ L_{21} & L_{22} & 0 \\ L_{31} & L_{32} & L_{33} \end{bmatrix}$$

Triangular matrices play an important role in linear algebra, because they simplify many computations. For example, consider the equations **Lx** = **c**, or

$$L_{11}x_1 = c_1$$

$$L_{21}x_1 + L_{22}x_2 = c_2$$

$$L_{31}x_1 + L_{32}x_2 + L_{33}x_3 = c_3$$

If we solve the equations forward, starting with the first equation, the computations are very easy, because each equation contains only one unknown at a time.

The solution would thus proceed as follows:

$$x_1 = c_1/L_{11}$$

$$x_2 = (c_2 - L_{21}x_1)/L_{22}$$

$$x_3 = (c_3 - L_{31}x_1 - L_{32}x_2)/L_{33}$$

This procedure is known as *forward substitution*. In a similar way, $\mathbf{Ux} = \mathbf{c}$, encountered in Gauss elimination, can easily be solved by *back substitution*, which starts with the last equation and proceeds backward through the equations.

The equations $\mathbf{LUx} = \mathbf{b}$, which are associated with LU decomposition, can also be solved quickly if we replace them with two sets of equivalent equations: $\mathbf{Ly} = \mathbf{b}$ and $\mathbf{Ux} = \mathbf{y}$. Now $\mathbf{Ly} = \mathbf{b}$ can be solved for $\mathbf{y}$ by forward substitution, followed by the solution of $\mathbf{Ux} = \mathbf{y}$ by means of back substitution.

The equations $\mathbf{Ix} = \mathbf{c}$, which are produced by Gauss-Jordan elimination, are equivalent to $\mathbf{x} = \mathbf{c}$ (recall the identity $\mathbf{Ix} = \mathbf{x}$), so that $\mathbf{c}$ is already the solution.

EXAMPLE 2.1

Determine whether the following matrix is singular:

$$\mathbf{A} = \begin{bmatrix} 2.1 & -0.6 & 1.1 \\ 3.2 & 4.7 & -0.8 \\ 3.1 & -6.5 & 4.1 \end{bmatrix}$$

Solution. Laplace's development of the determinant (see Appendix A2) about the first row of $\mathbf{A}$ yields

$$|\mathbf{A}| = 2.1 \begin{vmatrix} 4.7 & -0.8 \\ -6.5 & 4.1 \end{vmatrix} - (-0.6) \begin{vmatrix} 3.2 & -0.8 \\ 3.1 & 4.1 \end{vmatrix} + 1.1 \begin{vmatrix} 3.2 & 4.7 \\ 3.1 & -6.5 \end{vmatrix}$$

$$= 2.1(14.07) + 0.6(15.60) + 1.1(35.37) = 0$$

Since the determinant is zero, the matrix is singular. It can be verified that the singularity is due to the following row dependency: $(\text{row}\,3) = (3 \times \text{row}\,1) - (\text{row}\,2)$.

EXAMPLE 2.2

Solve the equations $\mathbf{Ax} = \mathbf{b}$, where

$$\mathbf{A} = \begin{bmatrix} 8 & -6 & 2 \\ -4 & 11 & -7 \\ 4 & -7 & 6 \end{bmatrix} \qquad \mathbf{b} = \begin{bmatrix} 28 \\ -40 \\ 33 \end{bmatrix}$$

knowing that the LU decomposition of the coefficient matrix is (you should verify this)

$$\mathbf{A} = \mathbf{LU} = \begin{bmatrix} 2 & 0 & 0 \\ -1 & 2 & 0 \\ 1 & -1 & 1 \end{bmatrix} \begin{bmatrix} 4 & -3 & 1 \\ 0 & 4 & -3 \\ 0 & 0 & 2 \end{bmatrix}$$

Solution. We first solve the equations $\mathbf{Ly} = \mathbf{b}$ by forward substitution:

$$
\begin{array}{ll}
2y_1 = 28 & y_1 = 28/2 = 14 \\
-y_1 + 2y_2 = -40 & y_2 = (-40 + y_1)/2 = (-40 + 14)/2 = -13 \\
y_1 - y_2 + y_3 = 33 & y_3 = 33 - y_1 + y_2 = 33 - 14 - 13 = 6
\end{array}
$$

The solution $\mathbf{x}$ is then obtained from $\mathbf{Ux} = \mathbf{y}$ by back substitution:

$$
\begin{array}{ll}
2x_3 = y_3 & x_3 = y_3/2 = 6/2 = 3 \\
4x_2 - 3x_3 = y_2 & x_2 = (y_2 + 3x_3)/4 = [-13 + 3(3)]/4 = -1 \\
4x_1 - 3x_2 + x_3 = y_1 & x_1 = (y_1 + 3x_2 - x_3)/4 = [14 + 3(-1) - 3]/4 = 2
\end{array}
$$

Hence the solution is $\mathbf{x} = \begin{bmatrix} 2 & -1 & 3 \end{bmatrix}^T$.

2.2 Gauss Elimination Method

Introduction

Gauss elimination is the most familiar method for solving simultaneous equations. It consists of two parts: the elimination phase and the back substitution phase. As indicated in Table 2.1, the function of the elimination phase is to transform the equations into the form $\mathbf{Ux} = \mathbf{c}$. The equations are then solved by back substitution. To illustrate the procedure, let us solve the following equations:

$$4x_1 - 2x_2 + x_3 = 11 \tag{a}$$

$$-2x_1 + 4x_2 - 2x_3 = -16 \tag{b}$$

$$x_1 - 2x_2 + 4x_3 = 17 \tag{c}$$

Elimination phase. The elimination phase uses only one of the elementary operations listed in Table 2.1—multiplying one equation (say, equation j) by a constant λ and subtracting it from another equation (equation i). The symbolic representation of this operation is

$$\text{Eq. } (i) \leftarrow \text{Eq. } (i) - \lambda \times \text{Eq. } (j) \tag{2.6}$$

The equation being subtracted, namely Eq. (j), is called the *pivot equation.*

We start the elimination by taking Eq. (a) to be the pivot equation and choosing the multipliers λ so as to eliminate x_1 from Eqs. (b) and (c):

$$\text{Eq. (b)} \leftarrow \text{Eq. (b)} - (-0.5) \times \text{Eq. (a)}$$

$$\text{Eq. (c)} \leftarrow \text{Eq. (c)} - 0.25 \times \text{Eq. (a)}$$

After this transformation, the equations become

$$4x_1 - 2x_2 + x_3 = 11 \tag{a}$$

$$3x_2 - 1.5x_3 = -10.5 \tag{b}$$

$$-1.5x_2 + 3.75x_3 = 14.25 \tag{c}$$

This completes the first pass. Now we pick (b) as the pivot equation and eliminate x_2 from (c):

$$\text{Eq. (c)} \leftarrow \text{Eq. (c)} - (-0.5) \times \text{Eq. (b)}$$

which yields the equations

$$4x_1 - 2x_2 + x_3 = 11 \tag{a}$$

$$3x_2 - 1.5x_3 = -10.5 \tag{b}$$

$$3x_3 = 9 \tag{c}$$

The elimination phase is now complete. The original equations have been replaced by equivalent equations that can be easily solved by back substitution.

As pointed out earlier, the augmented coefficient matrix is a more convenient instrument for performing the computations. Thus the original equations would be written as

$$\begin{bmatrix} 4 & -2 & 1 & | & 11 \\ -2 & 4 & -2 & | & -16 \\ 1 & -2 & 4 & | & 17 \end{bmatrix}$$

and the equivalent equations produced by the first and the second passes of Gauss elimination would appear as

$$\begin{bmatrix} 4 & -2 & 1 & | & 11.00 \\ 0 & 3 & -1.5 & | & -10.50 \\ 0 & -1.5 & 3.75 & | & 14.25 \end{bmatrix}$$

$$\begin{bmatrix} 4 & -2 & 1 & | & 11.0 \\ 0 & 3 & -1.5 & | & -10.5 \\ 0 & 0 & 3 & | & 9.0 \end{bmatrix}$$

It is important to note that the elementary row operation in Eq. (2.6) leaves the determinant of the coefficient matrix unchanged. This is fortunate, because the determinant of a triangular matrix is very easy to compute—it is the product of the diagonal elements (you can verify this quite easily). In other words,

$$|\mathbf{A}| = |\mathbf{U}| = U_{11} \times U_{22} \times \cdots \times U_{nn} \tag{2.7}$$

Back substitution phase. The unknowns can now be computed by back substitution in the manner described previously. Solving Eqs. (c), (b), and (a) in that order, we get

$$x_3 = 9/3 = 3$$

$$x_2 = (-10.5 + 1.5x_3)/3 = [-10.5 + 1.5(3)]/3 = -2$$

$$x_1 = (11 + 2x_2 - x_3)/4 = [11 + 2(-2) - 3]/4 = 1$$

Algorithm for Gauss Elimination Method

Elimination phase. Let us look at the equations at some instant during the elimination phase. Assume that the first k rows of **A** have already been transformed to upper triangular form. Therefore, the current pivot equation is the kth equation, and all the equations below it are still to be transformed. This situation is depicted by the following augmented coefficient matrix. Note that the components of **A** are not the coefficients of the original equations (except for the first row), because they have been altered by the elimination procedure. The same applies to the components of the constant vector **b**.

$$\left[\begin{array}{ccccccccc|c}
A_{11} & A_{12} & A_{13} & \cdots & A_{1k} & \cdots & A_{1j} & \cdots & A_{1n} & b_1 \\
0 & A_{22} & A_{23} & \cdots & A_{2k} & \cdots & A_{2j} & \cdots & A_{2n} & b_2 \\
0 & 0 & A_{33} & \cdots & A_{3k} & \cdots & A_{3j} & \cdots & A_{3n} & b_3 \\
\vdots & \vdots & \vdots & & \vdots & & \vdots & & \vdots & \vdots \\
0 & 0 & 0 & \cdots & A_{kk} & \cdots & A_{kj} & \cdots & A_{kn} & b_k \\
\vdots & \vdots & \vdots & & \vdots & & \vdots & & \vdots & \vdots \\
0 & 0 & 0 & \cdots & A_{ik} & \cdots & A_{ij} & \cdots & A_{in} & b_i \\
\vdots & \vdots & \vdots & & \vdots & & \vdots & & \vdots & \vdots \\
0 & 0 & 0 & \cdots & A_{nk} & \cdots & A_{nj} & \cdots & A_{nn} & b_n
\end{array}\right]
\begin{array}{l}
\\ \\ \\ \\
\leftarrow \text{ pivot row} \\
\\
\leftarrow \text{ row being} \\
\quad \text{transformed} \\
\\
\end{array}$$

Let the ith row be a typical row below the pivot equation that is to be transformed, meaning that the element A_{ik} is to be eliminated. We can achieve this by multiplying the pivot row by $\lambda = A_{ik}/A_{kk}$ and subtracting it from the ith row. The corresponding changes in the ith row are

$$A_{ij} \leftarrow A_{ij} - \lambda A_{kj}, \quad j = k, k+1, \ldots, n \tag{2.8a}$$

$$b_i \leftarrow b_i - \lambda b_k \tag{2.8b}$$

To transform the entire coefficient matrix to upper triangular form, k and i in Eqs. (2.8) must have the ranges $k = 1, 2, \ldots, n-1$ (chooses the pivot row), $i = k + 1, k + 2 \ldots, n$ (chooses the row to be transformed). The algorithm for the elimination phase now almost writes itself:

```
for k in range(0,n-1):
    for i in range(k+1,n):
```

```
if a[i,k] != 0.0:
    lam = a[i,k]/a[k,k]
    a[i,k+1:n] = a[i,k+1:n] - lam*a[k,k+1:n]
    b[i] = b[i] - lam*b[k]
```

To avoid unnecessary operations, the preceding algorithm departs slightly from Eqs. (2.8) in the following ways:

- If A_{ik} happens to be zero, the transformation of row i is skipped.
- The index j in Eq. (2.8a) starts with $k + 1$ rather than k. Therefore, A_{ik} is not replaced by zero, but retains its original value. As the solution phase never accesses the lower triangular portion of the coefficient matrix anyway, its contents are irrelevant.

Back substitution phase. After Gauss elimination the augmented coefficient matrix has the form

$$\begin{bmatrix} \mathbf{A} \mid \mathbf{b} \end{bmatrix} = \begin{bmatrix} A_{11} & A_{12} & A_{13} & \cdots & A_{1n} & b_1 \\ 0 & A_{22} & A_{23} & \cdots & A_{2n} & b_2 \\ 0 & 0 & A_{33} & \cdots & A_{3n} & b_3 \\ \vdots & \vdots & \vdots & \ddots & \vdots & \vdots \\ 0 & 0 & 0 & \cdots & A_{nn} & b_n \end{bmatrix}$$

The last equation, $A_{nn}x_n = b_n$, is solved first, yielding

$$x_n = b_n/A_{nn} \tag{2.9}$$

Consider now the stage of back substitution where $x_n, x_{n-1}, \ldots, x_{k+1}$ have been already been computed (in that order), and we are about to determine x_k from the kth equation

$$A_{kk}x_k + A_{k,k+1}x_{k+1} + \cdots + A_{kn}x_n = b_k$$

The solution is

$$x_k = \left(b_k - \sum_{j=k+1}^{n} A_{kj}x_j \right) \frac{1}{A_{kk}}, \quad k = n-1, n-2, \ldots, 1 \tag{2.10}$$

The corresponding algorithm for back substitution is

```
for k in range(n-1,-1,-1):
    x[k]=(b[k] - dot(a[k,k+1:n],x[k+1:n]))/a[k,k]
```

Operation count. The execution time of an algorithm depends largely on the number of long operations (multiplications and divisions) performed. It can be shown that Gauss elimination contains approximately $n^3/3$ such operations (n is the number of equations) in the elimination phase, and $n^2/2$ operations in back substitution. These numbers show that most of the computation time goes into the elimination phase. Moreover, the time increases very rapidly with the number of equations.

∎ gaussElimin

The function gaussElimin combines the elimination and the back substitution phases. During back substitution b is overwritten by the solution vector x, so that b contains the solution upon exit.

```
## module gaussElimin
''' x = gaussElimin(a,b).
    Solves [a]{b} = {x} by Gauss elimination.
'''
import numpy as np

def gaussElimin(a,b):
    n = len(b)
  # Elimination Phase
    for k in range(0,n-1):
        for i in range(k+1,n):
            if a[i,k] != 0.0:
                lam = a [i,k]/a[k,k]
                a[i,k+1:n] = a[i,k+1:n] - lam*a[k,k+1:n]
                b[i] = b[i] - lam*b[k]
  # Back substitution
    for k in range(n-1,-1,-1):
        b[k] = (b[k] - np.dot(a[k,k+1:n],b[k+1:n]))/a[k,k]
    return b
```

Multiple Sets of Equations

As mentioned earlier, it is frequently necessary to solve the equations $\mathbf{A}\mathbf{x} = \mathbf{b}$ for several constant vectors. Let there be m such constant vectors, denoted by $\mathbf{b}_1, \mathbf{b}_2, \ldots, \mathbf{b}_m$, and let the corresponding solution vectors be $\mathbf{x}_1, \mathbf{x}_2, \ldots, \mathbf{x}_m$. We denote multiple sets of equations by $\mathbf{A}\mathbf{X} = \mathbf{B}$, where

$$\mathbf{X} = \begin{bmatrix} \mathbf{x}_1 & \mathbf{x}_2 & \cdots & \mathbf{x}_m \end{bmatrix} \qquad \mathbf{B} = \begin{bmatrix} \mathbf{b}_1 & \mathbf{b}_2 & \cdots & \mathbf{b}_m \end{bmatrix}$$

are $n \times m$ matrices whose columns consist of solution vectors and constant vectors, respectively.

An economical way to handle such equations during the elimination phase is to include all m constant vectors in the augmented coefficient matrix, so that they are transformed simultaneously with the coefficient matrix. The solutions are then obtained by back substitution in the usual manner, one vector at a time. It would be quite easy to make the corresponding changes in gaussElimin. However, the LU decomposition method, described in the next section, is more versatile in handling multiple constant vectors.

EXAMPLE 2.3

Use Gauss elimination to solve the equations $\mathbf{AX} = \mathbf{B}$, where

$$\mathbf{A} = \begin{bmatrix} 6 & -4 & 1 \\ -4 & 6 & -4 \\ 1 & -4 & 6 \end{bmatrix} \qquad \mathbf{B} = \begin{bmatrix} -14 & 22 \\ 36 & -18 \\ 6 & 7 \end{bmatrix}$$

Solution. The augmented coefficient matrix is

$$\left[\begin{array}{ccc|cc} 6 & -4 & 1 & -14 & 22 \\ -4 & 6 & -4 & 36 & -18 \\ 1 & -4 & 6 & 6 & 7 \end{array} \right]$$

The elimination phase consists of the following two passes:

$$\text{row } 2 \leftarrow \text{row } 2 + (2/3) \times \text{row } 1$$

$$\text{row } 3 \leftarrow \text{row } 3 - (1/6) \times \text{row } 1$$

$$\left[\begin{array}{ccc|cc} 6 & -4 & 1 & -14 & 22 \\ 0 & 10/3 & -10/3 & 80/3 & -10/3 \\ 0 & -10/3 & 35/6 & 25/3 & 10/3 \end{array} \right]$$

and

$$\text{row } 3 \leftarrow \text{row } 3 + \text{row } 2$$

$$\left[\begin{array}{ccc|cc} 6 & -4 & 1 & -14 & 22 \\ 0 & 10/3 & -10/3 & 80/3 & -10/3 \\ 0 & 0 & 5/2 & 35 & 0 \end{array} \right]$$

In the solution phase, we first compute $\mathbf{x}_1$ by back substitution:

$$X_{31} = \frac{35}{5/2} = 14$$

$$X_{21} = \frac{80/3 + (10/3)X_{31}}{10/3} = \frac{80/3 + (10/3)14}{10/3} = 22$$

$$X_{11} = \frac{-14 + 4X_{21} - X_{31}}{6} = \frac{-14 + 4(22) - 14}{6} = 10$$

Thus the first solution vector is

$$\mathbf{x}_1 = \begin{bmatrix} X_{11} & X_{21} & X_{31} \end{bmatrix}^T = \begin{bmatrix} 10 & 22 & 14 \end{bmatrix}^T$$

The second solution vector is computed next, also using back substitution:

$$X_{32} = 0$$

$$X_{22} = \frac{-10/3 + (10/3)X_{32}}{10/3} = \frac{-10/3 + 0}{10/3} = -1$$

$$X_{12} = \frac{22 + 4X_{22} - X_{32}}{6} = \frac{22 + 4(-1) - 0}{6} = 3$$

Therefore,

$$\mathbf{x}_2 = \begin{bmatrix} X_{12} & X_{22} & X_{32} \end{bmatrix}^T = \begin{bmatrix} 3 & -1 & 0 \end{bmatrix}^T$$

EXAMPLE 2.4

An $n \times n$ Vandermode matrix $\mathbf{A}$ is defined by

$$A_{ij} = v_i^{n-j}, \quad i = 1, 2, \ldots, n, \quad j = 1, 2, \ldots, n$$

where $\mathbf{v}$ is a vector. Use the function gaussElimin to compute the solution of $\mathbf{Ax} = \mathbf{b}$, where $\mathbf{A}$ is the 6×6 Vandermode matrix generated from the vector

$$\mathbf{v} = \begin{bmatrix} 1.0 & 1.2 & 1.4 & 1.6 & 1.8 & 2.0 \end{bmatrix}^T$$

and

$$\mathbf{b} = \begin{bmatrix} 0 & 1 & 0 & 1 & 0 & 1 \end{bmatrix}^T$$

Also evaluate the accuracy of the solution (Vandermode matrices tend to be ill conditioned).

Solution

```
#!/usr/bin/python
## example2_4
import numpy as np
from gaussElimin import *

def vandermode(v):
    n = len(v)
    a = np.zeros((n,n))
    for j in range(n):
        a[:,j] = v**(n-j-1)
    return a

v = np.array([1.0, 1.2, 1.4, 1.6, 1.8, 2.0])
b = np.array([0.0, 1.0, 0.0, 1.0, 0.0, 1.0])
a = vandermode(v)
aOrig = a.copy()     # Save original matrix
bOrig = b.copy()     # and the constant vector
x = gaussElimin(a,b)
det = np.prod(np.diagonal(a))
print('x =\n',x)
print('\ndet =',det)
print('\nCheck result: [a]{x} - b =\n',np.dot(aOrig,x) - bOrig)
input("\nPress return to exit")
```

The program produced the following results:

```
x =
[   416.66666667   -3125.00000004    9250.00000012  -13500.00000017
   9709.33333345   -2751.00000003]


det = -1.13246207999e-006


Check result: [a]{x} - b =
 [0.00000000e+00   3.63797881e-12   0.00000000e+00   1.45519152e-11
  0.00000000e+00   5.82076609e-11]
```

As the determinant is quite small relative to the elements of **A** (you may want to print **A** to verify this), we expect a detectable roundoff error. Inspection of **x** leads us to suspect that the exact solution is

$$\mathbf{x} = \begin{bmatrix} 1250/3 & -3125 & 9250 & -13500 & 29128/3 & -2751 \end{bmatrix}^{T}$$

in which case the numerical solution would be accurate to about 10 decimal places. Another way to gauge the accuracy of the solution is to compute $\mathbf{Ax} - \mathbf{b}$ (the result should be **0**). The printout indicates that the solution is indeed accurate to at least 10 decimal places.

2.3   LU Decomposition Methods

Introduction

It is possible to show that any square matrix **A** can be expressed as a product of a lower triangular matrix **L** and an upper triangular matrix **U**:

$$\mathbf{A} = \mathbf{LU} \tag{2.11}$$

The process of computing **L** and **U** for a given **A** is known as *LU decomposition* or *LU factorization*. LU decomposition is not unique (the combinations of **L** and **U** for a prescribed **A** are endless), unless certain constraints are placed on **L** or **U**. These constraints distinguish one type of decomposition from another. Three commonly used decompositions are listed in Table 2.2.

Name	Constraints
Doolittle's decomposition	$L_{ii} = 1, \quad i = 1, 2, \ldots, n$
Crout's decomposition	$U_{ii} = 1, \quad i = 1, 2, \ldots, n$
Choleski's decomposition	$\mathbf{L} = \mathbf{U}^{T}$

Table 2.2. Three Commonly Used Decompositions

After decomposing **A**, it is easy to solve the equations $\mathbf{Ax} = \mathbf{b}$, as pointed out in Section 2.1. We first rewrite the equations as $\mathbf{LUx} = \mathbf{b}$. After using the notation

$\mathbf{Ux} = \mathbf{y}$, the equations become

$$\mathbf{Ly} = \mathbf{b},$$

which can be solved for $\mathbf{y}$ by forward substitution. Then

$$\mathbf{Ux} = \mathbf{y}$$

will yield $\mathbf{x}$ by the back substitution process.

The advantage of LU decomposition over the Gauss elimination method is that once $\mathbf{A}$ is decomposed, we can solve $\mathbf{Ax} = \mathbf{b}$ for as many constant vectors $\mathbf{b}$ as we please. The cost of each additional solution is relatively small, because the forward and back substitution operations are much less time consuming than the decomposition process.

Doolittle's Decomposition Method

Decomposition phase. Doolittle's decomposition is closely related to Gauss elimination. To illustrate the relationship, consider a 3×3 matrix $\mathbf{A}$ and assume that there exist triangular matrices

$$\mathbf{L} = \begin{bmatrix} 1 & 0 & 0 \\ L_{21} & 1 & 0 \\ L_{31} & L_{32} & 1 \end{bmatrix} \qquad \mathbf{U} = \begin{bmatrix} U_{11} & U_{12} & U_{13} \\ 0 & U_{22} & U_{23} \\ 0 & 0 & U_{33} \end{bmatrix}$$

such that $\mathbf{A} = \mathbf{LU}$. After completing the multiplication on the right-hand side, we get

$$\mathbf{A} = \begin{bmatrix} U_{11} & U_{12} & U_{13} \\ U_{11}L_{21} & U_{12}L_{21} + U_{22} & U_{13}L_{21} + U_{23} \\ U_{11}L_{31} & U_{12}L_{31} + U_{22}L_{32} & U_{13}L_{31} + U_{23}L_{32} + U_{33} \end{bmatrix} \qquad (2.12)$$

Let us now apply Gauss elimination to Eq. (2.12). The first pass of the elimination procedure consists of choosing the first row as the pivot row and applying the elementary operations

$$\text{row } 2 \leftarrow \text{row } 2 - L_{21} \times \text{row } 1 \text{ (eliminates} A_{21})$$

$$\text{row } 3 \leftarrow \text{row } 3 - L_{31} \times \text{row } 1 \text{ (eliminates} A_{31})$$

The result is

$$\mathbf{A}' = \begin{bmatrix} U_{11} & U_{12} & U_{13} \\ 0 & U_{22} & U_{23} \\ 0 & U_{22}L_{32} & U_{23}L_{32} + U_{33} \end{bmatrix}$$

In the next pass we take the second row as the pivot row, and use the operation

$$\text{row } 3 \leftarrow \text{row } 3 - L_{32} \times \text{row } 2 \text{ (eliminates} A_{32})$$

ending up with

$$\mathbf{A''} = \mathbf{U} = \begin{bmatrix} U_{11} & U_{12} & U_{13} \\ 0 & U_{22} & U_{23} \\ 0 & 0 & U_{33} \end{bmatrix}$$

The foregoing illustration reveals two important features of Doolittle's decomposition:

1. The matrix $\mathbf{U}$ is identical to the upper triangular matrix that results from Gauss elimination.
2. The off-diagonal elements of $\mathbf{L}$ are the pivot equation multipliers used during Gauss elimination; that is, L_{ij} is the multiplier that eliminated A_{ij}.

It is usual practice to store the multipliers in the lower triangular portion of the coefficient matrix, replacing the coefficients as they are eliminated (L_{ij} replacing A_{ij}). The diagonal elements of $\mathbf{L}$ do not have to be stored, because it is understood that each of them is unity. The final form of the coefficient matrix would thus be the following mixture of $\mathbf{L}$ and $\mathbf{U}$:

$$[\mathbf{L} \backslash \mathbf{U}] = \begin{bmatrix} U_{11} & U_{12} & U_{13} \\ L_{21} & U_{22} & U_{23} \\ L_{31} & L_{32} & U_{33} \end{bmatrix} \tag{2.13}$$

The algorithm for Doolittle's decomposition is thus identical to the Gauss elimination procedure in gaussElimin, except that each multiplier λ is now stored in the lower triangular portion of $\mathbf{A}$:

```
for k in range(0,n-1):
    for i in range(k+1,n):
        if a[i,k] != 0.0:
            lam = a[i,k]/a[k,k]
            a[i,k+1:n] = a[i,k+1:n] - lam*a[k,k+1:n]
            a[i,k] = lam
```

Solution phase. Consider now the procedure for the solution of $\mathbf{Ly} = \mathbf{b}$ by forward substitution. The scalar form of the equations is (recall that $L_{ii} = 1$)

$$y_1 = b_1$$

$$L_{21} y_1 + y_2 = b_2$$

$$\vdots$$

$$L_{k1} y_1 + L_{k2} y_2 + \cdots + L_{k,k-1} y_{k-1} + y_k = b_k$$

$$\vdots$$

Solving the kth equation for y_k yields

$$y_k = b_k - \sum_{j=1}^{k-1} L_{kj} y_j, \quad k = 2, 3, \ldots, n \tag{2.14}$$

Therefore, the forward substitution algorithm is

```
y[0] = b[0]
for k in range(1,n):
    y[k] = b[k] - dot(a[k,0:k],y[0:k])
```

The back substitution phase for solving $\mathbf{Ux} = \mathbf{y}$ is identical to what was used in the Gauss elimination method.

■ LUdecomp

This module contains both the decomposition and solution phases. The decomposition phase returns the matrix [$\mathbf{L}\backslash\mathbf{U}$] shown in Eq. (2.13). In the solution phase, the contents of $\mathbf{b}$ are replaced by $\mathbf{y}$ during forward substitution Similarly, the back substitution overwrites $\mathbf{y}$ with the solution $\mathbf{x}$.

```
## module LUdecomp
''' a = LUdecomp(a)
    LUdecomposition: [L][U] = [a]

    x = LUsolve(a,b)
    Solution phase: solves [L][U]{x} = {b}
'''
import numpy as np

def LUdecomp(a):
    n = len(a)
    for k in range(0,n-1):
        for i in range(k+1,n):
            if a[i,k] != 0.0:
                lam = a [i,k]/a[k,k]
                a[i,k+1:n] = a[i,k+1:n] - lam*a[k,k+1:n]
                a[i,k] = lam
    return a

def LUsolve(a,b):
    n = len(a)
    for k in range(1,n):
        b[k] = b[k] - np.dot(a[k,0:k],b[0:k])
    b[n-1] = b[n-1]/a[n-1,n-1]
    for k in range(n-2,-1,-1):
        b[k] = (b[k] - np.dot(a[k,k+1:n],b[k+1:n]))/a[k,k]
    return b
```

Choleski's Decomposition Method

Choleski's decomposition $\mathbf{A} = \mathbf{LL}^T$ has two limitations:

1. Since $\mathbf{LL}^T$ is always a symmetric matrix, Choleski's decomposition requires $\mathbf{A}$ to be *symmetric*.
2. The decomposition process involves taking square roots of certain combinations of the elements of $\mathbf{A}$. It can be shown that to avoid square roots of negative numbers $\mathbf{A}$ must be *positive definite*.

Choleski's decomposition contains approximately $n^3/6$ long operations plus n square root computations. This is about half the number of operations required in LU decomposition. The relative efficiency of Choleski's decomposition is due to its exploitation of symmetry.

Let us start by looking at Choleski's decomposition

$$\mathbf{A} = \mathbf{LL}^T \tag{2.15}$$

of a 3×3 matrix:

$$\begin{bmatrix} A_{11} & A_{12} & A_{13} \\ A_{21} & A_{22} & A_{23} \\ A_{31} & A_{32} & A_{33} \end{bmatrix} = \begin{bmatrix} L_{11} & 0 & 0 \\ L_{21} & L_{22} & 0 \\ L_{31} & L_{32} & L_{33} \end{bmatrix} \begin{bmatrix} L_{11} & L_{21} & L_{31} \\ 0 & L_{22} & L_{32} \\ 0 & 0 & L_{33} \end{bmatrix}$$

After completing the matrix multiplication on the right-hand side, we get

$$\begin{bmatrix} A_{11} & A_{12} & A_{13} \\ A_{21} & A_{22} & A_{23} \\ A_{31} & A_{32} & A_{33} \end{bmatrix} = \begin{bmatrix} L_{11}^2 & L_{11}L_{21} & L_{11}L_{31} \\ L_{11}L_{21} & L_{21}^2 + L_{22}^2 & L_{21}L_{31} + L_{22}L_{32} \\ L_{11}L_{31} & L_{21}L_{31} + L_{22}L_{32} & L_{31}^2 + L_{32}^2 + L_{33}^2 \end{bmatrix} \tag{2.16}$$

Note that the right-hand-side matrix is symmetric, as pointed out earlier. Equating the matrices $\mathbf{A}$ and $\mathbf{LL}^T$ element by element, we obtain six equations (because of symmetry only lower or upper triangular elements have to be considered) in the six unknown components of $\mathbf{L}$. By solving these equations in a certain order, it is possible to have only one unknown in each equation.

Consider the lower triangular portion of each matrix in Eq. (2.16) (the upper triangular portion would do as well). By equating the elements in the first column, starting with the first row and proceeding downward, we can compute L_{11}, L_{21}, and L_{31} in that order:

$$A_{11} = L_{11}^2 \qquad L_{11} = \sqrt{A_{11}}$$
$$A_{21} = L_{11}L_{21} \qquad L_{21} = A_{21}/L_{11}$$
$$A_{31} = L_{11}L_{31} \qquad L_{31} = A_{31}/L_{11}$$

The second column, starting with second row, yields L_{22} and L_{32}:

$$A_{22} = L_{21}^2 + L_{22}^2 \qquad L_{22} = \sqrt{A_{22} - L_{21}^2}$$
$$A_{32} = L_{21}L_{31} + L_{22}L_{32} \qquad L_{32} = (A_{32} - L_{21}L_{31})/L_{22}$$

Finally the third column, third row gives us L_{33}:

$$A_{33} = L_{31}^2 + L_{32}^2 + L_{33}^2 \qquad L_{33} = \sqrt{A_{33} - L_{31}^2 - L_{32}^2}$$

We can now extrapolate the results for an $n \times n$ matrix. We observe that a typical element in the lower triangular portion of $\mathbf{LL}^T$ is of the form

$$(\mathbf{LL}^T)_{ij} = L_{i1}L_{j1} + L_{i2}L_{j2} + \cdots + L_{ij}L_{jj} = \sum_{k=1}^{j} L_{ik}L_{jk}, \quad i \geq j$$

Equating this term to the corresponding element of $\mathbf{A}$ yields

$$A_{ij} = \sum_{k=1}^{j} L_{ik}L_{jk}, \quad i = j, j+1, \ldots, n, \quad j = 1, 2, \ldots, n \qquad (2.17)$$

The range of indices shown limits the elements to the lower triangular part. For the first column ($j = 1$), we obtain from Eq. (2.17)

$$L_{11} = \sqrt{A_{11}} \qquad L_{i1} = A_{i1}/L_{11}, \quad i = 2, 3, \ldots, n \qquad (2.18)$$

Proceeding to other columns, we observe that the unknown in Eq. (2.17) is L_{ij} (the other elements of $\mathbf{L}$ appearing in the equation have already been computed). Taking the term containing L_{ij} outside the summation in Eq. (2.17), we obtain

$$A_{ij} = \sum_{k=1}^{j-1} L_{ik}L_{jk} + L_{ij}L_{jj}$$

If $i = j$ (a diagonal term), the solution is

$$L_{jj} = \sqrt{A_{jj} - \sum_{k=1}^{j-1} L_{jk}^2}, \quad j = 2, 3, \ldots, n \qquad (2.19)$$

For a nondiagonal term we get

$$L_{ij} = \left(A_{ij} - \sum_{k=1}^{j-1} L_{ik}L_{jk} \right) / L_{jj}, \quad j = 2, 3, \ldots, n-1, \quad i = j+1, j+2, \ldots, n. \qquad (2.20)$$

■ choleski

Before presenting the algorithm for Choleski's decomposition, we make a useful observation: A_{ij} appears only in the formula for L_{ij}. Therefore, once L_{ij} has been computed, A_{ij} is no longer needed. This makes it possible to write the elements of $\mathbf{L}$ over the lower triangular portion of $\mathbf{A}$ as they are computed. The elements above the leading diagonal of $\mathbf{A}$ will remain untouched. The function listed next implements Choleski's decomposition. If a negative diagonal term is encountered during decomposition, an error message is printed and the program is terminated.

After the coefficient matrix **A** has been decomposed, the solution of **Ax** = **b** can be obtained by the usual forward and back substitution operations. The function choleskiSol (given here without derivation) carries out the solution phase.

```
## module choleski
''' L = choleski(a)
    Choleski decomposition: [L][L]transpose = [a]

    x = choleskiSol(L,b)
    Solution phase of Choleski's decomposition method
'''
import numpy as np
import math
import error

def choleski(a):
    n = len(a)
    for k in range(n):
        try:
            a[k,k] = math.sqrt(a[k,k]   \
                  - np.dot(a[k,0:k],a[k,0:k]))
        except ValueError:
            error.err('Matrix is not positive definite')
        for i in range(k+1,n):
            a[i,k] = (a[i,k] - np.dot(a[i,0:k],a[k,0:k]))/a[k,k]
    for k in range(1,n): a[0:k,k] = 0.0
    return a

def choleskiSol(L,b):
    n = len(b)
  # Solution of [L]{y} = {b}
    for k in range(n):
        b[k] = (b[k] - np.dot(L[k,0:k],b[0:k]))/L[k,k]
  # Solution of [L_transpose]{x} = {y}
    for k in range(n-1,-1,-1):
        b[k] = (b[k] - np.dot(L[k+1:n,k],b[k+1:n]))/L[k,k]
    return b
```

Other Methods

Crout's decomposition. Recall that the various decompositions **A** = **LU** are characterized by the constraints placed on the elements of **L** or **U**. In Doolittle's decomposition the diagonal elements of **L** were set to 1. An equally viable method is Crout's decomposition, where the 1's lie on the diagonal of **U**. There is little difference in the performance of the two methods.

Gauss-Jordan Elimination. The Gauss-Jordan method is essentially Gauss elimination taken to its limit. In the Gauss elimination method only the equations that lie below the pivot equation are transformed. In the Gauss-Jordan method the elimination is also carried out on equations above the pivot equation, resulting in a diagonal coefficient matrix. The main disadvantage of Gauss-Jordan elimination is that it involves about $n^3/2$ long operations, which is 1.5 times the number required in Gauss elimination.

EXAMPLE 2.5
Use Doolittle's decomposition method to solve the equations **Ax** = **b**, where

$$\mathbf{A} = \begin{bmatrix} 1 & 4 & 1 \\ 1 & 6 & -1 \\ 2 & -1 & 2 \end{bmatrix} \qquad \mathbf{b} = \begin{bmatrix} 7 \\ 13 \\ 5 \end{bmatrix}$$

Solution. We first decompose **A** by Gauss elimination. The first pass consists of the elementary operations

$$\text{row 2} \leftarrow \text{row 2} - 1 \times \text{row 1} \; (\text{eliminates } A_{21})$$

$$\text{row 3} \leftarrow \text{row 3} - 2 \times \text{row 1} \; (\text{eliminates } A_{31})$$

Storing the multipliers $L_{21} = 1$ and $L_{31} = 2$ in place of the eliminated terms, we obtain

$$\mathbf{A}' = \begin{bmatrix} 1 & 4 & 1 \\ 1 & 2 & -2 \\ 2 & -9 & 0 \end{bmatrix}$$

The second pass of Gauss elimination uses the operation

$$\text{row 3} \leftarrow \text{row 3} - (-4.5) \times \text{row 2} \; (\text{eliminates } A_{32})$$

Storing the multiplier $L_{32} = -4.5$ in place of A_{32}, we get

$$\mathbf{A}'' = [\mathbf{L} \backslash \mathbf{U}] = \begin{bmatrix} 1 & 4 & 1 \\ 1 & 2 & -2 \\ 2 & -4.5 & -9 \end{bmatrix}$$

The decomposition is now complete, with

$$\mathbf{L} = \begin{bmatrix} 1 & 0 & 0 \\ 1 & 1 & 0 \\ 2 & -4.5 & 1 \end{bmatrix} \qquad \mathbf{U} = \begin{bmatrix} 1 & 4 & 1 \\ 0 & 2 & -2 \\ 0 & 0 & -9 \end{bmatrix}$$

Solution of $\mathbf{Ly} = \mathbf{b}$ by forward substitution comes next. The augmented coefficient form of the equations is

$$\left[\mathbf{L} \mid \mathbf{b}\right] = \begin{bmatrix} 1 & 0 & 0 & 7 \\ 1 & 1 & 0 & 13 \\ 2 & -4.5 & 1 & 5 \end{bmatrix}$$

The solution is

$$y_1 = 7$$

$$y_2 = 13 - y_1 = 13 - 7 = 6$$

$$y_3 = 5 - 2y_1 + 4.5y_2 = 5 - 2(7) + 4.5(6) = 18$$

Finally, the equations $\mathbf{Ux} = \mathbf{y}$, or

$$\left[\mathbf{U} \mid \mathbf{y}\right] = \begin{bmatrix} 1 & 4 & 1 & 7 \\ 0 & 2 & -2 & 6 \\ 0 & 0 & -9 & 18 \end{bmatrix}$$

are solved by back substitution. This yields

$$x_3 = \frac{18}{-9} = -2$$

$$x_2 = \frac{6 + 2x_3}{2} = \frac{6 + 2(-2)}{2} = 1$$

$$x_1 = 7 - 4x_2 - x_3 = 7 - 4(1) - (-2) = 5$$

EXAMPLE 2.6
Compute Choleski's decomposition of the matrix

$$\mathbf{A} = \begin{bmatrix} 4 & -2 & 2 \\ -2 & 2 & -4 \\ 2 & -4 & 11 \end{bmatrix}$$

Solution. First we note that $\mathbf{A}$ is symmetric. Therefore, Choleski's decomposition is applicable, provided that the matrix is also positive definite. An a priori test for positive definiteness is not needed, because the decomposition algorithm contains its own test: If a square root of a negative number is encountered, the matrix is not positive definite and the decomposition fails.

Substituting the given matrix for $\mathbf{A}$ in Eq. (2.16) we obtain

$$\begin{bmatrix} 4 & -2 & 2 \\ -2 & 2 & -4 \\ 2 & -4 & 11 \end{bmatrix} = \begin{bmatrix} L_{11}^2 & L_{11}L_{21} & L_{11}L_{31} \\ L_{11}L_{21} & L_{21}^2 + L_{22}^2 & L_{21}L_{31} + L_{22}L_{32} \\ L_{11}L_{31} & L_{21}L_{31} + L_{22}L_{32} & L_{31}^2 + L_{32}^2 + L_{33}^2 \end{bmatrix}$$

Equating the elements in the lower (or upper) triangular portions yields

$$L_{11} = \sqrt{4} = 2$$

$$L_{21} = -2/L_{11} = -2/2 = -1$$

$$L_{31} = 2/L_{11} = 2/2 = 1$$

$$L_{22} = \sqrt{2 - L_{21}^2} = \sqrt{2 - 1^2} = 1$$

$$L_{32} = \frac{-4 - L_{21}L_{31}}{L_{22}} = \frac{-4 - (-1)(1)}{1} = -3$$

$$L_{33} = \sqrt{11 - L_{31}^2 - L_{32}^2} = \sqrt{11 - (1)^2 - (-3)^2} = 1$$

Therefore,

$$\mathbf{L} = \begin{bmatrix} 2 & 0 & 0 \\ -1 & 1 & 0 \\ 1 & -3 & 1 \end{bmatrix}$$

The result can easily be verified by performing the multiplication $\mathbf{LL}^T$.

EXAMPLE 2.7
Write a program that solves $\mathbf{AX} = \mathbf{B}$ with Doolittle's decomposition method and computes $|\mathbf{A}|$. Use the functions LUdecomp and LUsolve. Test the program with

$$\mathbf{A} = \begin{bmatrix} 3 & -1 & 4 \\ -2 & 0 & 5 \\ 7 & 2 & -2 \end{bmatrix} \qquad \mathbf{B} = \begin{bmatrix} 6 & -4 \\ 3 & 2 \\ 7 & -5 \end{bmatrix}$$

Solution

```
#!/usr/bin/python
## example2_7
import numpy as np
from LUdecomp import *

a = np.array([[ 3.0, -1.0,  4.0], \
              [-2.0,  0.0,  5.0], \
              [ 7.0,  2.0, -2.0]])
b = np.array([[ 6.0,  3.0,  7.0], \
              [-4.0,  2.0, -5.0]])
a = LUdecomp(a)                    # Decompose [a]
det = np.prod(np.diagonal(a))
print("\nDeterminant =",det)
for i in range(len(b)):           # Back-substitute one
    x = LUsolve(a,b[i])           # constant vector at a time
    print("x",i+1,"=",x)
input("\nPress return to exit")
```

The output of the program is

```
Determinant = -77.0
x 1 = [ 1.   1.   1.]
x 2 = [ -1.00000000e+00   1.00000000e+00   2.30695693e-17]
```

EXAMPLE 2.8

Solve the equations $\mathbf{Ax} = \mathbf{b}$ by Choleski's decomposition, where

$$\mathbf{A} = \begin{bmatrix} 1.44 & -0.36 & 5.52 & 0.00 \\ -0.36 & 10.33 & -7.78 & 0.00 \\ 5.52 & -7.78 & 28.40 & 9.00 \\ 0.00 & 0.00 & 9.00 & 61.00 \end{bmatrix} \qquad \mathbf{b} = \begin{bmatrix} 0.04 \\ -2.15 \\ 0 \\ 0.88 \end{bmatrix}$$

Also check the solution.

Solution

```
#!/usr/bin/python
## example2_8
import numpy as np
from choleski import *

a = np.array([[ 1.44, -0.36,  5.52,  0.0], \
              [-0.36, 10.33, -7.78,  0.0], \
              [ 5.52, -7.78, 28.40,  9.0], \
              [ 0.0,   0.0,   9.0,  61.0]])
b = np.array([0.04, -2.15, 0.0, 0.88])
aOrig = a.copy()
L = choleski(a)
x = choleskiSol(L,b)
print("x =",x)
print('\nCheck: A*x =\n',np.dot(aOrig,x))
input("\nPress return to exit")
```

The output is

```
x = [ 3.09212567 -0.73871706 -0.8475723   0.13947788]

Check: A*x =
[4.00000000e-02 -2.15000000e+00 -3.55271368e-15  8.80000000e-01]
```

PROBLEM SET 2.1

1. By evaluating the determinant, classify the following matrices as singular, ill conditioned, or well conditioned:

(a) $\mathbf{A} = \begin{bmatrix} 1 & 2 & 3 \\ 2 & 3 & 4 \\ 3 & 4 & 5 \end{bmatrix}$

(b) $\mathbf{A} = \begin{bmatrix} 2.11 & -0.80 & 1.72 \\ -1.84 & 3.03 & 1.29 \\ -1.57 & 5.25 & 4.30 \end{bmatrix}$

(c) $\mathbf{A} = \begin{bmatrix} 2 & -1 & 0 \\ -1 & 2 & -1 \\ 0 & -1 & 2 \end{bmatrix}$

(d) $\mathbf{A} = \begin{bmatrix} 4 & 3 & -1 \\ 7 & -2 & 3 \\ 5 & -18 & 13 \end{bmatrix}$

2. Given the LU decomposition $\mathbf{A} = \mathbf{LU}$, determine $\mathbf{A}$ and $|\mathbf{A}|$:

(a) $\mathbf{L} = \begin{bmatrix} 1 & 0 & 0 \\ 1 & 1 & 0 \\ 1 & 5/3 & 1 \end{bmatrix}$ $\mathbf{U} = \begin{bmatrix} 1 & 2 & 4 \\ 0 & 3 & 21 \\ 0 & 0 & 0 \end{bmatrix}$

(b) $\mathbf{L} = \begin{bmatrix} 2 & 0 & 0 \\ -1 & 1 & 0 \\ 1 & -3 & 1 \end{bmatrix}$ $\mathbf{U} = \begin{bmatrix} 2 & -1 & 1 \\ 0 & 1 & -3 \\ 0 & 0 & 1 \end{bmatrix}$

3. Use the results of LU decomposition

$$\mathbf{A} = \mathbf{LU} = \begin{bmatrix} 1 & 0 & 0 \\ 3/2 & 1 & 0 \\ 1/2 & 11/13 & 1 \end{bmatrix} \begin{bmatrix} 2 & -3 & -1 \\ 0 & 13/2 & -7/2 \\ 0 & 0 & 32/13 \end{bmatrix}$$

to solve $\mathbf{Ax} = \mathbf{b}$, where $\mathbf{b}^T = \begin{bmatrix} 1 & -1 & 2 \end{bmatrix}$.

4. Use Gauss elimination to solve the equations $\mathbf{Ax} = \mathbf{b}$, where

$$\mathbf{A} = \begin{bmatrix} 2 & -3 & -1 \\ 3 & 2 & -5 \\ 2 & 4 & -1 \end{bmatrix} \qquad \mathbf{b} = \begin{bmatrix} 3 \\ -9 \\ -5 \end{bmatrix}$$

5. Solve the equations $\mathbf{AX} = \mathbf{B}$ by Gauss elimination, where

$$\mathbf{A} = \begin{bmatrix} 2 & 0 & -1 & 0 \\ 0 & 1 & 2 & 0 \\ -1 & 2 & 0 & 1 \\ 0 & 0 & 1 & -2 \end{bmatrix} \qquad \mathbf{B} = \begin{bmatrix} 1 & 0 \\ 0 & 0 \\ 0 & 1 \\ 0 & 0 \end{bmatrix}$$

6. Solve the equations $\mathbf{Ax} = \mathbf{b}$ by Gauss elimination, where

$$\mathbf{A} = \begin{bmatrix} 0 & 0 & 2 & 1 & 2 \\ 0 & 1 & 0 & 2 & -1 \\ 1 & 2 & 0 & -2 & 0 \\ 0 & 0 & 0 & -1 & 1 \\ 0 & 1 & -1 & 1 & -1 \end{bmatrix} \qquad \mathbf{b} = \begin{bmatrix} 1 \\ 1 \\ -4 \\ -2 \\ -1 \end{bmatrix}$$

Hint: Reorder the equations before solving.

7. Find **L** and **U** so that

$$\mathbf{A} = \mathbf{LU} = \begin{bmatrix} 4 & -1 & 0 \\ -1 & 4 & -1 \\ 0 & -1 & 4 \end{bmatrix}$$

using (a) Doolittle's decomposition; (b) Choleski's decomposition.

8. Use Doolittle's decomposition method to solve **Ax** = **b**, where

$$\mathbf{A} = \begin{bmatrix} -3 & 6 & -4 \\ 9 & -8 & 24 \\ -12 & 24 & -26 \end{bmatrix} \qquad \mathbf{b} = \begin{bmatrix} -3 \\ 65 \\ -42 \end{bmatrix}$$

9. Solve the equations **AX** = **b** by Doolittle's decomposition method, where

$$\mathbf{A} = \begin{bmatrix} 2.34 & -4.10 & 1.78 \\ -1.98 & 3.47 & -2.22 \\ 2.36 & -15.17 & 6.18 \end{bmatrix} \qquad \mathbf{b} = \begin{bmatrix} 0.02 \\ -0.73 \\ -6.63 \end{bmatrix}$$

10. Solve the equations **AX** = **B** by Doolittle's decomposition method, where

$$\mathbf{A} = \begin{bmatrix} 4 & -3 & 6 \\ 8 & -3 & 10 \\ -4 & 12 & -10 \end{bmatrix} \qquad \mathbf{B} = \begin{bmatrix} 1 & 0 \\ 0 & 1 \\ 0 & 0 \end{bmatrix}$$

11. Solve the equations **Ax** = **b** by Choleski's decomposition method, where

$$\mathbf{A} = \begin{bmatrix} 1 & 1 & 1 \\ 1 & 2 & 2 \\ 1 & 2 & 3 \end{bmatrix} \qquad \mathbf{b} = \begin{bmatrix} 1 \\ 3/2 \\ 3 \end{bmatrix}$$

12. Solve the equations

$$\begin{bmatrix} 4 & -2 & -3 \\ 12 & 4 & -10 \\ -16 & 28 & 18 \end{bmatrix} \begin{bmatrix} x_1 \\ x_2 \\ x_3 \end{bmatrix} = \begin{bmatrix} 1.1 \\ 0 \\ -2.3 \end{bmatrix}$$

by Doolittle's decomposition method.

13. Determine **L** that results from Choleski's decomposition of the diagonal matrix

$$\mathbf{A} = \begin{bmatrix} \alpha_1 & 0 & 0 & \cdots \\ 0 & \alpha_2 & 0 & \cdots \\ 0 & 0 & \alpha_3 & \cdots \\ \vdots & \vdots & \vdots & \ddots \end{bmatrix}$$

14. ■ Modify the function gaussElimin so that it will work with m constant vectors. Test the program by solving **AX** = **B**, where

$$\mathbf{A} = \begin{bmatrix} 2 & -1 & 0 \\ -1 & 2 & -1 \\ 0 & -1 & 2 \end{bmatrix} \qquad \mathbf{B} = \begin{bmatrix} 1 & 0 & 0 \\ 0 & 1 & 0 \\ 0 & 0 & 1 \end{bmatrix}$$

15. ■ A well-known example of an ill-conditioned matrix is the *Hilbert matrix*:

$$\mathbf{A} = \begin{bmatrix} 1 & 1/2 & 1/3 & \cdots \\ 1/2 & 1/3 & 1/4 & \cdots \\ 1/3 & 1/4 & 1/5 & \cdots \\ \vdots & \vdots & \vdots & \ddots \end{bmatrix}$$

Write a program that specializes in solving the equations $\mathbf{Ax} = \mathbf{b}$ by Doolittle's decomposition method, where $\mathbf{A}$ is the Hilbert matrix of arbitrary size $n \times n$, and

$$b_i = \sum_{j=1}^{n} A_{ij}$$

The program should have no input apart from n. By running the program, determine the largest n for which the solution is within six significant figures of the exact solution

$$\mathbf{x} = \begin{bmatrix} 1 & 1 & 1 & \cdots \end{bmatrix}^T$$

16. Derive the forward and back substitution algorithms for the solution phase of Choleski's method. Compare them with the function `choleskiSol`.

17. ■ Determine the coefficients of the polynomial $y = a_0 + a_1 x + a_2 x^2 + a_3 x^3$ that passes through the points $(0, 10)$, $(1, 35)$, $(3, 31)$, and $(4, 2)$.

18. ■ Determine the fourth-degree polynomial $y(x)$ that passes through the points $(0, -1)$, $(1, 1)$, $(3, 3)$, $(5, 2)$, and $(6, -2)$.

19. ■ Find the fourth-degree polynomial $y(x)$ that passes through the points $(0, 1)$, $(0.75, -0.25)$, and $(1, 1)$ and has zero curvature at $(0, 1)$ and $(1, 1)$.

20. ■ Solve the equations $\mathbf{Ax} = \mathbf{b}$, where

$$\mathbf{A} = \begin{bmatrix} 3.50 & 2.77 & -0.76 & 1.80 \\ -1.80 & 2.68 & 3.44 & -0.09 \\ 0.27 & 5.07 & 6.90 & 1.61 \\ 1.71 & 5.45 & 2.68 & 1.71 \end{bmatrix} \quad \mathbf{b} = \begin{bmatrix} 7.31 \\ 4.23 \\ 13.85 \\ 11.55 \end{bmatrix}$$

By computing $|\mathbf{A}|$ and $\mathbf{Ax}$ comment on the accuracy of the solution.

21. Compute the condition number of the matrix

$$\mathbf{A} = \begin{bmatrix} 1 & -1 & -1 \\ 0 & 1 & -2 \\ 0 & 0 & 1 \end{bmatrix}$$

based on (a) the euclidean norm and (b) the infinity norm. You may use the function `inv(A)` in `numpy.linalg` to determine the inverse of $\mathbf{A}$.

22. ■ Write a function that returns the condition number of a matrix based on the euclidean norm. Test the function by computing the condition number of the ill-conditioned matrix:

$$\mathbf{A} = \begin{bmatrix} 1 & 4 & 9 & 16 \\ 4 & 9 & 16 & 25 \\ 9 & 16 & 25 & 36 \\ 16 & 25 & 36 & 49 \end{bmatrix}$$

Use the function inv(A) in numpy.linalg to determine the inverse of **A**.

23. ■ Test the function gaussElimin by solving $\mathbf{Ax} = \mathbf{b}$, where **A** is a $n \times n$ random matrix and $b_i = \sum_{j=1}^{n} A_{ij}$ (sum of the elements in the ith row of **A**). A random matrix can be generated with the rand function in the numpy.random module:

```
from numpy.random import ran

a = rand(n,n)
```

The solution should be $x = \begin{bmatrix} 1 & 1 & \cdots & 1 \end{bmatrix}^T$. Run the program with $n = 200$ or bigger.

24. ■ The function gaussElimin also works with complex numbers. Use it to solve $\mathbf{Ax} = \mathbf{b}$, where

$$\mathbf{A} = \begin{bmatrix} 5+i & 5+2i & -5+3i & 6-3i \\ 5+2i & 7-2i & 8-i & -1+3i \\ -5+3i & 8-i & -3-3i & 2+2i \\ 6-3i & -1+3i & 2+2i & 8+14i \end{bmatrix}$$

$$\mathbf{b} = \begin{bmatrix} 15-35i & 2+10i & -2-34i & 8+14i \end{bmatrix}^T$$

Note that Python uses j to denote $\sqrt{-1}$.

25. ■

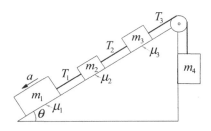

The four blocks of different masses m_i are connected by ropes of negligible mass. Three of the blocks lie on a inclined plane, the coefficients of friction between the blocks and the plane being μ_i. The equations of motion for the blocks can be shown to be

$$T_1 + m_1 a = m_1 g(\sin\theta - \mu_1 \cos\theta)$$

$$-T_1 + T_2 + m_2 a = m_2 g(\sin\theta - \mu_2 \cos\theta)$$

$$-T_2 + T_3 + m_3 a = m_3 g(\sin\theta - \mu_3 \cos\theta)$$

$$-T_3 + m_4 a = -m_4 g$$

where T_i denotes the tensile forces in the ropes and a is the acceleration of the system. Determine a and T_i if $\theta = 45°$, $g = 9.82$ m/s^2 and

$$\mathbf{m} = \begin{bmatrix} 10 & 4 & 5 & 6 \end{bmatrix}^T \text{ kg}$$

$$\boldsymbol{\mu} = \begin{bmatrix} 0.25 & 0.3 & 0.2 \end{bmatrix}^T$$

2.4 Symmetric and Banded Coefficient Matrices

Introduction

Engineering problems often lead to coefficient matrices that are *sparsely populated,* meaning that most elements of the matrix are zero. If all the nonzero terms are clustered about the leading diagonal, then the matrix is said to be *banded.* An example of a banded matrix is

$$\mathbf{A} = \begin{bmatrix} X & X & 0 & 0 & 0 \\ X & X & X & 0 & 0 \\ 0 & X & X & X & 0 \\ 0 & 0 & X & X & X \\ 0 & 0 & 0 & X & X \end{bmatrix}$$

where X's denote the nonzero elements that form the populated band (some of these elements may be zero). All the elements lying outside the band are zero. This banded matrix has a bandwidth of three, because there are at most three nonzero elements in each row (or column). Such a matrix is called *tridiagonal.*

If a banded matrix is decomposed in the form $\mathbf{A} = \mathbf{LU}$, both $\mathbf{L}$ and $\mathbf{U}$ retain the banded structure of $\mathbf{A}$. For example, if we decomposed the matrix shown above, we would get

$$\mathbf{L} = \begin{bmatrix} X & 0 & 0 & 0 & 0 \\ X & X & 0 & 0 & 0 \\ 0 & X & X & 0 & 0 \\ 0 & 0 & X & X & 0 \\ 0 & 0 & 0 & X & X \end{bmatrix} \quad \mathbf{U} = \begin{bmatrix} X & X & 0 & 0 & 0 \\ 0 & X & X & 0 & 0 \\ 0 & 0 & X & X & 0 \\ 0 & 0 & 0 & X & X \\ 0 & 0 & 0 & 0 & X \end{bmatrix}$$

The banded structure of a coefficient matrix can be exploited to save storage and computation time. If the coefficient matrix is also symmetric, further economies are possible. In this section I show how the methods of solution discussed previously can be adapted for banded and symmetric coefficient matrices.

Tridiagonal Coefficient Matrix

Consider the solution of $\mathbf{Ax} = \mathbf{b}$ by Doolittle's decomposition, where $\mathbf{A}$ is the $n \times n$ tridiagonal matrix

$$\mathbf{A} = \begin{bmatrix} d_1 & e_1 & 0 & 0 & \cdots & 0 \\ c_1 & d_2 & e_2 & 0 & \cdots & 0 \\ 0 & c_2 & d_3 & e_3 & \cdots & 0 \\ 0 & 0 & c_3 & d_4 & \cdots & 0 \\ \vdots & \vdots & \vdots & \vdots & \ddots & \vdots \\ 0 & 0 & \cdots & 0 & c_{n-1} & d_n \end{bmatrix}$$

As the notation implies, we are storing the non-zero elements of $\mathbf{A}$ in the vectors

$$\mathbf{c} = \begin{bmatrix} c_1 \\ c_2 \\ \vdots \\ c_{n-1} \end{bmatrix} \qquad \mathbf{d} = \begin{bmatrix} d_1 \\ d_2 \\ \vdots \\ d_{n-1} \\ d_n \end{bmatrix} \qquad \mathbf{e} = \begin{bmatrix} e_1 \\ e_2 \\ \vdots \\ e_{n-1} \end{bmatrix}$$

The resulting saving of storage can be significant. For example, a 100×100 tridiagonal matrix, containing 10,000 elements, can be stored in only $99 + 100 + 99 = 298$ locations, which represents a compression ratio of about 33:1.

Let us now apply LU decomposition to the coefficient matrix. We reduce row k by getting rid of c_{k-1} with the elementary operation

$$\text{row } k \leftarrow \text{row } k - (c_{k-1}/d_{k-1}) \times \text{row } (k-1), \quad k = 2, 3, \ldots, n$$

The corresponding change in d_k is

$$d_k \leftarrow d_k - (c_{k-1}/d_{k-1})e_{k-1} \tag{2.21}$$

whereas e_k is not affected. To finish up with Doolittle's decomposition of the form $[\mathbf{L} \backslash \mathbf{U}]$, we store the multiplier $\lambda = c_{k-1}/d_{k-1}$ in the location previously occupied by c_{k-1}:

$$c_{k-1} \leftarrow c_{k-1}/d_{k-1} \tag{2.22}$$

Thus the decomposition algorithm is

```
for k in range(1,n):
    lam = c[k-1]/d[k-1]
    d[k] = d[k] - lam*e[k-1]
    c[k-1] = lam
```

Next we look at the solution phase (i.e., solution of the $\mathbf{Ly} = \mathbf{b}$), followed by $\mathbf{Ux} = \mathbf{y}$. The equations $\mathbf{Ly} = \mathbf{b}$ can be portrayed by the augmented coefficient matrix

$$\left[\mathbf{L}\,\middle|\,\mathbf{b}\right] = \begin{bmatrix} 1 & 0 & 0 & 0 & \cdots & 0 & b_1 \\ c_1 & 1 & 0 & 0 & \cdots & 0 & b_2 \\ 0 & c_2 & 1 & 0 & \cdots & 0 & b_3 \\ 0 & 0 & c_3 & 1 & \cdots & 0 & b_4 \\ \vdots & \vdots & \vdots & \vdots & \cdots & \vdots & \vdots \\ 0 & 0 & \cdots & 0 & c_{n-1} & 1 & b_n \end{bmatrix}$$

Note that the original contents of $\mathbf{c}$ were destroyed and replaced by the multipliers during the decomposition. The solution algorithm for $\mathbf{y}$ by forward substitution is

```
y[0] = b[0]
for k in range(1,n):
    y[k] = b[k] - c[k-1]*y[k-1]
```

The augmented coefficient matrix representing $\mathbf{Ux} = \mathbf{y}$ is

$$\left[\mathbf{U}\,\middle|\,\mathbf{y}\right] = \begin{bmatrix} d_1 & e_1 & 0 & \cdots & 0 & 0 & y_1 \\ 0 & d_2 & e_2 & \cdots & 0 & 0 & y_2 \\ 0 & 0 & d_3 & \cdots & 0 & 0 & y_3 \\ \vdots & \vdots & \vdots & & \vdots & \vdots & \vdots \\ 0 & 0 & 0 & \cdots & d_{n-1} & e_{n-1} & y_{n-1} \\ 0 & 0 & 0 & \cdots & 0 & d_n & y_n \end{bmatrix}$$

Note again that the contents of $\mathbf{d}$ were altered from the original values during the decomposition phase (but $\mathbf{e}$ was unchanged). The solution for $\mathbf{x}$ is obtained by back substitution using the algorithm

```
x[n-1] = y[n-1]/d[n-1]
for k in range(n-2,-1,-1):
    x[k] = (y[k] - e[k]*x[k+1])/d[k]
end do
```

■ LUdecomp3

This module contains the functions LUdecomp3 and LUsolve3 for the decomposition and solution phases of a tridiagonal matrix. In LUsolve3, the vector $\mathbf{y}$ writes over the constant vector $\mathbf{b}$ during forward substitution. Similarly, the solution vector $\mathbf{x}$ overwrites $\mathbf{y}$ in the back substitution process. In other words, $\mathbf{b}$ contains the solution upon exit from LUsolve3.

```
## module LUdecomp3
''' c,d,e = LUdecomp3(c,d,e).
    LU decomposition of tridiagonal matrix [c\d\e]. On output
    {c},{d} and {e} are the diagonals of the decomposed matrix.
```

```
    x = LUsolve(c,d,e,b).
    Solves [c\d\e]{x} = {b}, where {c}, {d} and {e} are the
    vectors returned from LUdecomp3.
'''

def LUdecomp3(c,d,e):
    n = len(d)
    for k in range(1,n):
        lam = c[k-1]/d[k-1]
        d[k] = d[k] - lam*e[k-1]
        c[k-1] = lam
    return c,d,e

def LUsolve3(c,d,e,b):
    n = len(d)
    for k in range(1,n):
        b[k] = b[k] - c[k-1]*b[k-1]
    b[n-1] = b[n-1]/d[n-1]
    for k in range(n-2,-1,-1):
        b[k] = (b[k] - e[k]*b[k+1])/d[k]
    return b
```

Symmetric Coefficient Matrices

More often than not, coefficient matrices that arise in engineering problems are symmetric as well as banded. Therefore, it is worthwhile to discover special properties of such matrices and to learn how to use them in the construction of efficient algorithms.

If the matrix $\mathbf{A}$ is symmetric, then the LU decomposition can be presented in the form

$$\mathbf{A} = \mathbf{LU} = \mathbf{LDL}^T \tag{2.23}$$

where $\mathbf{D}$ is a diagonal matrix. An example is Choleski's decomposition $\mathbf{A} = \mathbf{LL}^T$ that was discussed earlier (in this case $\mathbf{D} = \mathbf{I}$).

For Doolittle's decomposition we have

$$\mathbf{U} = \mathbf{DL}^T = \begin{bmatrix} D_1 & 0 & 0 & \cdots & 0 \\ 0 & D_2 & 0 & \cdots & 0 \\ 0 & 0 & D_3 & \cdots & 0 \\ \vdots & \vdots & \vdots & \cdots & \vdots \\ 0 & 0 & 0 & \cdots & D_n \end{bmatrix} \begin{bmatrix} 1 & L_{21} & L_{31} & \cdots & L_{n1} \\ 0 & 1 & L_{32} & \cdots & L_{n2} \\ 0 & 0 & 1 & \cdots & L_{n3} \\ \vdots & \vdots & \vdots & \cdots & \vdots \\ 0 & 0 & 0 & \cdots & 1 \end{bmatrix}$$

which gives

$$
\mathbf{U} =
\begin{bmatrix}
D_1 & D_1 L_{21} & D_1 L_{31} & \cdots & D_1 L_{n1} \\
0 & D_2 & D_2 L_{32} & \cdots & D_2 L_{n2} \\
0 & 0 & D_3 & \cdots & D_3 L_{3n} \\
\vdots & \vdots & \vdots & \cdots & \vdots \\
0 & 0 & 0 & \cdots & D_n
\end{bmatrix}
\tag{2.24}
$$

We now see that during decomposition of a symmetric matrix only $\mathbf{U}$ has to be stored, because $\mathbf{D}$ and $\mathbf{L}$ can be easily recovered from $\mathbf{U}$. Thus Gauss elimination, which results in an upper triangular matrix of the form shown in Eq. (2.24), is sufficient to decompose a symmetric matrix.

There is an alternative storage scheme that can be employed during $\mathbf{LU}$ decomposition. The idea is to arrive at the matrix

$$
\mathbf{U}^* =
\begin{bmatrix}
D_1 & L_{21} & L_{31} & \cdots & L_{n1} \\
0 & D_2 & L_{32} & \cdots & L_{n2} \\
0 & 0 & D_3 & \cdots & L_{n3} \\
\vdots & \vdots & \vdots & \ddots & \vdots \\
0 & 0 & 0 & \cdots & D_n
\end{bmatrix}
\tag{2.25}
$$

Here $\mathbf{U}$ can be recovered from $U_{ij} = D_i L_{ji}$. It turns out that this scheme leads to a computationally more efficient solution phase; therefore, we adopt it for symmetric, banded matrices.

Symmetric, Pentadiagonal Coefficient Matrices

We encounter pentadiagonal (bandwidth = 5) coefficient matrices in the solution of fourth-order, ordinary differential equations by finite differences. Often these matrices are symmetric, in which case an $n \times n$ coefficient matrix has the form

$$
\mathbf{A} =
\begin{bmatrix}
d_1 & e_1 & f_1 & 0 & 0 & 0 & \cdots & 0 \\
e_1 & d_2 & e_2 & f_2 & 0 & 0 & \cdots & 0 \\
f_1 & e_2 & d_3 & e_3 & f_3 & 0 & \cdots & 0 \\
0 & f_2 & e_3 & d_4 & e_4 & f_4 & \cdots & 0 \\
\vdots & \vdots & \vdots & \vdots & \vdots & \vdots & \ddots & \vdots \\
0 & \cdots & 0 & f_{n-4} & e_{n-3} & d_{n-2} & e_{n-2} & f_{n-2} \\
0 & \cdots & 0 & 0 & f_{n-3} & e_{n-2} & d_{n-1} & e_{n-1} \\
0 & \cdots & 0 & 0 & 0 & f_{n-2} & e_{n-1} & d_n
\end{bmatrix}
\tag{2.26}
$$

As in the case of tridiagonal matrices, we store the nonzero elements in the three vectors

$$
\mathbf{d} = \begin{bmatrix} d_1 \\ d_2 \\ \vdots \\ d_{n-2} \\ d_{n-1} \\ d_n \end{bmatrix}
\qquad
\mathbf{e} = \begin{bmatrix} e_1 \\ e_2 \\ \vdots \\ e_{n-2} \\ e_{n-1} \end{bmatrix}
\qquad
\mathbf{f} = \begin{bmatrix} f_1 \\ f_2 \\ \vdots \\ f_{n-2} \end{bmatrix}
$$

Let us now look at the solution of the equations $\mathbf{Ax} = \mathbf{b}$ by Doolittle's decomposition. The first step is to transform $\mathbf{A}$ to upper triangular form by Gauss elimination. If elimination has progressed to the stage where the kth row has become the pivot row, we have the following situation

$$
\mathbf{A} = \begin{bmatrix}
\ddots & \vdots & \vdots & \vdots & \vdots & \vdots & \vdots & \vdots & \\
\cdots & 0 & d_k & e_k & f_k & 0 & 0 & 0 & \cdots \\
\cdots & 0 & e_k & d_{k+1} & e_{k+1} & f_{k+1} & 0 & 0 & \cdots \\
\cdots & 0 & f_k & e_{k+1} & d_{k+2} & e_{k+2} & f_{k+2} & 0 & \cdots \\
\cdots & 0 & 0 & f_{k+1} & e_{k+2} & d_{k+3} & e_{k+3} & f_{k+3} & \cdots \\
& \vdots & \vdots & \vdots & \vdots & \vdots & \vdots & \vdots & \ddots
\end{bmatrix} \leftarrow
$$

The elements e_k and f_k below the pivot row (the kth row) are eliminated by the operations

$$
\text{row } (k+1) \leftarrow \text{row } (k+1) - (e_k/d_k) \times \text{row } k
$$
$$
\text{row } (k+2) \leftarrow \text{row } (k+2) - (f_k/d_k) \times \text{row } k
$$

The only terms (other than those being eliminated) that are changed by the operations are

$$
d_{k+1} \leftarrow d_{k+1} - (e_k/d_k)e_k
$$
$$
e_{k+1} \leftarrow e_{k+1} - (e_k/d_k)f_k \tag{2.27a}
$$
$$
d_{k+2} \leftarrow d_{k+2} - (f_k/d_k)f_k
$$

Storage of the multipliers in the *upper* triangular portion of the matrix results in

$$
e_k \leftarrow e_k/d_k \qquad f_k \leftarrow f_k/d_k \tag{2.27b}
$$

At the conclusion of the elimination phase the matrix has the form (do not confuse $\mathbf{d}$, $\mathbf{e}$, and $\mathbf{f}$ with the original contents of $\mathbf{A}$)

$$
\mathbf{U}^* = \begin{bmatrix}
d_1 & e_1 & f_1 & 0 & \cdots & 0 \\
0 & d_2 & e_2 & f_2 & \cdots & 0 \\
0 & 0 & d_3 & e_3 & \cdots & 0 \\
\vdots & \vdots & \vdots & \vdots & \cdots & \vdots \\
0 & 0 & \cdots & 0 & d_{n-1} & e_{n-1} \\
0 & 0 & \cdots & 0 & 0 & d_n
\end{bmatrix}
$$

Now comes the solution phase. The equations $\mathbf{Ly} = \mathbf{b}$ have the augmented coefficient matrix

$$
\begin{bmatrix} \mathbf{L} \mid \mathbf{b} \end{bmatrix} = \left[\begin{array}{cccccc|c}
1 & 0 & 0 & 0 & \cdots & 0 & b_1 \\
e_1 & 1 & 0 & 0 & \cdots & 0 & b_2 \\
f_1 & e_2 & 1 & 0 & \cdots & 0 & b_3 \\
0 & f_2 & e_3 & 1 & \cdots & 0 & b_4 \\
\vdots & \vdots & \vdots & \vdots & \cdots & \vdots & \vdots \\
0 & 0 & 0 & f_{n-2} & e_{n-1} & 1 & b_n
\end{array} \right]
$$

Solution by forward substitution yields

$$
y_1 = b_1
$$
$$
y_2 = b_2 - e_1 y_1 \tag{2.28}
$$
$$
\vdots
$$
$$
y_k = b_k - f_{k-2} y_{k-2} - e_{k-1} y_{k-1}, \quad k = 3, 4, \ldots, n
$$

The equations to be solved by back substitution, namely $\mathbf{Ux} = \mathbf{y}$, have the augmented coefficient matrix

$$
\begin{bmatrix} \mathbf{U} \mid \mathbf{y} \end{bmatrix} = \left[\begin{array}{cccccc|c}
d_1 & d_1 e_1 & d_1 f_1 & 0 & \cdots & 0 & y_1 \\
0 & d_2 & d_2 e_2 & d_2 f_2 & \cdots & 0 & y_2 \\
0 & 0 & d_3 & d_3 e_3 & \cdots & 0 & y_3 \\
\vdots & \vdots & \vdots & \vdots & \cdots & \vdots & \vdots \\
0 & 0 & \cdots & 0 & d_{n-1} & d_{n-1} e_{n-1} & y_{n-1} \\
0 & 0 & \cdots & 0 & 0 & d_n & y_n
\end{array} \right]
$$

the solution of which is obtained by back substitution:

$$
x_n = y_n / d_n
$$
$$
x_{n-1} = y_{n-1}/d_{n-1} - e_{n-1} x_n \vdots
$$
$$
x_k = y_k / d_k - e_k x_{k+1} - f_k x_{k+2}, \quad k = n - 2, n - 3, \ldots, 1
$$

■ LUdecomp5

The function LUdecomp5 decomposes a symmetric, pentadiagonal matrix **A** of the form $\mathbf{A} = [\mathbf{f}\backslash\mathbf{e}\backslash\mathbf{d}\backslash\mathbf{e}\backslash\mathbf{f}]$. The original vectors **d**, **e**, and **f** are destroyed and replaced by the vectors of the decomposed matrix. After decomposition, the solution of $\mathbf{Ax} = \mathbf{b}$ can be obtained by LUsolve5. During forward substitution, the original **b** is replaced by **y**. Similarly, **y** is written over by **x** in the back substitution phase, so that **b** contains the solution vector upon exit from LUsolve5.

```
## module LUdecomp5
''' d,e,f = LUdecomp5(d,e,f).
    LU decomposition of symmetric pentadiagonal matrix [a], where
    {f}, {e} and {d} are the diagonals of [a]. On output
    {d},{e} and {f} are the diagonals of the decomposed matrix.

    x = LUsolve5(d,e,f,b).
    Solves [a]{x} = {b}, where {d}, {e} and {f} are the vectors
    returned from LUdecomp5.
    '''
def LUdecomp5(d,e,f):
    n = len(d)
    for k in range(n-2):
        lam = e[k]/d[k]
        d[k+1] = d[k+1] - lam*e[k]
        e[k+1] = e[k+1] - lam*f[k]
        e[k] = lam
        lam = f[k]/d[k]
        d[k+2] = d[k+2] - lam*f[k]
        f[k] = lam
    lam = e[n-2]/d[n-2]
    d[n-1] = d[n-1] - lam*e[n-2]
    e[n-2] = lam
    return d,e,f

def LUsolve5(d,e,f,b):
    n = len(d)
    b[1] = b[1] - e[0]*b[0]
    for k in range(2,n):
        b[k] = b[k] - e[k-1]*b[k-1] - f[k-2]*b[k-2]
    b[n-1] = b[n-1]/d[n-1]
    b[n-2] = b[n-2]/d[n-2] - e[n-2]*b[n-1]
    for k in range(n-3,-1,-1):
        b[k] = b[k]/d[k] - e[k]*b[k+1] - f[k]*b[k+2]
    return b
```

EXAMPLE 2.9

As a result of Gauss elimination, a symmetric matrix **A** was transformed to the upper triangular form

$$\mathbf{U} = \begin{bmatrix} 4 & -2 & 1 & 0 \\ 0 & 3 & -3/2 & 1 \\ 0 & 0 & 3 & -3/2 \\ 0 & 0 & 0 & 35/12 \end{bmatrix}$$

Determine the original matrix **A**.

Solution. First we find **L** in the decomposition $\mathbf{A} = \mathbf{LU}$. Dividing each row of **U** by its diagonal element yields

$$\mathbf{L}^T = \begin{bmatrix} 1 & -1/2 & 1/4 & 0 \\ 0 & 1 & -1/2 & 1/3 \\ 0 & 0 & 1 & -1/2 \\ 0 & 0 & 0 & 1 \end{bmatrix}$$

Therefore, $\mathbf{A} = \mathbf{LU}$, or

$$\mathbf{A} = \begin{bmatrix} 1 & 0 & 0 & 0 \\ -1/2 & 1 & 0 & 0 \\ 1/4 & -1/2 & 1 & 0 \\ 0 & 1/3 & -1/2 & 1 \end{bmatrix} \begin{bmatrix} 4 & -2 & 1 & 0 \\ 0 & 3 & -3/2 & 1 \\ 0 & 0 & 3 & -3/2 \\ 0 & 0 & 0 & 35/12 \end{bmatrix}$$

$$= \begin{bmatrix} 4 & -2 & 1 & 0 \\ -2 & 4 & -2 & 1 \\ 1 & -2 & 4 & -2 \\ 0 & 1 & -2 & 4 \end{bmatrix}$$

EXAMPLE 2.10

Determine **L** and **D** that result from Doolittle's decomposition $\mathbf{A} = \mathbf{LDL}^T$ of the symmetric matrix

$$\mathbf{A} = \begin{bmatrix} 3 & -3 & 3 \\ -3 & 5 & 1 \\ 3 & 1 & 10 \end{bmatrix}$$

Solution. We use Gauss elimination, storing the multipliers in the *upper* triangular portion of **A**. At the completion of elimination, the matrix will have the form of **U*** in Eq. (2.25).

The terms to be eliminated in the first pass are A_{21} and A_{31} using the elementary operations

$$\text{row } 2 \leftarrow \text{row } 2 - (-1) \times \text{row } 1$$

$$\text{row } 3 \leftarrow \text{row } 3 - (1) \times \text{row } 1$$

Storing the multipliers (-1 and 1) in the locations occupied by A_{12} and A_{13}, we get

$$\mathbf{A}' = \begin{bmatrix} 3 & -1 & 1 \\ 0 & 2 & 4 \\ 0 & 4 & 7 \end{bmatrix}$$

The second pass is the operation

$$\text{row } 3 \leftarrow \text{row } 3 - 2 \times \text{row } 2$$

yields, after overwriting A_{23} with the multiplier 2

$$\mathbf{A}'' = \begin{bmatrix} \mathbf{0} \backslash \mathbf{D} \backslash \mathbf{L}^T \end{bmatrix} = \begin{bmatrix} 3 & -1 & 1 \\ 0 & 2 & 2 \\ 0 & 0 & -1 \end{bmatrix}$$

Hence

$$\mathbf{L} = \begin{bmatrix} 1 & 0 & 0 \\ -1 & 1 & 0 \\ 1 & 2 & 1 \end{bmatrix} \quad \mathbf{D} = \begin{bmatrix} 3 & 0 & 0 \\ 0 & 2 & 0 \\ 0 & 0 & -1 \end{bmatrix}$$

EXAMPLE 2.11

Use the functions LUdecmp3 and LUsolve3 to solve $\mathbf{Ax} = \mathbf{b}$, where

$$\mathbf{A} = \begin{bmatrix} 2 & -1 & 0 & 0 & 0 \\ -1 & 2 & -1 & 0 & 0 \\ 0 & -1 & 2 & -1 & 0 \\ 0 & 0 & -1 & 2 & -1 \\ 0 & 0 & 0 & -1 & 2 \end{bmatrix} \quad \mathbf{b} = \begin{bmatrix} 5 \\ -5 \\ 4 \\ -5 \\ 5 \end{bmatrix}$$

Solution

```
#!/usr/bin/python
## example2_11
import numpy as np
from LUdecomp3 import *

d = np.ones((5))*2.0
c = np.ones((4))*(-1.0)
b = np.array([5.0, -5.0, 4.0, -5.0, 5.0])
e = c.copy()
c,d,e = LUdecomp3(c,d,e)
x = LUsolve3(c,d,e,b)
print("\nx =\n",x)
input("\nPress return to exit")
```

The output is

```
x =
[ 2. -1.  1. -1.  2.]
```

2.5 Pivoting

Introduction

Sometimes the order in which the equations are presented to the solution algorithm has a profound effect on the results. For example, consider these equations:

$$2x_1 - x_2 = 1$$

$$-x_1 + 2x_2 - x_3 = 0$$

$$-x_2 + x_3 = 0$$

The corresponding augmented coefficient matrix is

$$\begin{bmatrix} \mathbf{A} \,\big|\, \mathbf{b} \end{bmatrix} = \begin{bmatrix} 2 & -1 & 0 & 1 \\ -1 & 2 & -1 & 0 \\ 0 & -1 & 1 & 0 \end{bmatrix} \tag{a}$$

Equations (a) are in the "right order" in the sense that we would have no trouble obtaining the correct solution $x_1 = x_2 = x_3 = 1$ by Gauss elimination or LU decomposition. Now suppose that we exchange the first and third equations, so that the augmented coefficient matrix becomes

$$\begin{bmatrix} \mathbf{A} \,\big|\, \mathbf{b} \end{bmatrix} = \begin{bmatrix} 0 & -1 & 1 & 0 \\ -1 & 2 & -1 & 0 \\ 2 & -1 & 0 & 1 \end{bmatrix} \tag{b}$$

Because we did not change the equations (only their order was altered), the solution is still $x_1 = x_2 = x_3 = 1$. However, Gauss elimination fails immediately because of the presence of the zero pivot element (the element A_{11}).

This example demonstrates that it is sometimes essential to reorder the equations during the elimination phase. The reordering, or *row pivoting*, is also required if the pivot element is not zero, but is very small in comparison to other elements in the pivot row, as demonstrated by the following set of equations:

$$\begin{bmatrix} \mathbf{A} \,\big|\, \mathbf{b} \end{bmatrix} = \begin{bmatrix} \varepsilon & -1 & 1 & 0 \\ -1 & 2 & -1 & 0 \\ 2 & -1 & 0 & 1 \end{bmatrix} \tag{c}$$

These equations are the same as Eqs. (b), except that the small number ε replaces the zero element in Eq. (b). Therefore, if we let $\varepsilon \to 0$, the solutions of Eqs. (b) and (c) should become identical. After the first phase of Gauss elimination, the augmented coefficient matrix becomes

$$\begin{bmatrix} \mathbf{A}' \,\big|\, \mathbf{b}' \end{bmatrix} = \begin{bmatrix} \varepsilon & -1 & 1 & 0 \\ 0 & 2 - 1/\varepsilon & -1 + 1/\varepsilon & 0 \\ 0 & -1 + 2/\varepsilon & -2/\varepsilon & 1 \end{bmatrix} \tag{d}$$

Because the computer works with a fixed word length, all numbers are rounded off to a finite number of significant figures. If ε is very small, then $1/\varepsilon$ is huge, and an element such as $2 - 1/\varepsilon$ is rounded to $-1/\varepsilon$. Therefore, for sufficiently small ε, Eqs. (d) are actually stored as

$$
\left[\, \mathbf{A}' \,\middle|\, \mathbf{b}' \,\right] = \left[\begin{array}{ccc|c}
\varepsilon & -1 & 1 & 0 \\
0 & -1/\varepsilon & 1/\varepsilon & 0 \\
0 & 2/\varepsilon & -2/\varepsilon & 1
\end{array}\right]
$$

Because the second and third equations obviously contradict each other, the solution process fails again. This problem would not arise if the first and second, or the first and the third equations were interchanged in Eqs. (c) before the elimination.

The last example illustrates the extreme case where ε was so small that roundoff errors resulted in total failure of the solution. If we were to make ε somewhat bigger so that the solution would not "bomb" any more, the roundoff errors might still be large enough to render the solution unreliable. Again, this difficulty could be avoided by pivoting.

Diagonal Dominance

An $n \times n$ matrix $\mathbf{A}$ is said to be *diagonally dominant* if each diagonal element is larger than the sum of the other elements in the same row (we are talking here about absolute values). Thus diagonal dominance requires that

$$
|A_{ii}| > \sum_{\substack{j=1 \\ j \neq i}}^{n} |A_{ij}| \quad (i = 1, 2, \ldots, n) \tag{2.30}
$$

For example, the matrix

$$
\begin{bmatrix}
-2 & 4 & -1 \\
1 & -1 & 3 \\
4 & -2 & 1
\end{bmatrix}
$$

is not diagonally dominant. However, if we rearrange the rows in the following manner

$$
\begin{bmatrix}
4 & -2 & 1 \\
-2 & 4 & -1 \\
1 & -1 & 3
\end{bmatrix}
$$

then we have diagonal dominance.

It can be shown that if the coefficient matrix of the equations $\mathbf{A}\mathbf{x} = \mathbf{b}$ is diagonally dominant, then the solution does not benefit from pivoting; that is, the equations are already arranged in the optimal order. It follows that the strategy of pivoting should be to reorder the equations so that the coefficient matrix is as close to diagonal dominance as possible. This is the principle behind scaled row pivoting, discussed next.

Gauss Elimination with Scaled Row Pivoting

Consider the solution of $\mathbf{Ax} = \mathbf{b}$ by Gauss elimination with row pivoting. Recall that pivoting aims at improving diagonal dominance of the coefficient matrix (i.e., making the pivot element as large as possible in comparison to other elements in the pivot row). The comparison is made easier if we establish an array $\mathbf{s}$ with the elements

$$s_i = \max_j |A_{ij}|, \quad i = 1, 2, \ldots, n \tag{2.31}$$

Thus s_i, called the *scale factor* of row i, contains the absolute value of the largest element in the ith row of $\mathbf{A}$. The vector $\mathbf{s}$ can be obtained with the algorithm

```
for i in range(n):
    s[i] = max(abs(a[i,:]))
```

The *relative size* of an element A_{ij} (that is, relative to the largest element in the ith row) is defined as the ratio

$$r_{ij} = \frac{|A_{ij}|}{s_i} \tag{2.32}$$

Suppose that the elimination phase has reached the stage where the kth row has become the pivot row. The augmented coefficient matrix at this point is shown in the following matrix:

$$\begin{bmatrix} A_{11} & A_{12} & A_{13} & A_{14} & \cdots & A_{1n} & b_1 \\ 0 & A_{22} & A_{23} & A_{24} & \cdots & A_{2n} & b_2 \\ 0 & 0 & A_{33} & A_{34} & \cdots & A_{3n} & b_3 \\ \vdots & \vdots & \vdots & \vdots & \cdots & \vdots & \vdots \\ 0 & \cdots & 0 & A_{kk} & \cdots & A_{kn} & b_k \\ \vdots & \cdots & \vdots & \vdots & \cdots & \vdots & \vdots \\ 0 & \cdots & 0 & A_{nk} & \cdots & A_{nn} & b_n \end{bmatrix} \leftarrow$$

We do not automatically accept A_{kk} as the next pivot element, but look in the kth column below A_{kk} for a "better" pivot. The best choice is the element A_{pk} that has the largest relative size; that is, we choose p such that

$$r_{pk} = \max_j (r_{jk}), \quad j \geq k$$

If we find such an element, then we interchange the rows k and p, and proceed with the elimination pass as usual. Note that the corresponding row interchange must also be carried out in the scale factor array $\mathbf{s}$. The algorithm that does all this is as follows:

```
for k in range(0,n-1):

    # Find row containing element with largest relative size
    p = argmax(abs(a[k:n,k])/s[k:n]) + k

    # If this element is very small, matrix is singular
```

```
    if abs(a[p,k]) < tol: error.err('Matrix is singular')

  # Check whether rows k and p must be interchanged
    if p != k:
      # Interchange rows if needed
        swap.swapRows(b,k,p)
        swap.swapRows(s,k,p)
        swap.swapRows(a,k,p)
  # Proceed with elimination
```

The Python statement argmax(v) returns the index of the largest element in the vector v. The algorithms for exchanging rows (and columns) are included in the module swap shown next.

■ swap

The function swapRows interchanges rows i and j of a matrix or vector **v**, whereas swapCols interchanges columns i and j of a matrix.

```
## module swap
'''  swapRows(v,i,j).
     Swaps rows i and j of a vector or matrix [v].

     swapCols(v,i,j).
     Swaps columns of matrix [v].
'''
def swapRows(v,i,j):
    if len(v.shape) == 1:
        v[i],v[j] = v[j],v[i]
    else:
        v[[i,j],:] = v[[j,i],:]

def swapCols(v,i,j):
    v[:,[i,j]] = v[:,[j,i]]
```

■ gaussPivot

The function gaussPivot performs Gauss elimination with row pivoting. Apart from row swapping, the elimination and solution phases are identical to gaussElimin in Section 2.2.

```
## module gaussPivot
'''  x = gaussPivot(a,b,tol=1.0e-12).
     Solves [a]{x} = {b} by Gauss elimination with
     scaled row pivoting
```

```
'''

import numpy as np
import swap
import error

def gaussPivot(a,b,tol=1.0e-12):
    n = len(b)

  # Set up scale factors
    s = np.zeros(n)
    for i in range(n):
        s[i] = max(np.abs(a[i,:]))

    for k in range(0,n-1):

      # Row interchange, if needed
        p = np.argmax(np.abs(a[k:n,k])/s[k:n]) + k
        if abs(a[p,k]) < tol: error.err('Matrix is singular')
        if p != k:
            swap.swapRows(b,k,p)
            swap.swapRows(s,k,p)
            swap.swapRows(a,k,p)

      # Elimination
        for i in range(k+1,n):
            if a[i,k] != 0.0:
                lam = a[i,k]/a[k,k]
                a[i,k+1:n] = a[i,k+1:n] - lam*a[k,k+1:n]
                b[i] = b[i] - lam*b[k]
    if abs(a[n-1,n-1]) < tol: error.err('Matrix is singular')

  # Back substitution
    b[n-1] = b[n-1]/a[n-1,n-1]
    for k in range(n-2,-1,-1):
        b[k] = (b[k] - np.dot(a[k,k+1:n],b[k+1:n]))/a[k,k]
    return b
```

■ LUpivot

The Gauss elimination algorithm can be changed to Doolittle's decomposition with minor changes. The most important of these changes is keeping a record of the row interchanges during the decomposition phase. In LUdecomp this record is kept in the array seq. Initially seq contains [0, 1, 2, ...]. Whenever two rows are interchanged, the corresponding interchange is also carried out in seq. Thus seq shows the order in which the original rows have been rearranged. This information is passed on to the solution phase (LUsolve), which rearranges the elements of the constant vector in the same order before proceeding to forward and back substitutions.

```
## module LUpivot
''' a,seq = LUdecomp(a,tol=1.0e-9).
    LU decomposition of matrix [a] using scaled row pivoting.
    The returned matrix [a] = contains [U] in the upper
    triangle and the nondiagonal terms of [L] in the lower triangle.
    Note that [L][U] is a row-wise permutation of the original [a];
    the permutations are recorded in the vector {seq}.

    x = LUsolve(a,b,seq).
    Solves [L][U]{x} = {b}, where the matrix [a] = and the
    permutation vector {seq} are returned from LUdecomp.
'''
import numpy as np
import swap
import error

def LUdecomp(a,tol=1.0e-9):
    n = len(a)
    seq = np.array(range(n))

  # Set up scale factors
    s = np.zeros((n))
    for i in range(n):
        s[i] = max(abs(a[i,:]))

    for k in range(0,n-1):

      # Row interchange, if needed
        p = np.argmax(np.abs(a[k:n,k])/s[k:n]) + k
        if abs(a[p,k]) <  tol: error.err('Matrix is singular')
        if p != k:
            swap.swapRows(s,k,p)
            swap.swapRows(a,k,p)
            swap.swapRows(seq,k,p)

      # Elimination
        for i in range(k+1,n):
            if a[i,k] != 0.0:
                lam = a[i,k]/a[k,k]
```

```
                    a[i,k+1:n] = a[i,k+1:n] - lam*a[k,k+1:n]
                    a[i,k] = lam
        return a,seq

def LUsolve(a,b,seq):
    n = len(a)

  # Rearrange constant vector; store it in [x]
    x = b.copy()
    for i in range(n):
        x[i] = b[seq[i]]

  # Solution
    for k in range(1,n):
        x[k] = x[k] - np.dot(a[k,0:k],x[0:k])
    x[n-1] = x[n-1]/a[n-1,n-1]
    for k in range(n-2,-1,-1):
        x[k] = (x[k] - np.dot(a[k,k+1:n],x[k+1:n]))/a[k,k]
    return x
```

When to Pivot

Pivoting has two drawbacks. One is the increased cost of computation; the other is the destruction of symmetry and banded structure of the coefficient matrix. The latter is of particular concern in engineering computing, where the coefficient matrices are frequently banded and symmetric, a property that is used in the solution, as seen in the previous section. Fortunately, these matrices are often diagonally dominant as well, so that they would not benefit from pivoting anyway.

There are no infallible rules for determining when pivoting should be used. Experience indicates that pivoting is likely to be counterproductive if the coefficient matrix is banded. Positive definite and, to a lesser degree, symmetric matrices also seldom gain from pivoting. And we should not forget that pivoting is not the only means of controlling roundoff errors—there is also double precision arithmetic.

It should be strongly emphasized that these rules of thumb are only meant for equations that stem from real engineering problems. It is not difficult to concoct "textbook" examples that do not conform to these rules.

EXAMPLE 2.12

Employ Gauss elimination with scaled row pivoting to solve the equations $\mathbf{Ax} = \mathbf{b}$, where

$$
\mathbf{A} = \begin{bmatrix} 2 & -2 & 6 \\ -2 & 4 & 3 \\ -1 & 8 & 4 \end{bmatrix} \qquad \mathbf{b} = \begin{bmatrix} 16 \\ 0 \\ -1 \end{bmatrix}
$$

Solution. The augmented coefficient matrix and the scale factor array are

$$\left[\mathbf{A}\,\middle|\,\mathbf{b}\right] = \begin{bmatrix} 2 & -2 & 6 & 16 \\ -2 & 4 & 3 & 0 \\ -1 & 8 & 4 & -1 \end{bmatrix} \qquad \mathbf{s} = \begin{bmatrix} 6 \\ 4 \\ 8 \end{bmatrix}$$

Note that $\mathbf{s}$ contains the absolute value of the biggest element in each row of $\mathbf{A}$. At this stage, all the elements in the first column of $\mathbf{A}$ are potential pivots. To determine the best pivot element, we calculate the relative sizes of the elements in the first column:

$$\begin{bmatrix} r_{11} \\ r_{21} \\ r_{31} \end{bmatrix} = \begin{bmatrix} |A_{11}|/s_1 \\ |A_{21}|/s_2 \\ |A_{31}|/s_3 \end{bmatrix} = \begin{bmatrix} 1/3 \\ 1/2 \\ 1/8 \end{bmatrix}$$

Since r_{21} is the biggest element, we conclude that A_{21} makes the best pivot element. Therefore, we exchange rows 1 and 2 of the augmented coefficient matrix and the scale factor array, obtaining

$$\left[\mathbf{A}\,\middle|\,\mathbf{b}\right] = \begin{bmatrix} -2 & 4 & 3 & 0 \\ 2 & -2 & 6 & 16 \\ -1 & 8 & 4 & -1 \end{bmatrix} \leftarrow \qquad \mathbf{s} = \begin{bmatrix} 4 \\ 6 \\ 8 \end{bmatrix}$$

Now the first pass of Gauss elimination is carried out (the arrow points to the pivot row), yielding

$$\left[\mathbf{A}'\,\middle|\,\mathbf{b}'\right] = \begin{bmatrix} -2 & 4 & 3 & 0 \\ 0 & 2 & 9 & 16 \\ 0 & 6 & 5/2 & -1 \end{bmatrix} \qquad \mathbf{s} = \begin{bmatrix} 4 \\ 6 \\ 8 \end{bmatrix}$$

The potential pivot elements for the next elimination pass are A'_{22} and A'_{32}. We determine the "winner" from

$$\begin{bmatrix} * \\ r_{22} \\ r_{32} \end{bmatrix} = \begin{bmatrix} * \\ |A_{22}|/s_2 \\ |A_{32}|/s_3 \end{bmatrix} = \begin{bmatrix} * \\ 1/3 \\ 3/4 \end{bmatrix}$$

Note that r_{12} is irrelevant, because row 1 already acted as the pivot row. Therefore, it is excluded from further consideration. Because r_{32} is bigger than r_{22}, the third row is the better pivot row. After interchanging rows 2 and 3, we have

$$\left[\mathbf{A}'\,\middle|\,\mathbf{b}'\right] = \begin{bmatrix} -2 & 4 & 3 & 0 \\ 0 & 6 & 5/2 & -1 \\ 0 & 2 & 9 & 16 \end{bmatrix} \leftarrow \qquad \mathbf{s} = \begin{bmatrix} 4 \\ 8 \\ 6 \end{bmatrix}$$

The second elimination pass now yields

$$\left[\mathbf{A}''\,\middle|\,\mathbf{b}''\right] = \left[\mathbf{U}\,\middle|\,\mathbf{c}\right] = \begin{bmatrix} -2 & 4 & 3 & 0 \\ 0 & 6 & 5/2 & -1 \\ 0 & 0 & 49/6 & 49/3 \end{bmatrix}$$

This completes the elimination phase. It should be noted that $\mathbf{U}$ is the matrix that would result from LU decomposition of the following row-wise permutation of $\mathbf{A}$ (the ordering of rows is the same as achieved by pivoting):

$$\begin{bmatrix} -2 & 4 & 3 \\ -1 & 8 & 4 \\ 2 & -2 & 6 \end{bmatrix}$$

Because the solution of $\mathbf{U}\mathbf{x} = \mathbf{c}$ by back substitution is not affected by pivoting, we skip the details computation. The result is $\mathbf{x}^T = \begin{bmatrix} 1 & -1 & 2 \end{bmatrix}$.

Alternate Solution. It is not necessary to physically exchange equations during pivoting. We could accomplish Gauss elimination just as well by keeping the equations in place. The elimination would then proceed as follows (for the sake of brevity, the details of choosing the pivot equation are not repeated):

$$\begin{bmatrix} \mathbf{A} & | & \mathbf{b} \end{bmatrix} = \left[\begin{array}{ccc|c} 2 & -2 & 6 & 16 \\ -2 & 4 & 3 & 0 \\ -1 & 8 & 4 & -1 \end{array} \right] \leftarrow$$

$$\begin{bmatrix} \mathbf{A}' & | & \mathbf{b}' \end{bmatrix} = \left[\begin{array}{ccc|c} 0 & 2 & 9 & 16 \\ -2 & 4 & 3 & 0 \\ 0 & 6 & 5/2 & -1 \end{array} \right] \leftarrow$$

$$\begin{bmatrix} \mathbf{A}'' & | & \mathbf{b}'' \end{bmatrix} = \left[\begin{array}{ccc|c} 0 & 0 & 49/6 & 49/3 \\ -2 & 4 & 3 & 0 \\ 0 & 6 & 5/2 & -1 \end{array} \right]$$

Yet now the back substitution phase is a little more involved, because the order in which the equations must be solved has become scrambled. In hand computations this is not a problem, because we can determine the order by inspection. Unfortunately, "by inspection" does not work on a computer. To overcome this difficulty, we have to maintain an integer array $\mathbf{p}$ that keeps track of the row permutations during the elimination phase. The contents of $\mathbf{p}$ indicate the order in which the pivot rows were chosen. In this example, we would have at the end of Gauss elimination

$$\mathbf{p} = \begin{bmatrix} 2 \\ 3 \\ 1 \end{bmatrix}$$

showing that row 2 was the pivot row in the first elimination pass, followed by row 3 in the second pass. The equations are solved by back substitution in the reverse order: Equation 1 is solved first for x_3, then equation 3 is solved for x_2, and finally equation 2 yields x_1.

By dispensing with swapping of equations, the scheme outlined here is claimed to result in a faster algorithm than gaussPivot. This may be true if the programs are written in Fortran or C, but our tests show that in Python gaussPivot is about 30% faster than the in-place elimination scheme.

PROBLEM SET 2.2

1. Solve the equations $\mathbf{Ax} = \mathbf{b}$ by utilizing Doolittle's decomposition, where

$$\mathbf{A} = \begin{bmatrix} 3 & -3 & 3 \\ -3 & 5 & 1 \\ 3 & 1 & 5 \end{bmatrix} \qquad \mathbf{b} = \begin{bmatrix} 9 \\ -7 \\ 12 \end{bmatrix}$$

2. Use Doolittle's decomposition to solve $\mathbf{Ax} = \mathbf{b}$, where

$$\mathbf{A} = \begin{bmatrix} 4 & 8 & 20 \\ 8 & 13 & 16 \\ 20 & 16 & -91 \end{bmatrix} \qquad \mathbf{b} = \begin{bmatrix} 24 \\ 18 \\ -119 \end{bmatrix}$$

3. Determine $\mathbf{L}$ and $\mathbf{D}$ that result from Doolittle's decomposition of the symmetric matrix

$$\mathbf{A} = \begin{bmatrix} 2 & -2 & 0 & 0 & 0 \\ -2 & 5 & -6 & 0 & 0 \\ 0 & -6 & 16 & 12 & 0 \\ 0 & 0 & 12 & 39 & -6 \\ 0 & 0 & 0 & -6 & 14 \end{bmatrix}$$

4. Solve the tridiagonal equations $\mathbf{Ax} = \mathbf{b}$ by Doolittle's decomposition method, where

$$\mathbf{A} = \begin{bmatrix} 6 & 2 & 0 & 0 & 0 \\ -1 & 7 & 2 & 0 & 0 \\ 0 & -2 & 8 & 2 & 0 \\ 0 & 0 & 3 & 7 & -2 \\ 0 & 0 & 0 & 3 & 5 \end{bmatrix} \qquad \mathbf{b} = \begin{bmatrix} 2 \\ -3 \\ 4 \\ -3 \\ 1 \end{bmatrix}$$

5. Use Gauss elimination with scaled row pivoting to solve

$$\begin{bmatrix} 4 & -2 & 1 \\ -2 & 1 & -1 \\ -2 & 3 & 6 \end{bmatrix} \begin{bmatrix} x_1 \\ x_2 \\ x_3 \end{bmatrix} = \begin{bmatrix} 2 \\ -1 \\ 0 \end{bmatrix}$$

6. Solve $\mathbf{Ax} = \mathbf{b}$ by Gauss elimination with scaled row pivoting, where

$$\mathbf{A} = \begin{bmatrix} 2.34 & -4.10 & 1.78 \\ 1.98 & 3.47 & -2.22 \\ 2.36 & -15.17 & 6.81 \end{bmatrix} \qquad \mathbf{b} = \begin{bmatrix} 0.02 \\ -0.73 \\ -6.63 \end{bmatrix}$$

7. Solve the equations

$$\begin{bmatrix} 2 & -1 & 0 & 0 \\ 0 & 0 & -1 & 1 \\ 0 & -1 & 2 & -1 \\ -1 & 2 & -1 & 0 \end{bmatrix} \begin{bmatrix} x_1 \\ x_2 \\ x_3 \\ x_4 \end{bmatrix} = \begin{bmatrix} 1 \\ 0 \\ 0 \\ 0 \end{bmatrix}$$

by Gauss elimination with scaled row pivoting.

8. ■ Solve the equations

$$
\begin{bmatrix}
0 & 2 & 5 & -1 \\
2 & 1 & 3 & 0 \\
-2 & -1 & 3 & 1 \\
3 & 3 & -1 & 2
\end{bmatrix}
\begin{bmatrix}
x_1 \\
x_2 \\
x_3 \\
x_4
\end{bmatrix}
=
\begin{bmatrix}
-3 \\
3 \\
-2 \\
5
\end{bmatrix}
$$

9. ■ Solve the symmetric, tridiagonal equations

$$4x_1 - x_2 = 9$$

$$-x_{i-1} + 4x_i - x_{i+1} = 5, \quad i = 2, \ldots, n-1$$

$$-x_{n-1} + 4x_n = 5$$

with $n = 10$.

10. ■ Solve the equations $\mathbf{Ax} = \mathbf{b}$, where

$$
\mathbf{A} =
\begin{bmatrix}
1.3174 & 2.7250 & 2.7250 & 1.7181 \\
0.4002 & 0.8278 & 1.2272 & 2.5322 \\
0.8218 & 1.5608 & 0.3629 & 2.9210 \\
1.9664 & 2.0011 & 0.6532 & 1.9945
\end{bmatrix}
\qquad
\mathbf{b} =
\begin{bmatrix}
8.4855 \\
4.9874 \\
5.6665 \\
6.6152
\end{bmatrix}
$$

11. ■ Solve the following equations:

$$
\begin{bmatrix}
10 & -2 & -1 & 2 & 3 & 1 & -4 & 7 \\
5 & 11 & 3 & 10 & -3 & 3 & 3 & -4 \\
7 & 12 & 1 & 5 & 3 & -12 & 2 & 3 \\
8 & 7 & -2 & 1 & 3 & 2 & 2 & 4 \\
2 & -15 & -1 & 1 & 4 & -1 & 8 & 3 \\
4 & 2 & 9 & 1 & 12 & -1 & 4 & 1 \\
-1 & 4 & -7 & -1 & 1 & 1 & -1 & -3 \\
-1 & 3 & 4 & 1 & 3 & -4 & 7 & 6
\end{bmatrix}
\begin{bmatrix}
x_1 \\
x_2 \\
x_3 \\
x_4 \\
x_5 \\
x_6 \\
x_7 \\
x_8
\end{bmatrix}
=
\begin{bmatrix}
0 \\
12 \\
-5 \\
3 \\
-25 \\
-26 \\
9 \\
-7
\end{bmatrix}
$$

12. ■ The displacement formulation for the mass-spring system shown in Fig. (a) results in the following equilibrium equations of the masses

$$
\begin{bmatrix}
k_1 + k_2 + k_3 + k_5 & -k_3 & -k_5 \\
-k_3 & k_3 + k_4 & -k_4 \\
-k_5 & -k_4 & k_4 + k_5
\end{bmatrix}
\begin{bmatrix}
x_1 \\
x_2 \\
x_3
\end{bmatrix}
=
\begin{bmatrix}
W_1 \\
W_2 \\
W_3
\end{bmatrix}
$$

where k_i are the spring stiffnesses, W_i represent the weights of the masses, and x_i are the displacements of the masses from the undeformed configuration of the system. Write a program that solves these equations for given $\mathbf{k}$ and $\mathbf{W}$. Use the program to find the displacements if

$$k_1 = k_3 = k_4 = k \qquad k_2 = k_5 = 2k$$
$$W_1 = W_3 = 2W \qquad W_2 = W$$

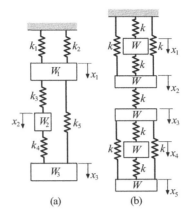

(a)   (b)

13. ■ The equlibrium equations of the mass-spring system in Fig. (b) are

$$\begin{bmatrix} 2 & -1 & 0 & 0 & 0 \\ -1 & 4 & -1 & 0 & 0 \\ 0 & -1 & 4 & -1 & -2 \\ 0 & 0 & -1 & 2 & -1 \\ 0 & 0 & -2 & -1 & 3 \end{bmatrix} \begin{bmatrix} x_1 \\ x_2 \\ x_3 \\ x_4 \\ x_5 \end{bmatrix} = \begin{bmatrix} W/k \\ W/k \\ W/k \\ W/k \\ W/k \end{bmatrix}$$

where k are the spring stiffnesses, W represent the weights of the masses, and x_i are the displacements of the masses from the undeformed configuration of the system. Determine the displacements.

14. ■

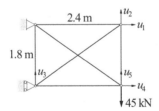

The displacement formulation for a plane truss is similar to that of a mass-spring system. The differences are (1) the stiffnesses of the members are $k_i = (EA/L)_i$, where E is the modulus of elasticity, A represents the cross-sectional area, and L is the length of the member; and (2) there are two components of displacement at each joint. For the statically indeterminate truss shown, the displacement formulation yields the symmetric equations $\mathbf{Ku} = \mathbf{p}$, where

$$\mathbf{K} = \begin{bmatrix} 27.58 & 7.004 & -7.004 & 0 & 0 \\ 7.004 & 29.57 & -5.253 & 0 & -24.32 \\ -7.004 & -5.253 & 29.57 & 0 & 0 \\ 0 & 0 & 0 & 27.58 & -7.004 \\ 0 & -24.32 & 0 & -7.004 & 29.57 \end{bmatrix} \text{MN/m}$$

$$\mathbf{p} = \begin{bmatrix} 0 & 0 & 0 & 0 & -45 \end{bmatrix}^T \text{kN}$$

Determine the displacements u_i of the joints.

15. ■

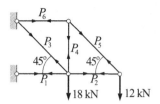

In the *force formulation* of a truss, the unknowns are the member forces P_i. For the statically determinate truss shown, force formulation is obtained by writing down the equilibrium equations of the joints

$$\begin{bmatrix} -1 & 1 & -1/\sqrt{2} & 0 & 0 & 0 \\ 0 & 0 & 1/\sqrt{2} & 1 & 0 & 0 \\ 0 & -1 & 0 & 0 & -1/\sqrt{2} & 0 \\ 0 & 0 & 0 & 0 & 1/\sqrt{2} & 0 \\ 0 & 0 & 0 & 0 & 1/\sqrt{2} & 1 \\ 0 & 0 & 0 & -1 & -1/\sqrt{2} & 0 \end{bmatrix} \begin{bmatrix} P_1 \\ P_2 \\ P_3 \\ P_4 \\ P_5 \\ P_6 \end{bmatrix} = \begin{bmatrix} 0 \\ 18 \\ 0 \\ 12 \\ 0 \\ 0 \end{bmatrix}$$

where the units of P_i are kN. (a) Solve the equations as they are with a computer program. (b) Rearrange the rows and columns so as to obtain a lower triangular coefficient matrix, and then solve the equations by back substitution using a calculator.

16. ■

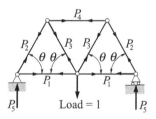

The force formulation of the symmetric truss shown results in the joint equilibrium equations

$$\begin{bmatrix} c & 1 & 0 & 0 & 0 \\ 0 & s & 0 & 0 & 1 \\ 0 & 0 & 2s & 0 & 0 \\ 0 & -c & c & 1 & 0 \\ 0 & s & s & 0 & 0 \end{bmatrix} \begin{bmatrix} P_1 \\ P_2 \\ P_3 \\ P_4 \\ P_5 \end{bmatrix} = \begin{bmatrix} 0 \\ 0 \\ 1 \\ 0 \\ 0 \end{bmatrix}$$

where $s = \sin \theta$, $c = \cos \theta$, and P_i are the unknown forces. Write a program that computes the forces, given the angle θ. Run the program with $\theta = 53°$.

17. ■

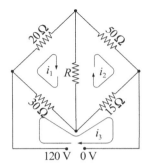

The electrical network shown can be viewed as consisting of three loops. Applying Kirchoff's law ($\sum$voltage drops $= \sum$voltage sources) to each loop yields the following equations for the loop currents i_1, i_2 and i_3:

$$(50 + R)i_1 - Ri_2 - 30i_3 = 0$$

$$-Ri_1 + (65 + R)i_2 - 15i_3 = 0$$

$$-30i_2 - 15i_2 + 45i_3 = 120$$

Compute the three loop currents for $R = 5\ \Omega$, $10\ \Omega$, and $20\ \Omega$.

18. ■

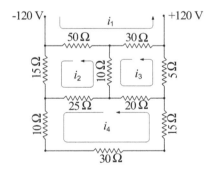

Determine the loop currents i_1 to i_4 in the electrical network shown.

19. ■ Consider the n simultaneous equations $\mathbf{Ax} = \mathbf{b}$, where

$$A_{ij} = (i + j)^2 \qquad b_i = \sum_{j=0}^{n-1} A_{ij}, \quad i = 0, 1, \ldots, n - 1, \quad j = 0, 1, \ldots, n - 1$$

Clearly, the solution is $\mathbf{x} = \begin{bmatrix} 1 & 1 & \cdots & 1 \end{bmatrix}^T$. Write a program that solves these equations for any given n (pivoting is recommended). Run the program with $n = 2, 3$ and 4, and comment on the results.

20. ■

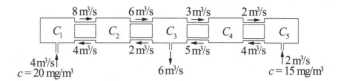

The diagram shows five mixing vessels connected by pipes. Water is pumped through the pipes at the steady rates shown on the diagram. The incoming water contains a chemical, the amount of which is specified by its concentration c (mg/m^3). Applying the principle of conservation of mass

mass of chemical flowing in = mass of chemical flowing out

to each vessel, we obtain the following simultaneous equations for the concentrations c_i within the vessels:

$$-8c_1 + 4c_2 = -80$$
$$8c_1 - 10c_2 + 2c_3 = 0$$
$$6c_2 - 11c_3 + 5c_4 = 0$$
$$3c_3 - 7c_4 + 4c_5 = 0$$
$$2c_4 - 4c_5 = -30$$

Note that the mass flow rate of the chemical is obtained by multiplying the volume flow rate of the water by the concentration. Verify the equations and determine the concentrations.

21. ■

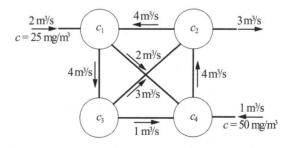

Four mixing tanks are connected by pipes. The fluid in the system is pumped through the pipes at the rates shown in the figure. The fluid entering the system contains a chemical of concentration c as indicated. Determine the concentration of the chemical in the four tanks, assuming a steady state.

22. ■ Solve the following equations:

$$7x_1 - 4x_2 + x_3 = 1$$

$$-4x_1 + 6x_2 - 4x_3 + x_4 = 1$$

$$x_{i-2} - 4x_{i-1} + 6x_i - 4x_{i+1} + x_{i+2} = 1 \quad (i = 3, 4, \ldots, 8)$$

$$x_7 - 4x_8 + 6x_9 - 4x_{10} = 1$$

$$x_8 - 4x_9 + 7x_{10} = 1$$

23. ■ Write a program that solves the complex tridiagonal equations $\mathbf{Ax} = \mathbf{b}$, where

$$
\mathbf{A} = \begin{bmatrix}
2 & -i & 0 & 0 & \cdots & 0 \\
-i & 2 & -i & 0 & \cdots & 0 \\
0 & -i & 2 & -i & \cdots & 0 \\
\vdots & \vdots & \ddots & \ddots & \ddots & \vdots \\
0 & 0 & \cdots & -i & 2 & -i \\
0 & 0 & 0 & \cdots & -i & 1
\end{bmatrix}
\qquad
\mathbf{b} = \begin{bmatrix}
100 + 100i \\
0 \\
0 \\
\vdots \\
0 \\
0
\end{bmatrix}
$$

The program should accommodate n equations, where n is arbitrary. Test it with $n = 10$.

*2.6   Matrix Inversion

Computing the inverse of a matrix and solving simultaneous equations are related tasks. The most economical way to invert an $n \times n$ matrix $\mathbf{A}$ is to solve the equations

$$\mathbf{AX} = \mathbf{I} \tag{2.33}$$

where $\mathbf{I}$ is the $n \times n$ identity matrix. The solution $\mathbf{X}$, also of size $n \times n$, will be the inverse of $\mathbf{A}$. The proof is simple: After we pre-multiply both sides of Eq. (2.33) by $\mathbf{A}^{-1}$ we have $\mathbf{A}^{-1}\mathbf{AX} = \mathbf{A}^{-1}\mathbf{I}$, which reduces to $\mathbf{X} = \mathbf{A}^{-1}$.

Inversion of large matrices should be avoided whenever possible because of its high cost. As seen from Eq. (2.33), inversion of $\mathbf{A}$ is equivalent to solving $\mathbf{Ax}_i = \mathbf{b}_i$ with $i = 1, 2, \ldots, n$, where $\mathbf{b}_i$ is the ith column of $\mathbf{I}$. Assuming that LU decomposition is employed in the solution, the solution phase (forward and back substitution) must be repeated n times, once for each $\mathbf{b}_i$. Since the cost of computation is proportional to n^3 for the decomposition phase and n^2 for each vector of the solution phase, the cost of inversion is considerably more expensive than the solution of $\mathbf{Ax} = \mathbf{b}$ (single constant vector $\mathbf{b}$).

Matrix inversion has another serious drawback—a banded matrix loses its structure during inversion. In other words, if $\mathbf{A}$ is banded or otherwise sparse, then $\mathbf{A}^{-1}$ is fully populated.

EXAMPLE 2.13
Write a function that inverts a matrix using LU decomposition with pivoting. Test the function by inverting

$$A = \begin{bmatrix} 0.6 & -0.4 & 1.0 \\ -0.3 & 0.2 & 0.5 \\ 0.6 & -1.0 & 0.5 \end{bmatrix}$$

Solution. The following function matInv uses the decomposition and solution procedures in the module LUpivot.

```
#!/usr/bin/python
## example2_13
import numpy as np
from LUpivot import *

def matInv(a):
    n = len(a[0])
    aInv = np.identity(n)
    a,seq = LUdecomp(a)
    for i in range(n):
        aInv[:,i] = LUsolve(a,aInv[:,i],seq)
    return aInv

a = np.array([[ 0.6, -0.4,   1.0],\
              [-0.3,  0.2,   0.5],\
              [ 0.6, -1.0,   0.5]])
aOrig = a.copy()   # Save original [a]
aInv = matInv(a)   # Invert [a] (original [a] is destroyed)
print("\naInv =\n",aInv)
print("\nCheck: a*aInv =\n", np.dot(aOrig,aInv))
input("\nPress return to exit")
```

The output is

```
aInv =
[[ 1.66666667 -2.22222222 -1.11111111]
 [ 1.25       -0.83333333 -1.66666667]
 [ 0.5         1.          0.        ]]

Check: a*aInv =
[[  1.00000000e+00  -4.44089210e-16  -1.11022302e-16]
 [  0.00000000e+00   1.00000000e+00   5.55111512e-17]
 [  0.00000000e+00  -3.33066907e-16   1.00000000e+00]]
```

EXAMPLE 2.14

Invert the matrix

$$
\mathbf{A} =
\begin{bmatrix}
2 & -1 & 0 & 0 & 0 & 0 \\
-1 & 2 & -1 & 0 & 0 & 0 \\
0 & -1 & 2 & -1 & 0 & 0 \\
0 & 0 & -1 & 2 & -1 & 0 \\
0 & 0 & 0 & -1 & 2 & -1 \\
0 & 0 & 0 & 0 & -1 & 5
\end{bmatrix}
$$

Solution. Because the matrix is tridiagonal, we solve $\mathbf{AX} = \mathbf{I}$ using the functions in the module $\texttt{LUdecomp3}$ (LU decomposition of tridiagonal matrices):

```python
#!/usr/bin/python
## example2_14
import numpy as np
from LUdecomp3 import *

n = 6
d = np.ones((n))*2.0
e = np.ones((n-1))*(-1.0)
c = e.copy()
d[n-1] = 5.0
aInv = np.identity(n)
c,d,e = LUdecomp3(c,d,e)
for i in range(n):
    aInv[:,i] = LUsolve3(c,d,e,aInv[:,i])
print("\nThe inverse matrix is:\n",aInv)
input("\nPress return to exit")
```

Running the program results in the following output:

```
The inverse matrix is:
[[ 0.84  0.68  0.52  0.36  0.2   0.04]
 [ 0.68  1.36  1.04  0.72  0.4   0.08]
 [ 0.52  1.04  1.56  1.08  0.6   0.12]
 [ 0.36  0.72  1.08  1.44  0.8   0.16]
 [ 0.2   0.4   0.6   0.8   1.    0.2 ]
 [ 0.04  0.08  0.12  0.16  0.2   0.24]]]
```

Note that $\mathbf{A}$ is tridiagonal, whereas $\mathbf{A}^{-1}$ is fully populated.

*2.7 Iterative Methods

Introduction

So far, we have discussed only direct methods of solution. The common character-istic of these methods is that they compute the solution with a finite number of op-erations. Moreover, if the computer were capable of infinite precision (no roundoff errors), the solution would be exact.

Iterative, or *indirect methods*, start with an initial guess of the solution **x** and then repeatedly improve the solution until the change in **x** becomes negligible. Because the required number of iterations can be large, the indirect methods are, in general, slower than their direct counterparts. However, iterative methods do have the follow-ing two advantages that make them attractive for certain problems:

1. It is feasible to store only the nonzero elements of the coefficient matrix. This makes it possible to deal with very large matrices that are sparse, but not neces-sarily banded. In many problems, there is no need to store the coefficient matrix at all.
2. Iterative procedures are self-correcting, meaning that roundoff errors (or even arithmetic mistakes) in one iterative cycle are corrected in subsequent cycles.

A serious drawback of iterative methods is that they do not always converge to the solution. It can be shown that convergence is guaranteed only if the coefficient matrix is diagonally dominant. The initial guess for **x** plays no role in determining whether convergence takes place—if the procedure converges for one starting vector, it would do so for any starting vector. The initial guess affects only the number of iterations that are required for convergence.

Gauss-Seidel Method

The equations $\mathbf{Ax} = \mathbf{b}$ are in scalar notation

$$\sum_{j=1}^{n} A_{ij} x_j = b_i, \quad i = 1, 2, \ldots, n$$

Extracting the term containing x_i from the summation sign yields

$$A_{ii} x_i + \sum_{\substack{j=1 \\ j \neq i}}^{n} A_{ij} x_j = b_i, \quad i = 1, 2, \ldots, n$$

Solving for x_i, we get

$$x_i = \frac{1}{A_{ii}} \left(b_i - \sum_{\substack{j=1 \\ j \neq i}}^{n} A_{ij} x_j \right), \quad i = 1, 2, \ldots, n$$

The last equation suggests the following iterative scheme:

$$x_i \leftarrow \frac{1}{A_{ii}} \left(b_i - \sum_{\substack{j=1 \\ j\neq i}}^{n} A_{ij}x_j \right), \quad i = 1, 2, \ldots, n \qquad (2.34)$$

We start by choosing the starting vector $\mathbf{x}$. If a good guess for the solution is not available, $\mathbf{x}$ can be chosen randomly. Equation (2.34) is then used to recompute each element of $\mathbf{x}$, always using the latest available values of x_j. This completes one iteration cycle. The procedure is repeated until the changes in $\mathbf{x}$ between successive iteration cycles become sufficiently small.

Convergence of the Gauss-Seidel method can be improved by a technique known as *relaxation*. The idea is to take the new value of x_i as a weighted average of its previous value and the value predicted by Eq. (2.34). The corresponding iterative formula is

$$x_i \leftarrow \frac{\omega}{A_{ii}} \left(b_i - \sum_{\substack{j=1 \\ j\neq i}}^{n} A_{ij}x_j \right) + (1-\omega)x_i, \quad i = 1, 2, \ldots, n \qquad (2.35)$$

where the weight ω is called the *relaxation factor*. It can be seen that if $\omega = 1$, no relaxation takes place, because Eqs. (2.34) and (2.35) produce the same result. If $\omega < 1$, Eq. (2.35) represents interpolation between the old x_i and the value given by Eq. (2.34). This is called *under-relaxation*. In cases where $\omega > 1$, we have extrapolation, or *over-relaxation*.

There is no practical method of determining the optimal value of ω beforehand; however, an estimate can be computed during run time. Let $\Delta x^{(k)} = \left| \mathbf{x}^{(k-1)} - \mathbf{x}^{(k)} \right|$ be the magnitude of the change in $\mathbf{x}$ during the kth iteration (carried out without relaxation [i.e., with $\omega = 1$]). If k is sufficiently large (say $k \geq 5$), it can be shown[2] that an approximation of the optimal value of ω is

$$\omega_{\text{opt}} \approx \frac{2}{1 + \sqrt{1 - \left(\Delta x^{(k+p)}/\Delta x^{(k)} \right)^{1/p}}} \qquad (2.36)$$

where p is a positive integer.

The essential elements of a Gauss-Seidel algorithm with relaxation are as follows:

> Carry out k iterations with $\omega = 1$ ($k = 10$ is reasonable).
> Record $\Delta x^{(k)}$.
> Perform additional p iterations.
> Record $\Delta x^{(k+p)}$.
> Compute ω_{opt} from Eq. (2.36).
> Perform all subsequent iterations with $\omega = \omega_{\text{opt}}$.

[2] See, for example, Terrence J. Akai, *Applied Numerical Methods for Engineers*, John Wiley & Sons (1994), p. 100.

■ gaussSeidel

The function gaussSeidel is an implementation of the Gauss-Seidel method with relaxation. It automatically computes ω_{opt} from Eq. (2.36) using $k = 10$ and $p = 1$. The user must provide the function iterEqs that computes the improved **x** from the iterative formulas in Eq. (2.35)—see Example 2.17. The function gaussSeidel returns the solution vector **x**, the number of iterations carried out, and the value of ω_{opt} used.

```
## module gaussSeidel
''' x,numIter,omega = gaussSeidel(iterEqs,x,tol = 1.0e-9)
    Gauss-Seidel method for solving [A]{x} = {b}.
    The matrix [A] should be sparse. User must supply the
    function iterEqs(x,omega) that returns the improved {x},
    given the current {x} ('omega' is the relaxation factor).
'''
import numpy as np
import math

def gaussSeidel(iterEqs,x,tol = 1.0e-9):
    omega = 1.0
    k = 10
    p = 1
    for i in range(1,501):
        xOld = x.copy()
        x = iterEqs(x,omega)
        dx = math.sqrt(np.dot(x-xOld,x-xOld))
        if dx < tol: return x,i,omega
      # Compute relaxation factor after k+p iterations
        if i == k: dx1 = dx
        if i == k + p:
            dx2 = dx
            omega = 2.0/(1.0 + math.sqrt(1.0 \
                    - (dx2/dx1)**(1.0/p)))
    print('Gauss-Seidel failed to converge')
```

Conjugate Gradient Method

Consider the problem of finding the vector **x** that minimizes the scalar function

$$f(\mathbf{x}) = \frac{1}{2}\mathbf{x}^T\mathbf{A}\mathbf{x} - \mathbf{b}^T\mathbf{x} \qquad (2.37)$$

where the matrix $\mathbf{A}$ is *symmetric* and *positive definite*. Because $f(\mathbf{x})$ is minimized when its gradient $\nabla f = \mathbf{A}\mathbf{x} - \mathbf{b}$ is zero, we see that minimization is equivalent to solving

$$\mathbf{A}\mathbf{x} = \mathbf{b} \tag{2.38}$$

Gradient methods accomplish the minimization by iteration, starting with an initial vector $\mathbf{x}_0$. Each iterative cycle k computes a refined solution

$$\mathbf{x}_{k+1} = \mathbf{x}_k + \alpha_k \mathbf{s}_k \tag{2.39}$$

The *step length* α_k is chosen so that $\mathbf{x}_{k+1}$ minimizes $f(\mathbf{x}_{k+1})$ in the *search direction* $\mathbf{s}_k$. That is, $\mathbf{x}_{k+1}$ must satisfy Eq. (2.38):

$$\mathbf{A}(\mathbf{x}_k + \alpha_k \mathbf{s}_k) = \mathbf{b} \tag{a}$$

Introducing the *residual*

$$\mathbf{r}_k = \mathbf{b} - \mathbf{A}\mathbf{x}_k \tag{2.40}$$

Eq. (a) becomes $\alpha \mathbf{A}\mathbf{s}_k = \mathbf{r}_k$. Pre-multiplying both sides by $\mathbf{s}_k^T$ and solving for α_k, we obtain

$$\alpha_k = \frac{\mathbf{s}_k^T \mathbf{r}_k}{\mathbf{s}_k^T \mathbf{A}\mathbf{s}_k} \tag{2.41}$$

We are still left with the problem of determining the search direction $\mathbf{s}_k$. Intuition tells us to choose $\mathbf{s}_k = -\nabla f = \mathbf{r}_k$, because this is the direction of the largest negative change in $f(\mathbf{x})$. The resulting procedure is known as the *method of steepest descent*. It is not a popular algorithm because its convergence can be slow. The more efficient conjugate gradient method uses the search direction

$$\mathbf{s}_{k+1} = \mathbf{r}_{k+1} + \beta_k \mathbf{s}_k \tag{2.42}$$

The constant β_k is chosen so that the two successive search directions are *conjugate* to each other, meaning

$$\mathbf{s}_{k+1}^T \mathbf{A}\mathbf{s}_k = 0 \tag{b}$$

The great attraction of conjugate gradients is that minimization in one conjugate direction does not undo previous minimizations (minimizations do not interfere with one another).

Substituting $\mathbf{s}_{k+1}$ from Eq. (2.42) into Eq. (b), we get

$$\left(\mathbf{r}_{k+1}^T + \beta_k \mathbf{s}_k^T \right) \mathbf{A}\mathbf{s}_k = 0$$

which yields

$$\beta_k = -\frac{\mathbf{r}_{k+1}^T \mathbf{A}\mathbf{s}_k}{\mathbf{s}_k^T \mathbf{A}\mathbf{s}_k} \tag{2.43}$$

Here is the outline of the conjugate gradient algorithm:

Choose $\mathbf{x}_0$ (any vector will do).
Let $\mathbf{r}_0 \leftarrow \mathbf{b} - \mathbf{A}\mathbf{x}_0$.
Let $\mathbf{s}_0 \leftarrow \mathbf{r}_0$ (lacking a previous search direction,
   choose the direction of steepest descent).
do with $k = 0, 1, 2, \ldots$:
$$\alpha_k \leftarrow \frac{\mathbf{s}_k^T \mathbf{r}_k}{\mathbf{s}_k^T \mathbf{A}\mathbf{s}_k}.$$
$\mathbf{x}_{k+1} \leftarrow \mathbf{x}_k + \alpha_k \mathbf{s}_k$.
$\mathbf{r}_{k+1} \leftarrow \mathbf{b} - \mathbf{A}\mathbf{x}_{k+1}$.
if $|\mathbf{r}_{k+1}| \leq \varepsilon$ exit loop (ε is the error tolerance).
$$\beta_k \leftarrow -\frac{\mathbf{r}_{k+1}^T \mathbf{A}\mathbf{s}_k}{\mathbf{s}_k^T \mathbf{A}\mathbf{s}_k}.$$
$\mathbf{s}_{k+1} \leftarrow \mathbf{r}_{k+1} + \beta_k \mathbf{s}_k$.

It can be shown that the residual vectors $\mathbf{r}_1, \mathbf{r}_2, \mathbf{r}_3, \ldots$ produced by the algorithm are mutually orthogonal; that is, $\mathbf{r}_i \cdot \mathbf{r}_j = 0$, $i \neq j$. Now suppose that we have carried out enough iterations to have computed the whole set of n residual vectors. The residual resulting from the next iteration must be a null vector ($\mathbf{r}_{n+1} = \mathbf{0}$), indicating that the solution has been obtained. It thus appears that the conjugate gradient algorithm is not an iterative method at all, because it reaches the exact solution after n computational cycles. In practice, however, convergence is usually achieved in less than n iterations.

The conjugate gradient method is not competitive with direct methods in the solution of small sets of equations. Its strength lies in the handling of large, sparse systems (where most elements of $\mathbf{A}$ are zero). It is important to note that $\mathbf{A}$ enters the algorithm only through its multiplication by a vector; that is, in the form $\mathbf{A}\mathbf{v}$, where $\mathbf{v}$ is a vector (either $\mathbf{x}_{k+1}$ or $\mathbf{s}_k$). If $\mathbf{A}$ is sparse, it is possible to write an efficient subroutine for the multiplication and pass it, rather than $\mathbf{A}$ itself, to the conjugate gradient algorithm.

■ conjGrad

The function conjGrad shown next implements the conjugate gradient algorithm. The maximum allowable number of iterations is set to n (the number of unknowns). Note that conjGrad calls the function Av that returns the product $\mathbf{A}\mathbf{v}$. This function must be supplied by the user (see Example 2.18). The user must also supply the starting vector $\mathbf{x}_0$ and the constant (right-hand-side) vector $\mathbf{b}$. The function returns the solution vector $\mathbf{x}$ and the number of iterations.

```
## module conjGrad
''' x, numIter = conjGrad(Av,x,b,tol=1.0e-9)
    Conjugate gradient method for solving [A]{x} = {b}.
```

> The matrix [A] should be sparse. User must supply
> the function Av(v) that returns the vector [A]{v}.
> ''''

```python
import numpy as np
import math

def conjGrad(Av,x,b,tol=1.0e-9):
    n = len(b)
    r = b - Av(x)
    s = r.copy()
    for i in range(n):
        u = Av(s)
        alpha = np.dot(s,r)/np.dot(s,u)
        x = x + alpha*s
        r = b - Av(x)
        if(math.sqrt(np.dot(r,r))) < tol:
            break
        else:
            beta = -np.dot(r,u)/np.dot(s,u)
            s = r + beta*s
    return x,i
```

EXAMPLE 2.15
Solve the equations

$$\begin{bmatrix} 4 & -1 & 1 \\ -1 & 4 & -2 \\ 1 & -2 & 4 \end{bmatrix} \begin{bmatrix} x_1 \\ x_2 \\ x_3 \end{bmatrix} = \begin{bmatrix} 12 \\ -1 \\ 5 \end{bmatrix}$$

by the Gauss-Seidel method without relaxation.

Solution. With the given data, the iteration formulas in Eq. (2.34) become

$$x_1 = \frac{1}{4}(12 + x_2 - x_3)$$

$$x_2 = \frac{1}{4}(-1 + x_1 + 2x_3)$$

$$x_3 = \frac{1}{4}(5 - x_1 + 2x_2)$$

Choosing the starting values $x_1 = x_2 = x_3 = 0$, the first iteration gives us

$$x_1 = \frac{1}{4}(12 + 0 - 0) = 3$$

$$x_2 = \frac{1}{4}[-1 + 3 + 2(0)] = 0.5$$

$$x_3 = \frac{1}{4}[5 - 3 + 2(0.5)] = 0.75$$

The second iteration yields

$$x_1 = \frac{1}{4}(12 + 0.5 - 0.75) = 2.9375$$

$$x_2 = \frac{1}{4}[-1 + 2.9375 + 2(0.75)] = 0.859\,38$$

$$x_3 = \frac{1}{4}[5 - 2.9375 + 2(0.859\,38)] = 0.945\,31$$

and the third iteration results in

$$x_1 = \frac{1}{4}(12 + 0.85938 - 0.94531) = 2.978\,52$$

$$x_2 = \frac{1}{4}[-1 + 2.97852 + 2(0.94531)] = 0.967\,29$$

$$x_3 = \frac{1}{4}[5 - 2.97852 + 2(0.96729)] = 0.989\,02$$

After five more iterations the results would agree with the exact solution $x_1 = 3$, $x_2 = x_3 = 1$ within five decimal places.

EXAMPLE 2.16

Solve the equations in Example 2.15 by the conjugate gradient method.

Solution. The conjugate gradient method should converge after three iterations. Choosing again for the starting vector

$$\mathbf{x}_0 = \begin{bmatrix} 0 & 0 & 0 \end{bmatrix}^T,$$

the computations outlined in the text proceed as follows:

First Iteration

$$\mathbf{r}_0 = \mathbf{b} - \mathbf{A}\mathbf{x}_0 = \begin{bmatrix} 12 \\ -1 \\ 5 \end{bmatrix} - \begin{bmatrix} 4 & -1 & 1 \\ -1 & 4 & -2 \\ 1 & -2 & 4 \end{bmatrix} \begin{bmatrix} 0 \\ 0 \\ 0 \end{bmatrix} = \begin{bmatrix} 12 \\ -1 \\ 5 \end{bmatrix}$$

$$\mathbf{s}_0 = \mathbf{r}_0 = \begin{bmatrix} 12 \\ -1 \\ 5 \end{bmatrix}$$

$$\mathbf{A}\mathbf{s}_0 = \begin{bmatrix} 4 & -1 & 1 \\ -1 & 4 & -2 \\ 1 & -2 & 4 \end{bmatrix} \begin{bmatrix} 12 \\ -1 \\ 5 \end{bmatrix} = \begin{bmatrix} 54 \\ -26 \\ 34 \end{bmatrix}$$

$$\alpha_0 = \frac{\mathbf{s}_0^T \mathbf{r}_0}{\mathbf{s}_0^T \mathbf{A}\mathbf{s}_0} = \frac{12^2 + (-1)^2 + 5^2}{12(54) + (-1)(-26) + 5(34)} = 0.201\,42$$

$$\mathbf{x}_1 = \mathbf{x}_0 + \alpha_0 \mathbf{s}_0 = \begin{bmatrix} 0 \\ 0 \\ 0 \end{bmatrix} + 0.201\,42 \begin{bmatrix} 12 \\ -1 \\ 5 \end{bmatrix} = \begin{bmatrix} 2.41\,704 \\ -0.201\,42 \\ 1.007\,10 \end{bmatrix}$$

Second Iteration

$$\mathbf{r}_1 = \mathbf{b} - \mathbf{Ax}_1 = \begin{bmatrix} 12 \\ -1 \\ 5 \end{bmatrix} - \begin{bmatrix} 4 & -1 & 1 \\ -1 & 4 & -2 \\ 1 & -2 & 4 \end{bmatrix} \begin{bmatrix} 2.417\,04 \\ -0.201\,42 \\ 1.007\,10 \end{bmatrix} = \begin{bmatrix} 1.123\,32 \\ 4.236\,92 \\ -1.848\,28 \end{bmatrix}$$

$$\beta_0 = -\frac{\mathbf{r}_1^T \mathbf{As}_0}{\mathbf{s}_0^T \mathbf{As}_0} = -\frac{1.123\,32(54) + 4.236\,92(-26) - 1.848\,28(34)}{12(54) + (-1)(-26) + 5(34)} = 0.133\,107$$

$$\mathbf{s}_1 = \mathbf{r}_1 + \beta_0 \mathbf{s}_0 = \begin{bmatrix} 1.123\,32 \\ 4.236\,92 \\ -1.848\,28 \end{bmatrix} + 0.133\,107 \begin{bmatrix} 12 \\ -1 \\ 5 \end{bmatrix} = \begin{bmatrix} 2.720\,76 \\ 4.103\,80 \\ -1.182\,68 \end{bmatrix}$$

$$\mathbf{As}_1 = \begin{bmatrix} 4 & -1 & 1 \\ -1 & 4 & -2 \\ 1 & -2 & 4 \end{bmatrix} \begin{bmatrix} 2.720\,76 \\ 4.103\,80 \\ -1.182\,68 \end{bmatrix} = \begin{bmatrix} 5.596\,56 \\ 16.059\,80 \\ -10.217\,60 \end{bmatrix}$$

$$\alpha_1 = \frac{\mathbf{s}_1^T \mathbf{r}_1}{\mathbf{s}_1^T \mathbf{As}_1}$$

$$= \frac{2.720\,76(1.123\,32) + 4.103\,80(4.236\,92) + (-1.182\,68)(-1.848\,28)}{2.720\,76(5.596\,56) + 4.103\,80(16.059\,80) + (-1.182\,68)(-10.217\,60)}$$

$$= 0.24276$$

$$\mathbf{x}_2 = \mathbf{x}_1 + \alpha_1 \mathbf{s}_1 = \begin{bmatrix} 2.417\,04 \\ -0.201\,42 \\ 1.007\,10 \end{bmatrix} + 0.24276 \begin{bmatrix} 2.720\,76 \\ 4.103\,80 \\ -1.182\,68 \end{bmatrix} = \begin{bmatrix} 3.077\,53 \\ 0.794\,82 \\ 0.719\,99 \end{bmatrix}$$

Third Iteration

$$\mathbf{r}_2 = \mathbf{b} - \mathbf{Ax}_2 = \begin{bmatrix} 12 \\ -1 \\ 5 \end{bmatrix} - \begin{bmatrix} 4 & -1 & 1 \\ -1 & 4 & -2 \\ 1 & -2 & 4 \end{bmatrix} \begin{bmatrix} 3.077\,53 \\ 0.794\,82 \\ 0.719\,99 \end{bmatrix} = \begin{bmatrix} -0.235\,29 \\ 0.338\,23 \\ 0.632\,15 \end{bmatrix}$$

$$\beta_1 = -\frac{\mathbf{r}_2^T \mathbf{As}_1}{\mathbf{s}_1^T \mathbf{As}_1}$$

$$= -\frac{(-0.235\,29)(5.596\,56) + 0.338\,23(16.059\,80) + 0.632\,15(-10.217\,60)}{2.720\,76(5.596\,56) + 4.103\,80(16.059\,80) + (-1.182\,68)(-10.217\,60)}$$

$$= 0.0251\,452$$

$$\mathbf{s}_2 = \mathbf{r}_2 + \beta_1 \mathbf{s}_1 = \begin{bmatrix} -0.235\,29 \\ 0.338\,23 \\ 0.632\,15 \end{bmatrix} + 0.025\,1452 \begin{bmatrix} 2.720\,76 \\ 4.103\,80 \\ -1.182\,68 \end{bmatrix} = \begin{bmatrix} -0.166\,876 \\ 0.441\,421 \\ 0.602\,411 \end{bmatrix}$$

$$\mathbf{As}_2 = \begin{bmatrix} 4 & -1 & 1 \\ -1 & 4 & -2 \\ 1 & -2 & 4 \end{bmatrix} \begin{bmatrix} -0.166\,876 \\ 0.441\,421 \\ 0.602\,411 \end{bmatrix} = \begin{bmatrix} -0.506\,514 \\ 0.727\,738 \\ 1.359\,930 \end{bmatrix}$$

$$\alpha_2 = \frac{\mathbf{r}_2^T \mathbf{s}_2}{\mathbf{s}_2^T \mathbf{As}_2}$$

$$= \frac{(-0.235\,29)(-0.166\,876) + 0.338\,23(0.441\,421) + 0.632\,15(0.602\,411)}{(-0.166\,876)(-0.506\,514) + 0.441\,421(0.727\,738) + 0.602\,411(1.359\,930)}$$

$$= 0.464\,80$$

$$\mathbf{x}_3 = \mathbf{x}_2 + \alpha_2 \mathbf{s}_2 = \begin{bmatrix} 3.077\,53 \\ 0.794\,82 \\ 0.719\,99 \end{bmatrix} + 0.464\,80 \begin{bmatrix} -0.166\,876 \\ 0.441\,421 \\ 0.602\,411 \end{bmatrix} = \begin{bmatrix} 2.999\,97 \\ 0.999\,99 \\ 0.999\,99 \end{bmatrix}$$

The solution $\mathbf{x}_3$ is correct to almost five decimal places. The small discrepancy may be caused by roundoff errors in the computations.

EXAMPLE 2.17
Write a computer program to solve the following n simultaneous equations by the Gauss-Seidel method with relaxation (the program should work with any value of n)[3]:

$$\begin{bmatrix} 2 & -1 & 0 & 0 & \cdots & 0 & 0 & 0 & 1 \\ -1 & 2 & -1 & 0 & \cdots & 0 & 0 & 0 & 0 \\ 0 & -1 & 2 & -1 & \cdots & 0 & 0 & 0 & 0 \\ \vdots & \vdots & \vdots & \vdots & & \vdots & \vdots & \vdots & \vdots \\ 0 & 0 & 0 & 0 & \cdots & -1 & 2 & -1 & 0 \\ 0 & 0 & 0 & 0 & \cdots & 0 & -1 & 2 & -1 \\ 1 & 0 & 0 & 0 & \cdots & 0 & 0 & -1 & 2 \end{bmatrix} \begin{bmatrix} x_1 \\ x_2 \\ x_3 \\ \vdots \\ x_{n-2} \\ x_{n-1} \\ x_n \end{bmatrix} = \begin{bmatrix} 0 \\ 0 \\ 0 \\ \vdots \\ 0 \\ 0 \\ 1 \end{bmatrix}$$

Run the program with $n = 20$. The exact solution can be shown to be $x_i = -n/4 + i/2$, $i = 1, 2, \ldots, n$.

Solution. In this case the iterative formulas in Eq. (2.35) are

$$x_1 = \omega(x_2 - x_n)/2 + (1 - \omega)x_1$$

$$x_i = \omega(x_{i-1} + x_{i+1})/2 + (1 - \omega)x_i, \quad i = 2, 3, \ldots, n-1 \qquad \text{(a)}$$

$$x_n = \omega(1 - x_1 + x_{n-1})/2 + (1 - \omega)x_n$$

These formulas are evaluated in the function `iterEqs`.

[3] Equations of this form are called *cyclic* tridiagonal. They occur in the finite difference formulation of second-order differential equations with periodic boundary conditions.

```
#!/usr/bin/python
## example2_17
import numpy as np
from gaussSeidel import *

def iterEqs(x,omega):
    n = len(x)
    x[0] = omega*(x[1] - x[n-1])/2.0 + (1.0 - omega)*x[0]
    for i in range(1,n-1):
        x[i] = omega*(x[i-1] + x[i+1])/2.0 + (1.0 - omega)*x[i]
    x[n-1] = omega*(1.0 - x[0] + x[n-2])/2.0 \
             + (1.0 - omega)*x[n-1]
    return x

n = eval(input("Number of equations ==> "))
x = np.zeros(n)
x,numIter,omega = gaussSeidel(iterEqs,x)
print("\nNumber of iterations =",numIter)
print("\nRelaxation factor =",omega)
print("\nThe solution is:\n",x)
input("\nPress return to exit")
```

The output from the program is

```
Number of equations ==> 20

Number of iterations = 259

Relaxation factor = 1.7054523107131399

The solution is:
 [ -4.50000000e+00 -4.00000000e+00 -3.50000000e+00 -3.00000000e+00
   -2.50000000e+00 -2.00000000e+00 -1.50000000e+00 -9.99999997e-01
   -4.99999998e-01  2.14047151e-09  5.00000002e-01  1.00000000e+00
    1.50000000e+00  2.00000000e+00  2.50000000e+00  3.00000000e+00
    3.50000000e+00  4.00000000e+00  4.50000000e+00  5.00000000e+00]
```

The convergence is very slow, because the coefficient matrix lacks diagonal dominance—substituting the elements of **A** into Eq. (2.30) produces an equality rather than the desired inequality. If we were to change each diagonal term of the coefficient from 2 to 4, **A** would be diagonally dominant and the solution would converge in only 17 iterations.

EXAMPLE 2.18

Solve Example 2.17 with the conjugate gradient method, also using $n = 20$.

Solution. The program shown next uses the function `conjGrad`. The solution vector **x** is initialized to zero in the program. The function `Ax(v)` returns the product $\mathbf{A} \cdot \mathbf{v}$, where **A** is the coefficient matrix and **v** is a vector. For the given **A**, the components of the vector **Ax(v)** are

$$(\mathbf{Ax})_1 = 2v_1 - v_2 + v_n$$

$$(\mathbf{Ax})_i = -v_{i-1} + 2v_i - v_{i+1}, \quad i = 2, 3, \ldots, n-1$$

$$(\mathbf{Ax})_n = -v_{n-1} + 2v_n + v_1$$

```python
#!/usr/bin/python
## example2_18
import numpy as np
from conjGrad import *

def Ax(v):
    n = len(v)
    Ax = np.zeros(n)
    Ax[0] = 2.0*v[0] - v[1]+v[n-1]
    Ax[1:n-1] = -v[0:n-2] + 2.0*v[1:n-1] -v[2:n]
    Ax[n-1] = -v[n-2] + 2.0*v[n-1] + v[0]
    return Ax

n = eval(input("Number of equations ==> "))
b = np.zeros(n)
b[n-1] = 1.0
x = np.zeros(n)
x,numIter = conjGrad(Ax,x,b)
print("\nThe solution is:\n",x)
print("\nNumber of iterations =",numIter)
input("\nPress return to exit")
```

Running the program results in

```
Number of equations ==> 20

The solution is:
[-4.5 -4. -3.5 -3. -2.5 -2. -1.5 -1. -0.5  0.   0.5  1.   1.5
   2.   2.5  3.   3.5  4.   4.5  5. ]

Number of iterations = 9
```

Note that convergence was reached in only 9 iterations, whereas 259 iterations were required in the Gauss-Seidel method.

PROBLEM SET 2.3

1. Let

$$
\mathbf{A} = \begin{bmatrix} 3 & -1 & 2 \\ 0 & 1 & 3 \\ -2 & 2 & -4 \end{bmatrix} \qquad \mathbf{B} = \begin{bmatrix} 0 & 1 & 3 \\ 3 & -1 & 2 \\ -2 & 2 & -4 \end{bmatrix}
$$

(note that $\mathbf{B}$ is obtained by interchanging the first two rows of $\mathbf{A}$). Knowing that

$$
\mathbf{A}^{-1} = \begin{bmatrix} 0.5 & 0 & 0.25 \\ 0.3 & 0.4 & 0.45 \\ -0.1 & 0.2 & -0.15 \end{bmatrix}
$$

determine $\mathbf{B}^{-1}$.

2. Invert the triangular matrices:

$$
\mathbf{A} = \begin{bmatrix} 2 & 4 & 3 \\ 0 & 6 & 5 \\ 0 & 0 & 2 \end{bmatrix} \qquad \mathbf{B} = \begin{bmatrix} 2 & 0 & 0 \\ 3 & 4 & 0 \\ 4 & 5 & 6 \end{bmatrix}
$$

3. Invert the triangular matrix:

$$
\mathbf{A} = \begin{bmatrix} 1 & 1/2 & 1/4 & 1/8 \\ 0 & 1 & 1/3 & 1/9 \\ 0 & 0 & 1 & 1/4 \\ 0 & 0 & 0 & 1 \end{bmatrix}
$$

4. Invert the following matrices:

(a) $\mathbf{A} = \begin{bmatrix} 1 & 2 & 4 \\ 1 & 3 & 9 \\ 1 & 4 & 16 \end{bmatrix}$ (b) $\mathbf{B} = \begin{bmatrix} 4 & -1 & 0 \\ -1 & 4 & -1 \\ 0 & -1 & 4 \end{bmatrix}$

5. Invert this matrix:

$$
\mathbf{A} = \begin{bmatrix} 4 & -2 & 1 \\ -2 & 1 & -1 \\ 1 & -2 & 4 \end{bmatrix}
$$

6. ■ Invert the following matrices with any method:

$$
\mathbf{A} = \begin{bmatrix} 5 & -3 & -1 & 0 \\ -2 & 1 & 1 & 1 \\ 3 & -5 & 1 & 2 \\ 0 & 8 & -4 & -3 \end{bmatrix} \qquad \mathbf{B} = \begin{bmatrix} 4 & -1 & 0 & 0 \\ -1 & 4 & -1 & 0 \\ 0 & -1 & 4 & -1 \\ 0 & 0 & -1 & 4 \end{bmatrix}
$$

7. ■ Invert the matrix by any method

$$
\mathbf{A} = \begin{bmatrix} 1 & 3 & -9 & 6 & 4 \\ 2 & -1 & 6 & 7 & 1 \\ 3 & 2 & -3 & 15 & 5 \\ 8 & -1 & 1 & 4 & 2 \\ 11 & 1 & -2 & 18 & 7 \end{bmatrix}
$$

and comment on the reliability of the result.

8. ■ The joint displacements **u** of the plane truss in Prob. 14, Problem Set 2.2 are related to the applied joint forces **p** by

$$\mathbf{K}\mathbf{u} = \mathbf{p} \tag{a}$$

where

$$\mathbf{K} = \begin{bmatrix} 27.580 & 7.004 & -7.004 & 0.000 & 0.000 \\ 7.004 & 29.570 & -5.253 & 0.000 & -24.320 \\ -7.004 & -5.253 & 29.570 & 0.000 & 0.000 \\ 0.000 & 0.000 & 0.000 & 27.580 & -7.004 \\ 0.000 & -24.320 & 0.000 & -7.004 & 29.570 \end{bmatrix} \text{MN/m}$$

is called the *stiffness matrix* of the truss. If Eq. (a) is inverted by multiplying each side by $\mathbf{K}^{-1}$, we obtain $\mathbf{u} = \mathbf{K}^{-1}\mathbf{p}$, where $\mathbf{K}^{-1}$ is known as the *flexibility matrix*. The physical meaning of the elements of the flexibility matrix is $K_{ij}^{-1} = $ displacements u_i ($i = 1, 2, \ldots 5$) produced by the unit load $p_j = 1$. Compute (a) the flexibility matrix of the truss; (b) the displacements of the joints due to the load $p_5 = -45$ kN (the load shown in Problem 14, Problem Set 2.2).

9. ■ Invert the matrices:

$$\mathbf{A} = \begin{bmatrix} 3 & -7 & 45 & 21 \\ 12 & 11 & 10 & 17 \\ 6 & 25 & -80 & -24 \\ 17 & 55 & -9 & 7 \end{bmatrix} \qquad \mathbf{B} = \begin{bmatrix} 1 & 1 & 1 & 1 \\ 1 & 2 & 2 & 2 \\ 2 & 3 & 4 & 4 \\ 4 & 5 & 6 & 7 \end{bmatrix}$$

10. ■ Write a program for inverting an $n \times n$ lower triangular matrix. The inversion procedure should contain only forward substitution. Test the program by inverting this matrix:

$$\mathbf{A} = \begin{bmatrix} 36 & 0 & 0 & 0 \\ 18 & 36 & 0 & 0 \\ 9 & 12 & 36 & 0 \\ 5 & 4 & 9 & 36 \end{bmatrix}$$

11. Use the Gauss-Seidel method to solve

$$\begin{bmatrix} -2 & 5 & 9 \\ 7 & 1 & 1 \\ -3 & 7 & -1 \end{bmatrix} \begin{bmatrix} x_1 \\ x_2 \\ x_3 \end{bmatrix} = \begin{bmatrix} 1 \\ 6 \\ -26 \end{bmatrix}$$

12. Solve the following equations with the Gauss-Seidel method:

$$\begin{bmatrix} 12 & -2 & 3 & 1 \\ -2 & 15 & 6 & -3 \\ 1 & 6 & 20 & -4 \\ 0 & -3 & 2 & 9 \end{bmatrix} \begin{bmatrix} x_1 \\ x_2 \\ x_3 \\ x_4 \end{bmatrix} = \begin{bmatrix} 0 \\ 0 \\ 20 \\ 0 \end{bmatrix}$$

13. Use the Gauss-Seidel method with relaxation to solve $\mathbf{Ax} = \mathbf{b}$, where

$$\mathbf{A} = \begin{bmatrix} 4 & -1 & 0 & 0 \\ -1 & 4 & -1 & 0 \\ 0 & -1 & 4 & -1 \\ 0 & 0 & -1 & 3 \end{bmatrix} \qquad \mathbf{b} = \begin{bmatrix} 15 \\ 10 \\ 10 \\ 10 \end{bmatrix}$$

Take $x_i = b_i/A_{ii}$ as the starting vector, and use $\omega = 1.1$ for the relaxation factor.

14. Solve the equations

$$\begin{bmatrix} 2 & -1 & 0 \\ -1 & 2 & -1 \\ 0 & -1 & 1 \end{bmatrix} \begin{bmatrix} x_1 \\ x_2 \\ x_3 \end{bmatrix} = \begin{bmatrix} 1 \\ 1 \\ 1 \end{bmatrix}$$

by the conjugate gradient method. Start with $\mathbf{x} = \mathbf{0}$.

15. Use the conjugate gradient method to solve

$$\begin{bmatrix} 3 & 0 & -1 \\ 0 & 4 & -2 \\ -1 & -2 & 5 \end{bmatrix} \begin{bmatrix} x_1 \\ x_2 \\ x_3 \end{bmatrix} = \begin{bmatrix} 4 \\ 10 \\ -10 \end{bmatrix}$$

starting with $\mathbf{x} = \mathbf{0}$.

16. ■ Write a program for solving $\mathbf{Ax} = \mathbf{b}$ by the Gauss-Seidel method based on the function `gaussSeidel`. Input should consist of the matrix $\mathbf{A}$ and the vector $\mathbf{b}$. Test the program with

$$\mathbf{A} = \begin{bmatrix} 3 & -2 & 1 & 0 & 0 & 1 \\ -2 & 4 & -2 & 1 & 0 & 0 \\ 1 & -2 & 4 & -2 & 1 & 0 \\ 0 & 1 & -2 & 4 & -2 & 1 \\ 0 & 0 & 1 & -2 & 4 & -2 \\ 1 & 0 & 0 & 1 & -2 & 3 \end{bmatrix} \qquad \mathbf{b} = \begin{bmatrix} 10 \\ -8 \\ 10 \\ 10 \\ -8 \\ 10 \end{bmatrix}$$

Note that $\mathbf{A}$ is not diagonally dominant, but this does not necessarily preclude convergence.

17. ■ Modify the program in Example 2.17 (Gauss-Seidel method) so that it will solve the following equations:

$$\begin{bmatrix} 4 & -1 & 0 & 0 & \cdots & 0 & 0 & 0 & 1 \\ -1 & 4 & -1 & 0 & \cdots & 0 & 0 & 0 & 0 \\ 0 & -1 & 4 & -1 & \cdots & 0 & 0 & 0 & 0 \\ \vdots & \vdots & \vdots & \vdots & \cdots & \vdots & \vdots & \vdots & \vdots \\ 0 & 0 & 0 & 0 & \cdots & -1 & 4 & -1 & 0 \\ 0 & 0 & 0 & 0 & \cdots & 0 & -1 & 4 & -1 \\ 1 & 0 & 0 & 0 & \cdots & 0 & 0 & -1 & 4 \end{bmatrix} \begin{bmatrix} x_1 \\ x_2 \\ x_3 \\ \vdots \\ x_{n-2} \\ x_{n-1} \\ x_n \end{bmatrix} = \begin{bmatrix} 0 \\ 0 \\ 0 \\ \vdots \\ 0 \\ 0 \\ 100 \end{bmatrix}$$

Run the program with $n = 20$ and compare the number of iterations with Example 2.17.

18. ■ Modify the program in Example 2.18 to solve the equations in Prob. 17 by the conjugate gradient method. Run the program with $n = 20$.

19. ■

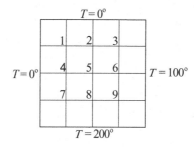

The edges of the square plate are kept at the temperatures shown. Assuming steady-state heat conduction, the differential equation governing the temperature T in the interior is

$$\frac{\partial^2 T}{\partial x^2} + \frac{\partial^2 T}{\partial y^2} = 0.$$

If this equation is approximated by finite differences using the mesh shown, we obtain the following algebraic equations for temperatures at the mesh points:

$$
\begin{bmatrix}
-4 & 1 & 0 & 1 & 0 & 0 & 0 & 0 & 0 \\
1 & -4 & 1 & 0 & 1 & 0 & 0 & 0 & 0 \\
0 & 1 & -4 & 0 & 0 & 1 & 0 & 0 & 0 \\
1 & 0 & 0 & -4 & 1 & 0 & 1 & 0 & 0 \\
0 & 1 & 0 & 1 & -4 & 1 & 0 & 1 & 0 \\
0 & 0 & 1 & 0 & 1 & -4 & 0 & 0 & 1 \\
0 & 0 & 0 & 1 & 0 & 0 & -4 & 1 & 0 \\
0 & 0 & 0 & 0 & 1 & 0 & 1 & -4 & 1 \\
0 & 0 & 0 & 0 & 0 & 1 & 0 & 1 & -4
\end{bmatrix}
\begin{bmatrix}
T_1 \\ T_2 \\ T_3 \\ T_4 \\ T_5 \\ T_6 \\ T_7 \\ T_8 \\ T_9
\end{bmatrix}
=
\begin{bmatrix}
0 \\ 0 \\ 100 \\ 0 \\ 0 \\ 100 \\ 200 \\ 200 \\ 300
\end{bmatrix}
$$

Solve these equations with the conjugate gradient method.

20. ■

The equilibrium equations of the blocks in the spring-block system are

$$3(x_2 - x_1) - 2x_1 = -80$$

$$3(x_3 - x_2) - 3(x_2 - x_1) = 0$$

$$3(x_4 - x_3) - 3(x_3 - x_2) = 0$$

$$3(x_5 - x_4) - 3(x_4 - x_3) = 60$$

$$-2x_5 - 3(x_5 - x_4) = 0$$

where x_i are the horizontal displacements of the blocks measured in mm. (a) Write a program that solves these equations by the Gauss-Seidel method without relaxation. Start with $\mathbf{x} = \mathbf{0}$ and iterate until four-figure accuracy after the decimal point is achieved. Also print the number of iterations required. (b) Solve the equations using the function `gaussSeidel` using the same convergence criterion as in part (a). Compare the number of iterations in parts (a) and (b).

21. ■ Solve the equations in Prob. 20 with the conjugate gradient method using the function `conjGrad`. Start with $\mathbf{x} = \mathbf{0}$ and iterate until four-figure accuracy after the decimal point is achieved.

2.8   Other Methods

A matrix can be decomposed in numerous ways, some of which are generally useful, whereas others find use in special applications. The most important of the latter are the QR factorization and the singular value decomposition.

The *QR decomposition* of a matrix $\mathbf{A}$ is

$$\mathbf{A} = \mathbf{QR}$$

where $\mathbf{Q}$ is an orthogonal matrix (recall that the matrix $\mathbf{Q}$ is orthogonal if $\mathbf{Q}^{-1} = \mathbf{Q}^T$) and $\mathbf{R}$ is an upper triangular matrix. Unlike LU factorization, QR decomposition does not require pivoting to sustain stability, but it does involve about twice as many operations. Because of its relative inefficiency, the QR factorization is not used as a general-purpose tool, but finds it niche in applications that put a premium on stability (e.g., solution of eigenvalue problems). The numpy module includes the function `qr` that does the factorization:

```
Q,R = numpy.linalg.qr(A)
```

The *singular value decomposition* (SVD) is a useful diagnostic tool for singular or ill-conditioned matrices. Here the factorization is

$$\mathbf{A} = \mathbf{U}\mathbf{\Lambda}\mathbf{V}^T$$

where $\mathbf{U}$ and $\mathbf{V}$ are orthogonal matrices and

$$\mathbf{\Lambda} = \begin{bmatrix} \lambda_1 & 0 & 0 & \cdots \\ 0 & \lambda_2 & 0 & \cdots \\ 0 & 0 & \lambda_3 & \cdots \\ \vdots & \vdots & \vdots & \ddots \end{bmatrix}$$

is a diagonal matrix. The λ's are called the *singular values* of the matrix $\mathbf{A}$. They can be shown to be positive or zero. If $\mathbf{A}$ is symmetric and positive definite, then the

λ's are the eigenvalues of **A**. A nice characteristic of the singular value decomposition is that it works even if **A** is singular or ill conditioned. The conditioning of **A** can be diagnosed from magnitudes of λ's: The matrix is singular if one or more of the λ's are zero, and it is ill conditioned if the condition number $\lambda_{max}/\lambda_{min}$ is very large. The singular value decomposition function that comes with the numpy module is svd:

```
U,lam,V = numpy.linalg.svd(A)
```

3 Interpolation and Curve Fitting

> Given the $n+1$ data points (x_i, y_i), $i = 0, 1, \ldots, n$, estimate $y(x)$.

3.1 Introduction

Discrete data sets, or tables of the form

x_0	x_1	x_2	$\cdots$	x_n
y_0	y_1	y_2	$\cdots$	y_n

are commonly involved in technical calculations. The source of the data may be experimental observations or numerical computations. There is a distinction between interpolation and curve fitting. In interpolation we construct a curve *through* the data points. In doing so, we make the implicit assumption that the data points are accurate and distinct. In contrast, curve fitting is applied to data that contain scatter (noise), usually caused by measurement errors. Here we want to find a smooth curve that *approximates* the data in some sense. Thus the curve does not necessarily hit the data points. The difference between interpolation and curve fitting is illustrated in Figure 3.1.

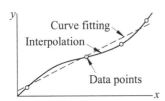

Figure 3.1. Interpolation vs. curve fitting.

3.2   Polynomial Interpolation

Lagrange's Method

The simplest form of an interpolant is a polynomial. It is always possible to construct a *unique* polynomial of degree n that passes through $n + 1$ distinct data points. One means of obtaining this polynomial is the formula of Lagrange,

$$P_n(x) = \sum_{i=0}^{n} y_i \ell_i(x) \tag{3.1a}$$

where the subscript n denotes the degree of the polynomial and

$$\ell_i(x) = \frac{x - x_0}{x_i - x_0} \cdot \frac{x - x_1}{x_i - x_1} \cdots \cdot \frac{x - x_{i-1}}{x_i - x_{i-1}} \cdot \frac{x - x_{i+1}}{x_i - x_{i+1}} \cdots \frac{x - x_n}{x_i - x_n}$$

$$= \prod_{\substack{j=0 \\ j \neq i}}^{n} \frac{x - x_i}{x_i - x_j}, \quad i = 0, 1, \dots, n \tag{3.1b}$$

are called the *cardinal functions*.

For example, if $n = 1$, the interpolant is the straight line $P_1(x) = y_0 \ell_0(x) + y_1 \ell_1(x)$, where

$$\ell_0(x) = \frac{x - x_1}{x_0 - x_1} \qquad \ell_1(x) = \frac{x - x_0}{x_1 - x_0}$$

With $n = 2$, interpolation is parabolic: $P_2(x) = y_0 \ell_0(x) + y_1 \ell_1(x) + y_2 \ell_2(x)$, where now

$$\ell_0(x) = \frac{(x - x_1)(x - x_2)}{(x_0 - x_1)(x_0 - x_2)}$$

$$\ell_1(x) = \frac{(x - x_0)(x - x_2)}{(x_1 - x_0)(x_1 - x_2)}$$

$$\ell_2(x) = \frac{(x - x_0)(x - x_1)}{(x_2 - x_0)(x_2 - x_1)}$$

The cardinal functions are polynomials of degree n and have the property

$$\ell_i(x_j) = \begin{cases} 0 \text{ if } i \neq j \\ 1 \text{ if } i = j \end{cases} = \delta_{ij} \tag{3.2}$$

where δ_{ij} is the Kronecker delta. This property is illustrated in Figure 3.2 for three-point interpolation ($n = 2$) with $x_0 = 0$, $x_1 = 2$, and $x_2 = 3$.

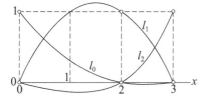

Figure 3.2. Example of quadratic cardinal functions.

To prove that the interpolating polynomial passes through the data points, we substitute $x = x_j$ into Eq. (3.1a) and then use Eq. (3.2). The result is

$$P_n(x_j) = \sum_{i=0}^{n} y_i \ell_i(x_j) = \sum_{i=0}^{n} y_i \delta_{ij} = y_j$$

It can be shown that the error in polynomial interpolation is

$$f(x) - P_n(x) = \frac{(x - x_0)(x - x_1) \cdots (x - x_n)}{(n + 1)!} f^{(n+1)}(\xi) \qquad (3.3)$$

where ξ lies somewhere in the interval (x_0, x_n); its value is otherwise unknown. It is instructive to note that the farther a data point is from x, the more it contributes to the error at x.

Newton's Method

Although Lagrange's method is conceptually simple, it does not lend itself to an efficient algorithm. A better computational procedure is obtained with Newton's method, where the interpolating polynomial is written in the form

$$P_n(x) = a_0 + (x - x_0)a_1 + (x - x_0)(x - x_1)a_2 + \cdots + (x - x_0)(x - x_1) \cdots (x - x_{n-1})a_n$$

This polynomial lends itself to an efficient evaluation procedure. Consider, for example, four data points ($n = 3$). Here the interpolating polynomial is

$$P_3(x) = a_0 + (x - x_0)a_1 + (x - x_0)(x - x_1)a_2 + (x - x_0)(x - x_1)(x - x_2)a_3$$
$$= a_0 + (x - x_0)\{a_1 + (x - x_1)[a_2 + (x - x_2)a_3]\}$$

which can be evaluated backward with the following recurrence relations:

$$P_0(x) = a_3$$
$$P_1(x) = a_2 + (x - x_2)P_0(x)$$
$$P_2(x) = a_1 + (x - x_1)P_1(x)$$
$$P_3(x) = a_0 + (x - x_0)P_2(x)$$

For arbitrary n we have

$$P_0(x) = a_n \qquad P_k(x) = a_{n-k} + (x - x_{n-k})P_{k-1}(x), \quad k = 1, 2, \ldots, n \qquad (3.4)$$

Denoting the x-coordinate array of the data points by xData and the degree of the polynomial by n, we have the following algorithm for computing $P_n(x)$:

```
p = a[n]
for k in range(1,n+1):
    p = a[n-k] + (x - xData[n-k])*p
```

The coefficients of P_n are determined by forcing the polynomial to pass through each data point: $y_i = P_n(x_i)$, $i = 0, 1, \ldots, n$. This yields these simultaneous equations:

$$y_0 = a_0$$

$$y_1 = a_0 + (x_1 - x_0)a_1$$

$$y_2 = a_0 + (x_2 - x_0)a_1 + (x_2 - x_0)(x_2 - x_1)a_2 \qquad \text{(a)}$$

$$\vdots$$

$$y_n = a_0 + (x_n - x_0)a_1 + \cdots + (x_n - x_0)(x_n - x_1)\cdots(x_n - x_{n-1})a_n$$

Introducing the *divided differences*

$$\nabla y_i = \frac{y_i - y_0}{x_i - x_0}, \quad i = 1, 2, \ldots, n$$

$$\nabla^2 y_i = \frac{\nabla y_i - \nabla y_1}{x_i - x_1}, \quad i = 2, 3, \ldots, n$$

$$\nabla^3 y_i = \frac{\nabla^2 y_i - \nabla^2 y_2}{x_i - x_2}, \quad i = 3, 4, \ldots n \qquad (3.5)$$

$$\vdots$$

$$\nabla^n y_n = \frac{\nabla^{n-1} y_n - \nabla^{n-1} y_{n-1}}{x_n - x_{n-1}}$$

the solution of Eqs. (a) is

$$a_0 = y_0 \qquad a_1 = \nabla y_1 \qquad a_2 = \nabla^2 y_2 \qquad \cdots \qquad a_n = \nabla^n y_n \qquad (3.6)$$

If the coefficients are computed by hand, it is convenient to work with the format in Table 3.1 (shown for $n = 4$).

x_0	y_0				
x_1	y_1	∇y_1			
x_2	y_2	∇y_2	$\nabla^2 y_2$		
x_3	y_3	∇y_3	$\nabla^2 y_3$	$\nabla^3 y_3$	
x_4	y_4	∇y_4	$\nabla^2 y_4$	$\nabla^3 y_4$	$\nabla^4 y_4$

Table 3.1. Tableau for Newton's method.

The diagonal terms (y_0, ∇y_1, $\nabla^2 y_2$, $\nabla^3 y_3$, and $\nabla^4 y_4$) in the table are the coefficients of the polynomial. If the data points are listed in a different order, the entries in the table will change, but the resultant polynomial will be the same—recall that a polynomial of degree n interpolating $n + 1$ distinct data points is unique.

Machine computations can be carried out within a one-dimensional array **a** employing the following algorithm (we use the notation $m = n + 1 =$ number of data points):

```
a = yData.copy()
for k in range(1,m):
    for i in range(k,m):
        a[i] = (a[i] - a[k-1])/(xData[i] - xData[k-1])
```

Initially, **a** contains the y-coordinates of the data, so that it is identical to the second column in Table 3.1. Each pass through the outer loop generates the entries in the next column, which overwrite the corresponding elements of **a**. Therefore, **a** ends up containing the diagonal terms of Table 3.1 (i.e., the coefficients of the polynomial).

■ newtonPoly

This module contains the two functions required for interpolation by Newton's method. Given the data point arrays xData and yData, the function coeffts returns the coefficient array **a**. After the coefficients are found, the interpolant $P_n(x)$ can be evaluated at any value of x with the function evalPoly.

```
## module newtonPoly
''' p = evalPoly(a,xData,x).
    Evaluates Newton's polynomial p at x. The coefficient
    vector {a} can be computed by the function 'coeffts'.

    a = coeffts(xData,yData).
    Computes the coefficients of Newton's polynomial.
'''
def evalPoly(a,xData,x):
    n = len(xData) - 1  # Degree of polynomial
    p = a[n]
    for k in range(1,n+1):
        p = a[n-k] + (x -xData[n-k])*p
    return p

def coeffts(xData,yData):
    m = len(xData)  # Number of data points
    a = yData.copy()
    for k in range(1,m):
        a[k:m] = (a[k:m] - a[k-1])/(xData[k:m] - xData[k-1])
    return a
```

Neville's Method

Newton's method of interpolation involved two steps: computation of the coefficients, followed by evaluation of the polynomial. It works well if the interpolation is carried out repeatedly at different values of x using the same polynomial. If only one point is to interpolated, a method that computes the interpolant in a single step, such as Neville's algorithm, is a more convenient choice.

Let $P_k[x_i, x_{i+1}, \ldots, x_{i+k}]$ denote the polynomial of degree k that passes through the $k + 1$ data points $(x_i, y_i), (x_{i+1}, y_{i+1}), \ldots, (x_{i+k}, y_{i+k})$. For a single data point, we have

$$P_0[x_i] = y_i \tag{3.7}$$

The interpolant based on two data points is

$$P_1[x_i, x_{i+1}] = \frac{(x - x_{i+1}) P_0[x_i] + (x_i - x) P_0[x_{i+1}]}{x_i - x_{i+1}}$$

It is easily verified that $P_1[x_i, x_{i+1}]$ passes through the two data points; that is, $P_1[x_i, x_{i+1}] = y_i$ when $x = x_i$, and $P_1[x_i, x_{i+1}] = y_{i+1}$ when $x = x_{i+1}$.

The three-point interpolant is

$$P_2[x_i, x_{i+1}, x_{i+2}] = \frac{(x - x_{i+2}) P_1[x_i, x_{i+1}] + (x_i - x) P_1[x_{i+1}, x_{i+2}]}{x_i - x_{i+2}}$$

To show that this interpolant does intersect the data points, we first substitute $x = x_i$, obtaining

$$P_2[x_i, x_{i+1}, x_{i+2}] = P_1[x_i, x_{i+1}] = y_i$$

Similarly, $x = x_{i+2}$ yields

$$P_2[x_i, x_{i+1}, x_{i+2}] = P_2[x_{i+1}, x_{i+2}] = y_{i+2}$$

Finally, when $x = x_{i+1}$ we have

$$P_1[x_i, x_{i+1}] = P_1[x_{i+1}, x_{i+2}] = y_{i+1}$$

so that

$$P_2[x_i, x_{i+1}, x_{i+2}] = \frac{(x_{i+1} - x_{i+2}) y_{i+1} + (x_i - x_{i+1}) y_{i+1}}{x_i - x_{i+2}} = y_{i+1}$$

Having established the pattern, we can now deduce the general recursive formula:

$$P_k[x_i, x_{i+1}, \ldots, x_{i+k}] \tag{3.8}$$
$$= \frac{(x - x_{i+k}) P_{k-1}[x_i, x_{i+1}, \ldots, x_{i+k-1}] + (x_i - x) P_{k-1}[x_{i+1}, x_{i+2}, \ldots, x_{i+k}]}{x_i - x_{i+k}}$$

Given the value of x, the computations can be carried out in the following tabular format (shown for four data points):

	$k = 0$	$k = 1$	$k = 2$	$k = 3$
x_0	$P_0[x_0] = y_0$	$P_1[x_0, x_1]$	$P_2[x_0, x_1, x_2]$	$P_3[x_0, x_1, x_2, x_3]$
x_1	$P_0[x_1] = y_1$	$P_1[x_1, x_2]$	$P_2[x_1, x_2, x_3]$	
x_2	$P_0[x_2] = y_2$	$P_1[x_2, x_3]$		
x_3	$P_0[x_3] = y_3$			

Table 3.2. Tableau for Neville's method.

Denoting the number of data points by m, the algorithm that computes the elements of the table is

```
y = yData.copy()
for k in range (1,m):
    for i in range(m-k):
        y[i] = ((x - xData[i+k])*y[i] + (xData[i] - x)*y[i+1])/ \
            (xData[i]-xData[i+k])
```

This algorithm works with the one-dimensional array **y**, which initially contains the y-values of the data (the second column in Table 3.2). Each pass through the outer loop computes the elements of **y** in the next column, which overwrite the previous entries. At the end of the procedure, **y** contains the diagonal terms of the table. The value of the interpolant (evaluated at x) that passes through all the data points is the first element of **y**.

■ neville

The following function implements Neville's method; it returns $P_n(x)$:

```
## module neville
''' p = neville(xData,yData,x).
    Evaluates the polynomial interpolant p(x) that passes
    through the specified data points by Neville's method.
'''
def neville(xData,yData,x):
    m = len(xData)    # number of data points
    y = yData.copy()
    for k in range(1,m):
        y[0:m-k] = ((x - xData[k:m])*y[0:m-k] +          \
                    (xData[0:m-k] - x)*y[1:m-k+1])/ \
                    (xData[0:m-k] - xData[k:m])
    return y[0]
```

Limitations of Polynomial Interpolation

Polynomial interpolation should be carried out with the fewest feasible number of data points. Linear interpolation, using the nearest two points, is often sufficient if the data points are closely spaced. Three to six nearest-neighbor points produce good results in most cases. An interpolant intersecting more than six points must be viewed with suspicion. The reason is that the data points that are far from the point of interest do not contribute to the accuracy of the interpolant. In fact, they can be detrimental.

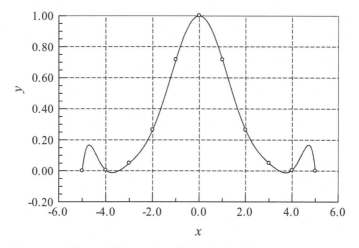

Figure 3.3. Polynomial interpolant displaying oscillations.

The danger of using too many points is illustrated in Figure 3.3. The 11 equally spaced data points are represented by the circles. The solid line is the interpolant, a polynomial of degree 10, that intersects all the points. As seen in the figure, a polynomial of such a high degree has a tendency to oscillate excessively between the data points. A much smoother result would be obtained by using a cubic interpolant spanning four nearest-neighbor points.

Polynomial extrapolation (interpolating outside the range of data points) is dangerous. As an example, consider Figure 3.4. Six data points are shown as circles. The fifth-degree interpolating polynomial is represented by the solid line. The interpolant looks fine within the range of data points, but drastically departs from the obvious trend when $x > 12$. Extrapolating y at $x = 14$, for example, would be absurd in this case.

If extrapolation cannot be avoided, the following measures can be useful:

- Plot the data and visually verify that the extrapolated value makes sense.
- Use a low-order polynomial based on nearest-neighbor data points. A linear or quadratic interpolant, for example, would yield a reasonable estimate of $y(14)$ for the data in Figure 3.4.

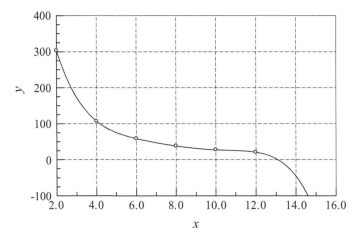

Figure 3.4. Extrapolation may not follow the trend of data.

- Work with a plot of log x vs. log y, which is usually much smoother than the x-y curve and thus safer to extrapolate. Frequently this plot is almost a straight line. This is illustrated in Figure 3.5, which represents the logarithmic plot of the data in Figure 3.4.

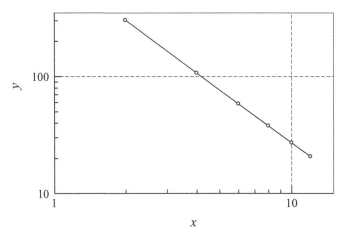

Figure 3.5. Logarithmic plot of the data in Figure 3.4.

EXAMPLE 3.1

Given the data points

x	0	2	3
y	7	11	28

use Lagrange's method to determine y at $x = 1$.

Solution

$$\ell_0 = \frac{(x - x_1)(x - x_2)}{(x_0 - x_1)(x_0 - x_2)} = \frac{(1 - 2)(1 - 3)}{(0 - 2)(0 - 3)} = \frac{1}{3}$$

$$\ell_1 = \frac{(x - x_0)(x - x_2)}{(x_1 - x_0)(x_1 - x_2)} = \frac{(1 - 0)(1 - 3)}{(2 - 0)(2 - 3)} = 1$$

$$\ell_2 = \frac{(x - x_0)(x - x_1)}{(x_2 - x_0)(x_2 - x_1)} = \frac{(1 - 0)(1 - 2)}{(3 - 0)(3 - 2)} = -\frac{1}{3}$$

$$y = y_0\ell_0 + y_1\ell_1 + y_2\ell_2 = \frac{7}{3} + 11 - \frac{28}{3} = 4$$

EXAMPLE 3.2

The data points

x	−2	1	4	−1	3	−4
y	−1	2	59	4	24	−53

lie on a polynomial. Determine the degree of this polynomial by constructing a divided difference table, similar to Table 3.1.

Solution

i	x_i	y_i	∇y_i	$\nabla^2 y_i$	$\nabla^3 y_i$	$\nabla^4 y_i$	$\nabla^5 y_i$
0	−2	−1					
1	1	2	1				
2	4	59	10	3			
3	−1	4	5	−2	1		
4	3	24	5	2	1	0	
5	−4	−53	26	−5	1	0	0

Here are a few sample calculations used in arriving at the figures in the table:

$$\nabla y_2 = \frac{y_2 - y_0}{x_2 - x_0} = \frac{59 - (-1)}{4 - (-2)} = 10$$

$$\nabla^2 y_2 = \frac{\nabla y_2 - \nabla y_1}{x_2 - x_1} = \frac{10 - 1}{4 - 1} = 3$$

$$\nabla^3 y_5 = \frac{\nabla^2 y_5 - \nabla^2 y_2}{x_5 - x_2} = \frac{-5 - 3}{-4 - 4} = 1$$

From the table we see that the last nonzero coefficient (last nonzero diagonal term) of Newton's polynomial is $\nabla^3 y_3$, which is the coefficient of the cubic term. Hence the polynomial is a cubic.

EXAMPLE 3.3

Given the data points

x	4.0	3.9	3.8	3.7
y	−0.06604	−0.02724	0.01282	0.05383

determine the root of $y(x) = 0$ by Neville's method.

Solution. This is an example of *inverse interpolation*, in which the roles of x and y are interchanged. Instead of computing y at a given x, we are finding x that corresponds to a given y (in this case, $y = 0$). Employing the format of Table 3.2 (with x and y interchanged, of course), we obtain

i	y_i	$P_0[\,] = x_i$	$P_1[\,,]$	$P_2[\,,\,]$	$P_3[\,,\,,]$
0	-0.06604	4.0	3.8298	3.8316	3.8317
1	-0.02724	3.9	3.8320	3.8318	
2	0.01282	3.8	3.8313		
3	0.05383	3.7			

The following are sample computations used in the table:

$$P_1[y_0, y_1] = \frac{(y - y_1)\, P_0[y_0] + (y_0 - y)\, P_0[y_1]}{y_0 - y_1}$$

$$= \frac{(0 + 0.02724)(4.0) + (-0.06604 - 0)(3.9)}{-0.06604 + 0.02724} = 3.8298$$

$$P_2[y_1, y_2, y_3] = \frac{(y - y_3)\, P_1[y_1, y_2] + (y_1 - y)\, P_1[y_2, y_3]}{y_1 - y_3}$$

$$= \frac{(0 - 0.05383)(3.8320) + (-0.02724 - 0)(3.8313)}{-0.02724 - 0.05383} = 3.8318$$

All the P's in the table are estimates of the root resulting from different orders of interpolation involving different data points. For example, $P_1[y_0, y_1]$ is the root obtained from linear interpolation based on the first two points, and $P_2[y_1, y_2, y_3]$ is the result from quadratic interpolation using the last three points. The root obtained from cubic interpolation over all four data points is $x = P_3[y_0, y_1, y_2, y_3] = 3.8317$.

EXAMPLE 3.4

The data points in the table lie on the plot of $f(x) = 4.8 \cos \dfrac{\pi x}{20}$. Interpolate this data by Newton's method at $x = 0, 0.5, 1.0, \ldots, 8.0$, and compare the results with the "exact" values $y_i = f(x_i)$.

x	0.15	2.30	3.15	4.85	6.25	7.95
y	4.79867	4.49013	4.2243	3.47313	2.66674	1.51909

Solution

```
#!/usr/bin/python
## example3_4
import numpy as np
import math
from newtonPoly import *

xData = np.array([0.15,2.3,3.15,4.85,6.25,7.95])
yData = np.array([4.79867,4.49013,4.2243,3.47313,2.66674,1.51909])
a = coeffts(xData,yData)
print(" x    yInterp   yExact")
```

```
print("----------------------")
for x in np.arange(0.0,8.1,0.5):
    y = evalPoly(a,xData,x)
    yExact = 4.8*math.cos(math.pi*x/20.0)
    print('{:3.1f} {:9.5f} {:9.5f}'.format(x,y,yExact))
input("\nPress return to exit")
```

The results are

x	yInterp	yExact
0.0	4.80003	4.80000
0.5	4.78518	4.78520
1.0	4.74088	4.74090
1.5	4.66736	4.66738
2.0	4.56507	4.56507
2.5	4.43462	4.43462
3.0	4.27683	4.27683
3.5	4.09267	4.09267
4.0	3.88327	3.88328
4.5	3.64994	3.64995
5.0	3.39411	3.39411
5.5	3.11735	3.11735
6.0	2.82137	2.82137
6.5	2.50799	2.50799
7.0	2.17915	2.17915
7.5	1.83687	1.83688
8.0	1.48329	1.48328

Rational Function Interpolation

Some data are better interpolated by *rational functions* rather than polynomials. A rational function $R(x)$ is the quotient of two polynomials:

$$R(x) = \frac{P_m(x)}{Q_n(x)} = \frac{a_1 x^m + a_2 x^{m-1} + \cdots + a_m x + a_{m+1}}{b_1 x^n + b_2 x^{n-1} + \cdots + b_n x + b_{n+1}}$$

Because $R(x)$ is a ratio, it can be scaled so that one of the coefficients (usually b_{n+1}) is unity. That leaves $m + n + 1$ undetermined coefficients that must be computed by forcing $R(x)$ through $m + n + 1$ data points.

A popular version of $R(x)$ is the so-called *diagonal rational function*, where the degree of the numerator is equal to that of the denominator ($m = n$) if $m + n$ is even, or less by one ($m = n - 1$) if $m + n$ is odd. The advantage of using the diagonal form is that the interpolation can be carried out with a Neville-type algorithm, similar to that outlined in Table 3.2. The recursive formula that is the basis of the algorithm is due to Stoer and Bulirsch.[1] It is somewhat more complex than Eq. (3.8) used in Neville's method:

$$R[x_i, x_{i+1}, \ldots, x_{i+k}] = R[x_{i+1}, x_{i+2}, \ldots, x_{i+k}] \tag{3.9a}$$
$$+ \frac{R[x_{i+1}, x_{i+2}, \ldots, x_{i+k}] - R[x_i, x_{i+1}, \ldots, x_{i+k-1}]}{S}$$

where

$$S = \frac{x - x_i}{x - x_{i+k}} \left(1 - \frac{R[x_{i+1}, x_{i+2}, \ldots, x_{i+k}] - R[x_i, x_{i+1}, \ldots, x_{i+k-1}]}{R[x_{i+1}, x_{i+2}, \ldots, x_{i+k}] - R[x_{i+1}, x_{i+2}, \ldots, x_{i+k-1}]} \right) - 1 \tag{3.9b}$$

In Eqs. (3.9) $R[x_i, x_{i+1}, \ldots, x_{i+k}]$ denotes the diagonal rational function that passes through the data points (x_i, y_i), (x_{i+1}, y_{i+1}), $\ldots$, (x_{i+k}, y_{i+k}). It is also understood that $R[x_i, x_{i+1}, \ldots, x_{i-1}] = 0$ (corresponding to the case $k = -1$) and $R[x_i] = y_i$ (the case $k = 0$).

The computations can be carried out in a tableau, similar to Table 3.2 used for Neville's method. Table 3.3 is an example of the tableau for four data points. We start by filling the column $k = -1$ with zeros and entering the values of y_i in the column $k = 0$. The remaining entries are computed by applying Eqs. (3.9).

	$k = -1$	$k = 0$	$k = 1$	$k = 2$	$k = 3$
x_1	0	$R[x_1] = y_1$	$R[x_1, x_2]$	$R[x_1, x_2, x_3]$	$R[x_1, x_2, x_3, x_4]$
x_2	0	$R[x_2] = y_2$	$R[x_2, x_3]$	$R[x_2, x_3, x_4]$	
x_3	0	$R[x_3] = y_3$	$R[x_3, x_4]$		
x_4	0	$R[x_4] = y_4$			

Table 3.3. A Tableau for Four Data Points.

■ rational

We managed to implement Neville's algorithm with the tableau "compressed" to a one-dimensional array. This will not work with the rational function interpolation, where the formula for computing an R in the kth column involves entries in columns $k - 1$ as well as $k - 2$. However, we can work with two one-dimensional arrays, one array (called r in the program) containing the latest values of R while the other array (rOld) saves the previous entries. Here is the algorithm for diagonal rational function interpolation:

```
## module rational
''' p = rational(xData,yData,x)
    Evaluates the diagonal rational function interpolant p(x)
    that passes through the data points
'''
import numpy as np
```

---

[1] Stoer, J., and Bulirsch, R., *Introduction to Numerical Analysis,* Springer, 1980.

```
def rational(xData,yData,x):
    m = len(xData)
    r = yData.copy()
    rOld = np.zeros(m)
    for k in range(m-1):
        for i in range(m-k-1):
            if abs(x - xData[i+k+1]) < 1.0e-9:
                return yData[i+k+1]
            else:
                c1 = r[i+1] - r[i]
                c2 = r[i+1] - rOld[i+1]
                c3 = (x - xData[i])/(x - xData[i+k+1])
                r[i] = r[i+1] + c1/(c3*(1.0 - c1/c2) - 1.0)
                rOld[i+1] = r[i+1]
    return r[0]
```

EXAMPLE 3.5
Given the data

x	0	0.6	0.8	0.95
y	0	1.3764	3.0777	12.7062

determine $y(0.5)$ by the diagonal rational function interpolation.

Solution. The plot of the data points indicates that y may have a pole at around $x = 1$. Such a function is a very poor candidate for polynomial interpolation, but can be readily represented by a rational function.

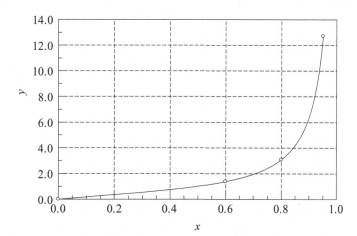

We set up our work in the format of Table 3.3. After completing the computations, the table looks like this:

	k = -1	k = 0	k = 1	k = 2	k = 3
i = 1	0	0	0	0	0.9544	1.0131
i = 2	0.6	0	1.3764	1.0784	1.0327	
i = 3	0.8	0	3.0777	1.2235		
i = 4	0.95	0	12.7062			

Let us now look at a few sample computations. We obtain, for example, $R[x_3, x_4]$ by substituting $i = 3$, $k = 1$ into Eqs. (3.9). This yields

$$S = \frac{x - x_3}{x - x_4}\left(1 - \frac{R[x_4] - R[x_3]}{R[x_4] - R[x_4, \ldots, x_3]}\right) - 1$$

$$= \frac{0.5 - 0.8}{0.5 - 0.95}\left(1 - \frac{12.7062 - 3.0777}{12.7062 - 0}\right) - 1 = -0.83852$$

$$R[x_3, x_4] = R[x_4] + \frac{R[x_4] - R[x_3]}{S}$$

$$= 12.7062 + \frac{12.7062 - 3.0777}{-0.83852} = 1.2235$$

The entry $R[x_2, x_3, x_4]$ is obtained with $i = 2$, $k = 2$. The result is

$$S = \frac{x - x_2}{x - x_4}\left(1 - \frac{R[x_3, x_4] - R[x_2, x_3]}{R[x_3, x_4] - R[x_3]}\right) - 1$$

$$= \frac{0.5 - 0.6}{0.5 - 0.95}\left(1 - \frac{1.2235 - 1.0784}{1.2235 - 3.0777}\right) - 1 = -0.76039$$

$$R[x_2, x_3, x_4] = R[x_3, x_4] + \frac{R[x_3, x_4] - R[x_2, x_3]}{S}$$

$$= 1.2235 + \frac{1.2235 - 1.0784}{-0.76039} = 1.0327$$

The interpolant at $x = 0.5$ based on all four data points is $R[x_1, x_2, x_3, x_4] = 1.0131$.

EXAMPLE 3.6
Interpolate the data shown at x-increments of 0.05 and plot the results. Use both the polynomial interpolation and the rational function interpolation.

x	0.1	0.2	0.5	0.6	0.8	1.2	1.5
y	-1.5342	-1.0811	-0.4445	-0.3085	-0.0868	0.2281	0.3824

Solution

```
#!/usr/bin/python
## example 3_6
import numpy as np
from rational import *
from neville import *
import matplotlib.pyplot as plt
```

```
xData = np.array([0.1,0.2,0.5,0.6,0.8,1.2,1.5])
yData = np.array([-1.5342,-1.0811,-0.4445,-0.3085, \
                  -0.0868,0.2281,0.3824])
x = np.arange(0.1,1.55,0.05)
n = len(x)
y = np.zeros((n,2))
for i in range(n):
    y[i,0] = rational(xData,yData,x[i])
    y[i,1] = neville(xData,yData,x[i])
plt.plot(xData,yData,'o',x,y[:,0],'-',x,y[:,1],'--')
plt.xlabel('x')
plt.legend(('Data','Rational','Neville'),loc = 0)
plt.show()
input("\nPress return to exit")
```

The output is shown next. In this case, the rational function interpolant is smoother, and thus superior, to the polynomial (Neville's) interpolant.

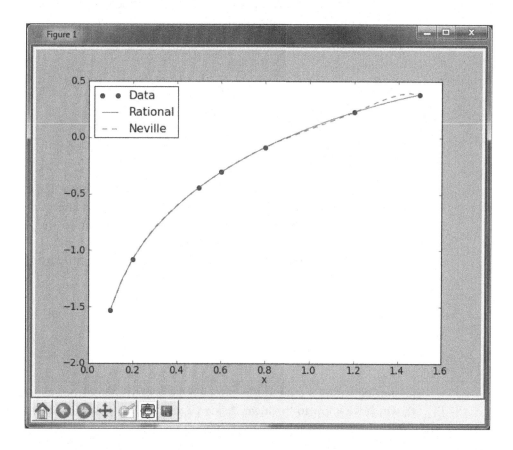

3.3 Interpolation with Cubic Spline

If there are more than a few data points, a cubic spline is hard to beat as a global interpolant. It is considerably "stiffer" than a polynomial in the sense that it has less tendency to oscillate between data points.

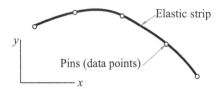

Elastic strip

Pins (data points)

Figure 3.6. Mechanical model of a natural cubic spline.

The mechanical model of a cubic spline is shown in Figure 3.6. It is a thin, elastic beam that is attached with pins to the data points. Because the beam is unloaded between the pins, each segment of the spline curve is a cubic polynomial—recall from beam theory that $d^4y/dx^4 = q/(EI)$, so that $y(x)$ is a cubic since $q = 0$. At the pins, the slope and bending moment (and hence the second derivative) are continuous. There is no bending moment at the two end pins; consequently, the second derivative of the spline is zero at the end points. Because these end conditions occur naturally in the beam model, the resulting curve is known as the *natural cubic spline*. The pins (i.e., the data points) are called the *knots* of the spline.

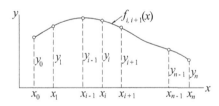

Figure 3.7. A cubic spline.

Figure 3.7 shows a cubic spline that spans $n + 1$ knots. We use the notation $f_{i,i+1}(x)$ for the cubic polynomial that spans the segment between knots i and $i + 1$. Note that the spline is a *piecewise cubic* curve, put together from the n cubics $f_{0,1}(x), f_{1,2}(x), \ldots, f_{n-1,n}(x)$, all of which have different coefficients.

Denoting the second derivative of the spline at knot i by k_i, continuity of second derivatives requires that

$$f''_{i-1,i}(x_i) = f''_{i,i+1}(x_i) = k_i \tag{a}$$

At this stage, each k is unknown, except for

$$k_0 = k_n = 0$$

The starting point for computing the coefficients of $f_{i,i+1}(x)$ is the expression for $f''_{i,i+1}(x)$, which we know to be linear. Using Lagrange's two-point interpolation, we can write

$$f''_{i,i+1}(x) = k_i \ell_i(x) + k_{i+1}\ell_{i+1}(x)$$

where

$$\ell_i(x) = \frac{x - x_{i+1}}{x_i - x_{i+1}} \qquad \ell_{1+1}(x) = \frac{x - x_i}{x_{i+1} - x_i}$$

Therefore,

$$f''_{i,i+1}(x) = \frac{k_i(x - x_{i+1}) - k_{i+1}(x - x_i)}{x_i - x_{i+1}} \qquad (b)$$

Integrating twice with respect to x, we obtain

$$f_{i,i+1}(x) = \frac{k_i(x - x_{i+1})^3 - k_{i+1}(x - x_i)^3}{6(x_i - x_{i+1})} + A(x - x_{i+1}) - B(x - x_i) \qquad (c)$$

where A and B are constants of integration. The terms arising from the integration would usually be written as $Cx + D$. By letting $C = A - B$ and $D = -Ax_{i+1} + Bx_i$, we end up with the last two terms of Eq. (c), which are more convenient to use in the computations that follow.

Imposing the condition $f_{i,i+1}(x_i) = y_i$, we get from Eq. (c)

$$\frac{k_i(x_i - x_{i+1})^3}{6(x_i - x_{i+1})} + A(x_i - x_{i+1}) = y_i$$

Therefore,

$$A = \frac{y_i}{x_i - x_{i+1}} - \frac{k_i}{6}(x_i - x_{i+1}) \qquad (d)$$

Similarly, $f_{i,i+1}(x_{i+1}) = y_{i+1}$ yields

$$B = \frac{y_{i+1}}{x_i - x_{i+1}} - \frac{k_{i+1}}{6}(x_i - x_{i+1}) \qquad (e)$$

Substituting Eqs. (d) and (e) into Eq. (c) results in

$$\begin{aligned} f_{i,i+1}(x) = \frac{k_i}{6}\left[\frac{(x - x_{i+1})^3}{x_i - x_{i+1}} - (x - x_{i+1})(x_i - x_{i+1})\right] \\ - \frac{k_{i+1}}{6}\left[\frac{(x - x_i)^3}{x_i - x_{i+1}} - (x - x_i)(x_i - x_{i+1})\right] \\ + \frac{y_i(x - x_{i+1}) - y_{i+1}(x - x_i)}{x_i - x_{i+1}} \end{aligned} \qquad (3.10)$$

The second derivatives k_i of the spline at the interior knots are obtained from the slope continuity conditions $f'_{i-1,i}(x_i) = f'_{i,i+1}(x_i)$, where $i = 1, 2, \ldots, n - 1$. After a little algebra, this results in the simultaneous equations

$$k_{i-1}(x_{i-1} - x_i) + 2k_i(x_{i-1} - x_{i+1}) + k_{i+1}(x_i - x_{i+1})$$

$$= 6\left(\frac{y_{i-1} - y_i}{x_{i-1} - x_i} - \frac{y_i - y_{i+1}}{x_i - x_{i+1}}\right), \quad i = 1, 2, \cdots, n - 1 \qquad (3.11)$$

Because Eqs. (3.11) have a tridiagonal coefficient matrix, they can be solved economically with the functions in module LUdecomp3 described in Section 2.4.

If the data points are evenly spaced at intervals h, then $x_{i-1} - x_i = x_i - x_{i+1} = -h$, and the Eqs. (3.11) simplify to

$$k_{i-1} + 4k_i + k_{i+1} = \frac{6}{h^2}(y_{i-1} - 2y_i + y_{i+1}), \quad i = 1, 2, \ldots, n-1 \qquad (3.12)$$

■ cubicSpline

The first stage of cubic spline interpolation is to set up Eqs. (3.11) and solve them for the unknown k's (recall that $k_0 = k_n = 0$). This task is carried out by the function curvatures. The second stage is the computation of the interpolant at x from Eq. (3.10). This step can be repeated any number of times with different values of x using the function evalSpline. The function findSegment embedded in evalSpline finds the segment of the spline that contains x using the method of bisection. It returns the segment number; that is, the value of the subscript i in Eq. (3.10).

```
## module cubicSpline
''' k = curvatures(xData,yData).
    Returns the curvatures of cubic spline at its knots.

    y = evalSpline(xData,yData,k,x).
    Evaluates cubic spline at x. The curvatures k can be
    computed with the function 'curvatures'.
'''
import numpy as np
from LUdecomp3 import *

def curvatures(xData,yData):
    n = len(xData) - 1
    c = np.zeros(n)
    d = np.ones(n+1)
    e = np.zeros(n)
    k = np.zeros(n+1)
    c[0:n-1] = xData[0:n-1] - xData[1:n]
    d[1:n] = 2.0*(xData[0:n-1] - xData[2:n+1])
    e[1:n] = xData[1:n] - xData[2:n+1]
    k[1:n] =6.0*(yData[0:n-1] - yData[1:n]) \
                /(xData[0:n-1] - xData[1:n]) \
            -6.0*(yData[1:n] - yData[2:n+1])   \
                /(xData[1:n] - xData[2:n+1])
    LUdecomp3(c,d,e)
    LUsolve3(c,d,e,k)
    return k

def evalSpline(xData,yData,k,x):
```

```
def findSegment(xData,x):
    iLeft = 0
    iRight = len(xData)- 1
    while 1:
        if (iRight-iLeft) <= 1: return iLeft
        i =(iLeft + iRight)/2
        if x < xData[i]: iRight = i
        else: iLeft = i

i = findSegment(xData,x)
h = xData[i] - xData[i+1]
y = ((x - xData[i+1])**3/h - (x - xData[i+1])*h)*k[i]/6.0 \
  - ((x - xData[i])**3/h - (x - xData[i])*h)*k[i+1]/6.0   \
  + (yData[i]*(x - xData[i+1])                            \
    - yData[i+1]*(x - xData[i]))/h
return y
```

EXAMPLE 3.7
Use a natural cubic spline to determine y at $x = 1.5$. The data points are

x	1	2	3	4	5
y	0	1	0	1	0

Solution. The five knots are equally spaced at $h = 1$. Recalling that the second derivative of a natural spline is zero at the first and last knot, we have $k_0 = k_4 = 0$. The second derivatives at the other knots are obtained from Eq. (3.12). Using $i = 1, 2, 3$ results in the simultaneous equations

$$0 + 4k_1 + k_2 = 6\,[0 - 2(1) + 0] = -12$$

$$k_1 + 4k_2 + k_3 = 6\,[1 - 2(0) + 1] = 12$$

$$k_2 + 4k_3 + 0 = 6\,[0 - 2(1) + 0] = -12$$

The solution is $k_1 = k_3 = -30/7$, $k_2 = 36/7$.

The point $x = 1.5$ lies in the segment between knots 0 and 1. The corresponding interpolant is obtained from Eq. (3.10) by setting $i = 0$. With $x_i - x_{i+1} = -h = -1$, we obtain from Eq. (3.10)

$$f_{0,1}(x) = -\frac{k_0}{6}\left[(x - x_1)^3 - (x - x_1)\right] + \frac{k_1}{6}\left[(x - x_0)^3 - (x - x_0)\right]$$
$$- \left[y_0(x - x_1) - y_1(x - x_0)\right]$$

Therefore,

$$y(1.5) = f_{0,1}(1.5)$$

$$= 0 + \frac{1}{6}\left(-\frac{30}{7}\right)\left[(1.5-1)^3 - (1.5-1)\right] - [0 - 1(1.5-1)]$$

$$= 0.7679$$

The plot of the interpolant, which in this case is made up of four cubic segments, is shown in the figure.

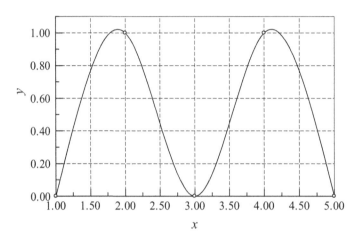

EXAMPLE 3.8

Sometimes it is preferable to replace one or both of the end conditions of the cubic spline with something other than the natural conditions. Use the end condition $f'_{0,1}(0) = 0$ (zero slope), rather than $f''_{0,1}(0) = 0$ (zero curvature), to determine the cubic spline interpolant at $x = 2.6$, given the data points

x	0	1	2	3
y	1	1	0.5	0

Solution. We must first modify Eqs. (3.12) to account for the new end condition. Setting $i = 0$ in Eq. (3.10) and differentiating, we get

$$f'_{0,1}(x) = \frac{k_0}{6}\left[3\frac{(x-x_1)^2}{x_0-x_1} - (x_0-x_1)\right] - \frac{k_1}{6}\left[3\frac{(x-x_0)^2}{x_0-x_1} - (x_0-x_1)\right] + \frac{y_0 - y_1}{x_0 - x_1}$$

Thus the end condition $f'_{0,1}(x_0) = 0$ yields

$$\frac{k_0}{3}(x_0 - x_1) + \frac{k_1}{6}(x_0 - x_1) + \frac{y_0 - y_1}{x_0 - x_1} = 0$$

or

$$2k_0 + k_1 = -6\frac{y_0 - y_1}{(x_0 - x_1)^2}$$

From the given data we see that $y_0 = y_1 = 1$, so that the last equation becomes

$$2k_0 + k_1 = 0 \qquad \text{(a)}$$

The other equations in Eq. (3.12) are unchanged. Knowing that $k_3 = 0$, they are

$$k_0 + 4k_1 + k_2 = 6[1 - 2(1) + 0.5] = -3 \qquad \text{(b)}$$

$$k_1 + 4k_2 = 6[1 - 2(0.5) + 0] = 0 \qquad \text{(c)}$$

The solution of Eqs. (a)–(c) is $k_0 = 0.4615$, $k_1 = -0.9231$, $k_2 = 0.2308$.

The interpolant can now be evaluated from Eq. (3.10). Substituting $i = 2$ and $x_i - x_{i+1} = -1$, we obtain

$$f_{2,3}(x) = \frac{k_2}{6}\left[-(x - x_3)^3 + (x - x_3)\right] - \frac{k_3}{6}\left[-(x - x_2)^3 + (x - x_2)\right]$$
$$-y_2(x - x_3) + y_3(x - x_2)$$

Therefore,

$$y(2.6) = f_{2,3}(2.6) = \frac{0.2308}{6}\left[-(-0.4)^3 + (-0.4)\right] - 0 - 0.5(-0.4) + 0$$
$$= 0.1871$$

EXAMPLE 3.9
Use the module cubicSpline to write a program that interpolates between given data points with a natural cubic spline. The program must be able to evaluate the interpolant for more than one value of x. As a test, use data points specified in Example 3.4 and compute the interpolant at $x = 1.5$ and $x = 4.5$ (because of symmetry, these values should be equal).

Solution

```
#!/usr/bin/python
## example3_9
import numpy as np
from cubicSpline import *

xData = np.array([1,2,3,4,5],float)
yData = np.array([0,1,0,1,0],float)
k = curvatures(xData,yData)
while True:
    try: x = eval(input("\nx ==> "))
    except SyntaxError: break
    print("y =",evalSpline(xData,yData,k,x))
input("Done. Press return to exit")
```

Running the program produces the following result:

```
x ==> 1.5
y = 0.767857142857

x ==> 4.5
y = 0.767857142857

x ==>
Done. Press return to exit
```

PROBLEM SET 3.1

1. Given the data points

x	−1.2	0.3	1.1
y	−5.76	−5.61	−3.69

   determine y at $x = 0$ using (a) Neville's method and (b) Lagrange's method.

2. Find the zero of $y(x)$ from the following data:

x	0	0.5	1	1.5	2	2.5	3
y	1.8421	2.4694	2.4921	1.9047	0.8509	−0.4112	−1.5727

   Use Lagrange's interpolation over (a) three; and (b) four nearest-neighbor data points. *Hint*: After finishing part (a), part (b) can be computed with a relatively small effort.

3. The function $y(x)$ represented by the data in Prob. 2 has a maximum at $x = 0.7692$. Compute this maximum by Neville's interpolation over four nearest-neighbor data points.

4. Use Neville's method to compute y at $x = \pi/4$ from the data points

x	0	0.5	1	1.5	2
y	−1.00	1.75	4.00	5.75	7.00

5. Given the data

x	0	0.5	1	1.5	2
y	−0.7854	0.6529	1.7390	2.2071	1.9425

   find y at $x = \pi/4$ and at $\pi/2$. Use the method that you consider to be most convenient.

6. The points

x	−2	1	4	−1	3	−4
y	−1	2	59	4	24	−53

   lie on a polynomial. Use the divided difference table of Newton's method to determine the degree of the polynomial.

7. Use Newton's method to find the polynomial that fits the following points:

x	−3	2	−1	3	1
y	0	5	−4	12	0

8. Use Neville's method to determine the equation of the quadratic that passes through the points

x	−1	1	3
y	17	−7	−15

9. Density of air ρ varies with elevation h in the following manner:

h (km)	0	3	6
ρ (kg/m^3)	1.225	0.905	0.652

   Express $\rho(h)$ as a quadratic function using Lagrange's method.

10. Determine the natural cubic spline that passes through the data points

x	0	1	2
y	0	2	1

   Note that the interpolant consists of two cubics, one valid in $0 \le x \le 1$, the other in $1 \le x \le 2$. Verify that these cubics have the same first and second derivatives at $x = 1$.

11. Given the data points

x	1	2	3	4	5
y	13	15	12	9	13

   determine the natural cubic spline interpolant at $x = 3.4$.

12. Compute the zero of the function $y(x)$ from the following data:

x	0.2	0.4	0.6	0.8	1.0
y	1.150	0.855	0.377	−0.266	−1.049

   Use inverse interpolation with the natural cubic spline. *Hint*: Reorder the data so that the values of y are in ascending order.

13. Solve Example 3.8 with a cubic spline that has constant second derivatives within its first and last segments (the end segments are parabolic). The end conditions for this spline are $k_0 = k_1$ and $k_{n-1} = k_n$.

14. ■ Write a computer program for interpolation by Neville's method. The program must be able to compute the interpolant at several user-specified values of x. Test the program by determining y at $x = 1.1$, 1.2, and 1.3 from the following data:

x	−2.0	−0.1	−1.5	0.5
y	2.2796	1.0025	1.6467	1.0635

x	−0.6	2.2	1.0	1.8
y	1.0920	2.6291	1.2661	1.9896

(Answer: $y = 1.3262$, 1.3938, 1.4639)

15. ■ The specific heat c_p of aluminum depends on temperature T as follows[2]:

T (°C)	−250	−200	−100	0	100	300
c_p (kJ/kg·K)	0.0163	0.318	0.699	0.870	0.941	1.04

Plot the polynomial and the rational function interpolants from $T = -250°$ to 500°. Comment on the results.

16. ■ Using the data

x	0	0.0204	0.1055	0.241	0.582	0.712	0.981
y	0.385	1.04	1.79	2.63	4.39	4.99	5.27

plot the rational function interpolant from $x = 0$ to $x = 1$.

17. ■ The table shows the drag coefficient c_D of a sphere as a function of Reynold's number Re.[3] Use a natural cubic spline to find c_D at $Re = 5, 50, 500,$ and $5,000$. *Hint*: Use a log-log scale.

Re	0.2	2	20	200	2000	20 000
c_D	103	13.9	2.72	0.800	0.401	0.433

18. ■ Solve Prob. 17 using a polynomial interpolant intersecting four nearest-neighbor data points (do not use a log scale).

19. ■ The kinematic viscosity μ_k of water varies with temperature T in the following manner:

T (°C)	0	21.1	37.8	54.4	71.1	87.8	100
μ_k (10^{-3} m^2/s)	1.79	1.13	0.696	0.519	0.338	0.321	0.296

Interpolate μ_k at $T = 10°, 30°, 60°,$ and $90°$C.

20. ■ The table shows how the relative density ρ of air varies with altitude h. Determine the relative density of air at 10.5 km.

h (km)	0	1.525	3.050	4.575	6.10	7.625	9.150
ρ	1	0.8617	0.7385	0.6292	0.5328	0.4481	0.3741

21. ■ The vibrational amplitude of a driveshaft is measured at various speeds. The results are

Speed (rpm)	0	400	800	1200	1600
Amplitude (mm)	0	0.072	0.233	0.712	3.400

Use rational function interpolation to plot amplitude vs. speed from 0 to 2,500 rpm. From the plot estimate the speed of the shaft at resonance.

---

[2] Source: Black, Z.B. and Hartley, J.G., *Thermodynamics*, Harper & Row, 1985.
[3] Source: Kreith, F., *Principles of Heat Transfer*, Harper & Row, 1973.

3.4   Least-Squares Fit

Overview

If the data are obtained from experiments, they typically contain a significant amount of random noise caused by measurement errors. The task of curve fitting is to find a smooth curve that fits the data points "on the average." This curve should have a simple form (e.g., a low-order polynomial), so as to not reproduce the noise.

Let

$$f(x) = f(x; a_0, a_1, \ldots, a_m)$$

be the function that is to be fitted to the $n + 1$ data points (x_i, y_i), $i = 0, 1, \ldots, n$. The notation implies that we have a function of x that contains $m + 1$ variable parameters $a_0, a_1, \ldots, a_m$, where $m < n$. The form of $f(x)$ is determined beforehand, usually from the theory associated with the experiment from which the data are obtained. The only means of adjusting the fit are the parameters. For example, if the data represent the displacements y_i of an over-damped mass-spring system at time t_i, the theory suggests the choice $f(t) = a_0 t e^{-a_1 t}$. Thus curve fitting consists of two steps: choosing the form of $f(x)$, followed by computation of the parameters that produce the best fit to the data.

This brings us to the question: What is meant by the "best" fit? If the noise is confined to the y-coordinate, the most commonly used measure is the *least-squares fit*, which minimizes the function

$$S(a_0, a_1, \ldots, a_m) = \sum_{i=0}^{n} \left[y_i - f(x_i) \right]^2 \tag{3.13}$$

with respect to each a_j. Therefore, the optimal values of the parameters are given by the solution of the equations

$$\frac{\partial S}{\partial a_k} = 0, \quad k = 0, 1, \ldots, m. \tag{3.14}$$

The terms $r_i = y_i - f(x_i)$ in Eq. (3.13) are called *residuals*; they represent the discrepancy between the data points and the fitting function at x_i. The function S to be minimized is thus the sum of the squares of the residuals. Equations (3.14) are generally nonlinear in a_j and may thus be difficult to solve. Often the fitting function is chosen as a linear combination of specified functions $f_j(x)$,

$$f(x) = a_0 f_0(x) + a_1 f_1(x) + \cdots + a_m f_m(x)$$

in which case Eqs. (3.14) are linear. If the fitting function is a polynomial, we have $f_0(x) = 1, f_1(x) = x, f_2(x) = x^2$, and so on.

The spread of the data about the fitting curve is quantified by the *standard deviation*, defined as

$$\sigma = \sqrt{\frac{S}{n - m}} \tag{3.15}$$

Note that if $n = m$, we have *interpolation*, not curve fitting. In that case both the numerator and the denominator in Eq. (3.15) are zero, so that σ is indeterminate.

Fitting a Straight Line

Fitting a straight line

$$f(x) = a + bx \tag{3.16}$$

to data is also known as *linear regression*. In this case the function to be minimized is

$$S(a, b) = \sum_{i=0}^{n} \left[y_i - f(x_i) \right]^2 = \sum_{i=0}^{n} (y_i - a - bx_i)^2$$

Equations (3.14) now become

$$\frac{\partial S}{\partial a} = \sum_{i=0}^{n} -2(y_i - a - bx_i) = 2 \left[a\,(n+1) + b\sum_{i=0}^{n} x_i - \sum_{i=0}^{n} y_i \right] = 0$$

$$\frac{\partial S}{\partial b} = \sum_{i=0}^{n} -2(y_i - a - bx_i)x_i = 2 \left(a\sum_{i=0}^{n} x_i + b\sum_{i=0}^{n} x_i^2 - \sum_{i=0}^{n} x_i y_i \right) = 0$$

Dividing both equations by $2\,(n+1)$ and rearranging terms, we get

$$a + \bar{x}b = \bar{y} \qquad \bar{x}a + \left(\frac{1}{n+1}\sum_{i=0}^{n} x_i^2 \right) b = \frac{1}{n+1}\sum_{i=0}^{n} x_i y_i$$

where

$$\bar{x} = \frac{1}{n+1}\sum_{i=0}^{n} x_i \qquad \bar{y} = \frac{1}{n+1}\sum_{i=0}^{n} y_i \tag{3.17}$$

are the mean values of the x and y data. The solution for the parameters is

$$a = \frac{\bar{y}\sum x_i^2 - \bar{x}\sum x_i y_i}{\sum x_i^2 - n\bar{x}^2} \qquad b = \frac{\sum x_i y_i - \bar{x}\sum y_i}{\sum x_i^2 - n\bar{x}^2} \tag{3.18}$$

These expressions are susceptible to roundoff errors (the two terms in each numerator as well as in each denominator can be roughly equal). Therefore it is better to compute the parameters from

$$b = \frac{\sum y_i(x_i - \bar{x})}{\sum x_i(x_i - \bar{x})} \qquad a = \bar{y} - \bar{x}b \tag{3.19}$$

which are equivalent to Eqs. (3.18), but are much less affected by rounding off.

Fitting Linear Forms

Consider the least-squares fit of the *linear form*

$$f(x) = a_0 f_0(x) + a_1 f_1(x) + \ldots + a_m f_m(x) = \sum_{j=0}^{m} a_j f_j(x) \tag{3.20}$$

where each $f_j(x)$ is a predetermined function of x, called a *basis function*. Substitution in Eq. (3.13) yields

$$S = \sum_{i=0}^{n} \left[y_i - \sum_{j=0}^{m} a_j f_j(x_i) \right]^2$$

Thus Eqs. (3.14) are

$$\frac{\partial S}{\partial a_k} = -2 \left\{ \sum_{i=0}^{n} \left[y_i - \sum_{j=0}^{m} a_j f_j(x_i) \right] f_k(x_i) \right\} = 0, \quad k = 0, 1, \ldots, m$$

Dropping the constant (-2) and interchanging the order of summation, we get

$$\sum_{j=0}^{m} \left[\sum_{i=0}^{n} f_j(x_i) f_k(x_i) \right] a_j = \sum_{i=0}^{n} f_k(x_i) y_i, \quad k = 0, 1, \ldots, m$$

In matrix notation these equations are

$$\mathbf{Aa} = \mathbf{b} \tag{3.21a}$$

where

$$A_{kj} = \sum_{i=0}^{n} f_j(x_i) f_k(x_i) \qquad b_k = \sum_{i=0}^{n} f_k(x_i) y_i \tag{3.21b}$$

Equation (3.21a), known as the *normal equations* of the least-squares fit, can be solved with the methods discussed in Chapter 2. Note that the coefficient matrix is symmetric (i.e., $A_{kj} = A_{jk}$).

Polynomial Fit

A commonly used linear form is a polynomial. If the degree of the polynomial is m, we have $f(x) = \sum_{j=0}^{m} a_j x^j$. Here the basis functions are

$$f_j(x) = x^j \qquad (j = 0, 1, \ldots, m) \tag{3.22}$$

so that Eqs. (3.21b) become

$$A_{kj} = \sum_{i=0}^{n} x_i^{j+k} \qquad b_k = \sum_{i=0}^{n} x_i^k y_i$$

or

$$\mathbf{A} = \begin{bmatrix} n & \sum x_i & \sum x_i^2 & \cdots & \sum x_i^m \\ \sum x_i & \sum x_i^2 & \sum x_i^3 & \cdots & \sum x_i^{m+1} \\ \vdots & \vdots & \vdots & \ddots & \vdots \\ \sum x_i^{m-1} & \sum x_i^m & \sum x_i^{m+1} & \cdots & \sum x_i^{2m} \end{bmatrix} \qquad \mathbf{b} = \begin{bmatrix} \sum y_i \\ \sum x_i y_i \\ \vdots \\ \sum x_i^m y_i \end{bmatrix} \tag{3.23}$$

where $\sum$ stands for $\sum_{i=0}^{n}$. The normal equations become progressively ill conditioned with increasing m. Fortunately, this is of little practical consequence, because only low-order polynomials are useful in curve fitting. Polynomials of high order are not recommended, because they tend to reproduce the noise inherent in the data.

■ polyFit

The function `polyFit` in this module sets up and solves the normal equations for the coefficients of a polynomial of degree m. It returns the coefficients of the polynomial. To facilitate computations, the terms n, $\sum x_i$, $\sum x_i^2$, ..., $\sum x_i^{2m}$ that make up the coefficient matrix in Eq. (3.23) are first stored in the vector **s** and then inserted into **A**. The normal equations are then solved by Gauss elimination with pivoting. After the solution is found, the standard deviation σ can be computed with the function `std-Dev`. The polynomial evaluation in `stdDev` is carried out by the embedded function `evalPoly`—see Section 4.7 for an explanation of the algorithm.

```
## module polyFit
''' c = polyFit(xData,yData,m).
    Returns coefficients of the polynomial
    p(x) = c[0] + c[1]x + c[2]x^2 +...+ c[m]x^m
    that fits the specified data in the least
    squares sense.

    sigma = stdDev(c,xData,yData).
    Computes the std. deviation between p(x)
    and the data.
'''
import numpy as np
import math
from gaussPivot import *

def polyFit(xData,yData,m):
    a = np.zeros((m+1,m+1))
    b = np.zeros(m+1)
    s = np.zeros(2*m+1)
    for i in range(len(xData)):
        temp = yData[i]
        for j in range(m+1):
            b[j] = b[j] + temp
            temp = temp*xData[i]
        temp = 1.0
        for j in range(2*m+1):
            s[j] = s[j] + temp
            temp = temp*xData[i]
    for i in range(m+1):
        for j in range(m+1):
```

```
               a[i,j] = s[i+j]
       return gaussPivot(a,b)


def stdDev(c,xData,yData):

    def evalPoly(c,x):
        m = len(c) - 1
        p = c[m]
        for j in range(m):
            p = p*x + c[m-j-1]
        return p

    n = len(xData) - 1
    m = len(c) - 1
    sigma = 0.0
    for i in range(n+1):
        p = evalPoly(c,xData[i])
        sigma = sigma + (yData[i] - p)**2
    sigma = math.sqrt(sigma/(n - m))
    return sigma
```

▪ plotPoly

The function `plotPoly` listed next is handy for plotting the data points and the fitting polynomial.

```
## module plotPoly
''' plotPoly(xData,yData,coeff,xlab='x',ylab='y')
    Plots data points and the fitting
    polynomial defined by its coefficient
    array coeff = [a0, a1. ...]
    xlab and ylab are optional axis labels
'''
import numpy as np
import matplotlib.pyplot as plt

def plotPoly(xData,yData,coeff,xlab='x',ylab='y'):
    m = len(coeff)
    x1 = min(xData)
    x2 = max(xData)
    dx = (x2 - x1)/20.0
    x = np.arange(x1,x2 + dx/10.0,dx)
    y = np.zeros((len(x)))*1.0
    for i in range(m):
```

```
    y = y + coeff[i]*x**i
plt.plot(xData,yData,'o',x,y,'-')
plt.xlabel(xlab); plt.ylabel(ylab)
plt.grid (True)
plt.show()
```

Weighting of Data

There are occasions when our confidence in the accuracy of data varies from point to point. For example, the instrument taking the measurements may be more sensitive in a certain range of data. Sometimes the data represent the results of several experiments, each carried out under different conditions. Under these circumstances we may want to assign a confidence factor, or *weight*, to each data point and minimize the sum of the squares of the *weighted residuals* $r_i = W_i [y_i - f(x_i)]$, where W_i are the weights. Hence the function to be minimized is

$$S(a_0, a_1, \ldots, a_m) = \sum_{i=0}^{n} W_i^2 [y_i - f(x_i)]^2 \qquad (3.24)$$

This procedure forces the fitting function $f(x)$ closer to the data points that have higher weights.

Weighted linear regression. If the fitting function is the straight line $f(x) = a + bx$, Eq. (3.24) becomes

$$S(a, b) = \sum_{i=0}^{n} W_i^2 (y_i - a - bx_i)^2 \qquad (3.25)$$

The conditions for minimizing S are

$$\frac{\partial S}{\partial a} = -2 \sum_{i=0}^{n} W_i^2 (y_i - a - bx_i) = 0$$

$$\frac{\partial S}{\partial b} = -2 \sum_{i=0}^{n} W_i^2 (y_i - a - bx_i) x_i = 0$$

or

$$a \sum_{i=0}^{n} W_i^2 + b \sum_{i=0}^{n} W_i^2 x_i = \sum_{i=0}^{n} W_i^2 y_i \qquad (3.26a)$$

$$a \sum_{i=0}^{n} W_i^2 x_i + b \sum_{i=0}^{n} W_i^2 x_i^2 = \sum_{i=0}^{n} W_i^2 x_i y_i \qquad (3.26b)$$

Dividing the Eq. (3.26a) by $\sum W_i^2$ and introducing the *weighted averages,*

$$\hat{x} = \frac{\sum W_i^2 x_i}{\sum W_i^2} \qquad \hat{y} = \frac{\sum W_i^2 y_i}{\sum W_i^2} \tag{3.27}$$

we obtain

$$a = \hat{y} - b\hat{x}. \tag{3.28a}$$

Substituting into Eq. (3.26b) and solving for b yields, after some algebra,

$$b = \frac{\sum W_i^2 y_i (x_i - \hat{x})}{\sum W_i^2 x_i (x_i - \hat{x})} \tag{3.28b}$$

Note that Eqs. (3.28) are quite similar to Eqs. (3.19) of unweighted data.

Fitting exponential functions. A special application of weighted linear regression arises in fitting various exponential functions to data. Consider as an example the fitting function

$$f(x) = ae^{bx}$$

Normally, the least-squares fit would lead to equations that are nonlinear in a and b. But if we fit $\ln y$ rather than y, the problem is transformed to linear regression: fitting the function

$$F(x) = \ln f(x) = \ln a + bx$$

to the data points $(x_i, \ln y_i)$, $i = 0, 1, \ldots, n$. This simplification comes at a price: The least-squares fit to the logarithm of the data is not quite the same as the least-squares fit to the original data. The residuals of the logarithmic fit are

$$R_i = \ln y_i - F(x_i) = \ln y_i - (\ln a + bx_i) \tag{3.29a}$$

whereas the residuals used in fitting the original data are

$$r_i = y_i - f(x_i) = y_i - ae^{bx_i} \tag{3.29b}$$

This discrepancy can be largely eliminated by weighting the logarithmic fit. From Eq. (3.29b) we obtain $\ln(r_i - y_i) = \ln(ae^{bx_i}) = \ln a + bx_i$, so that Eq. (3.29a) can be written as

$$R_i = \ln y_i - \ln(r_i - y_i) = \ln\left(1 - \frac{r_i}{y_i}\right)$$

If the residuals r_i are sufficiently small $(r_i << y_i)$, we can use the approximation $\ln(1 - r_i/y_i) \approx r_i/y_i$, so that

$$R_i \approx r_i/y_i$$

We can now see that by minimizing $\sum R_i^2$, we have inadvertently introduced the weights $1/y_i$. This effect can be negated if we apply the weights $W_i = y_i$ when fitting $F(x)$ to $(\ln y_i, x_i)$. That is, minimizing

$$S = \sum_{i=0}^{n} y_i^2 R_i^2 \tag{3.30}$$

is a good approximation to minimizing $\sum r_i^2$.

Other examples that also benefit from the weights $W_i = y_i$ are given in Table 3.4.

$f(x)$	$F(x)$	Data to be fitted by $F(x)$
axe^{bx}	$\ln\left[f(x)/x\right] = \ln a + bx$	$\left[x_i, \ln(y_i/x_i)\right]$
ax^b	$\ln f(x) = \ln a + b\ln(x)$	$(\ln x_i, \ln y_i)$

Table 3.4. Fitting exponential functions.

EXAMPLE 3.10

Fit a straight line to the data shown and compute the standard deviation.

x	0.0	1.0	2.0	2.5	3.0
y	2.9	3.7	4.1	4.4	5.0

Solution. The averages of the data are

$$\bar{x} = \frac{1}{5}\sum x_i = \frac{0.0 + 1.0 + 2.0 + 2.5 + 3.0}{5} = 1.7$$

$$\bar{y} = \frac{1}{5}\sum y_i = \frac{2.9 + 3.7 + 4.1 + 4.4 + 5.0}{5} = 4.02$$

The intercept a and slope b of the interpolant can now be determined from Eq. (3.19):

$$b = \frac{\sum y_i(x_i - \bar{x})}{\sum x_i(x_i - \bar{x})}$$

$$= \frac{2.9(-1.7) + 3.7(-0.7) + 4.1(0.3) + 4.4(0.8) + 5.0(1.3)}{0.0(-1.7) + 1.0(-0.7) + 2.0(0.3) + 2.5(0.8) + 3.0(1.3)}$$

$$= \frac{3.73}{5.8} = 0.6431$$

$$a = \bar{y} - \bar{x}b = 4.02 - 1.7(0.6431) = 2.927$$

Therefore, the regression line is $f(x) = 2.927 + 0.6431x$, which is shown in the figure together with the data points.

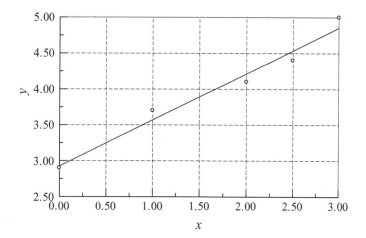

We start the evaluation of the standard deviation by computing the residuals:

x	0.000	1.000	2.000	2.500	3.000
y	2.900	3.700	4.100	4.400	5.000
$f(x)$	2.927	3.570	4.213	4.535	4.856
$y - f(x)$	-0.027	0.130	-0.113	-0.135	0.144

The sum of the squares of the residuals is

$$S = \sum \left[y_i - f(x_i) \right]^2$$
$$= (-0.027)^2 + (0.130)^2 + (-0.113)^2 + (-0.135)^2 + (0.144)^2 = 0.06936$$

so that the standard deviation in Eq. (3.15) becomes

$$\sigma = \sqrt{\frac{S}{5-2}} = \sqrt{\frac{0.06936}{3}} = 0.1520$$

EXAMPLE 3.11
Determine the parameters a and b so that $f(x) = ae^{bx}$ fits the following data in the least-squares sense.

x	1.2	2.8	4.3	5.4	6.8	7.9
y	7.5	16.1	38.9	67.0	146.6	266.2

Use two different methods: (1) fit $\ln y_i$; and (2) fit $\ln y_i$ with weights $W_i = y_i$. Compute the standard deviation in each case.

Solution of Part (1). The problem is to fit the function $\ln(ae^{bx}) = \ln a + bx$ to the data

x	1.2	2.8	4.3	5.4	6.8	7.9
$z = \ln y$	2.015	2.779	3.661	4.205	4.988	5.584

We are now dealing with linear regression, where the parameters to be found are $A = \ln a$ and b. Following the steps in Example 3.8, we get (skipping some of the arithmetic details),

$$\bar{x} = \frac{1}{6}\sum x_i = 4.733 \qquad \bar{z} = \frac{1}{6}\sum z_i = 3.872$$

$$b = \frac{\sum z_i(x_i - \bar{x})}{\sum x_i(x_i - \bar{x})} = \frac{16.716}{31.153} = 0.5366 \qquad A = \bar{z} - \bar{x}b = 1.3323$$

Therefore, $a = e^A = 3.790$ and the fitting function becomes $f(x) = 3.790e^{0.5366}$. The plots of $f(x)$ and the data points are shown in the figure.

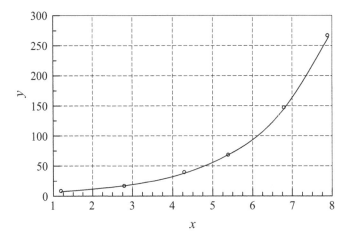

Here is the computation of standard deviation:

x	1.20	2.80	4.30	5.40	6.80	7.90
y	7.50	16.10	38.90	67.00	146.60	266.20
$f(x)$	7.21	17.02	38.07	68.69	145.60	262.72
$y - f(x)$	0.29	−0.92	0.83	−1.69	1.00	3.48

$$S = \sum [y_i - f(x_i)]^2 = 17.59$$

$$\sigma = \sqrt{\frac{S}{6-2}} = 2.10$$

As pointed out earlier, this is an approximate solution of the stated problem, because we did not fit y_i, but $\ln y_i$. Judging by the plot, the fit seems to be quite good.

Solution of Part (2). We again fit $\ln(ae^{bx}) = \ln a + bx$ to $z = \ln y$, but this time using the weights $W_i = y_i$. From Eqs. (3.27) the weighted averages of the data are (recall that we fit $z = \ln y$)

$$\hat{x} = \frac{\sum y_i^2 x_i}{\sum y_i^2} = \frac{737.5 \times 10^3}{98.67 \times 10^3} = 7.474$$

$$\hat{z} = \frac{\sum y_i^2 z_i}{\sum y_i^2} = \frac{528.2 \times 10^3}{98.67 \times 10^3} = 5.353$$

and Eqs. (3.28) yield for the parameters

$$b = \frac{\sum y_i^2 z_i(x_i - \hat{x})}{\sum y_i^2 x_i(x_i - \hat{x})} = \frac{35.39 \times 10^3}{65.05 \times 10^3} = 0.5440$$

$$\ln a = \hat{z} - b\hat{x} = 5.353 - 0.5440(7.474) = 1.287$$

Therefore,

$$a = e^{\ln a} = e^{1.287} = 3.622$$

so that the fitting function is $f(x) = 3.622e^{0.5440x}$. As expected, this result is somewhat different from that obtained in Part (1).

The computations of the residuals and the standard deviation are as follows:

x	1.20	2.80	4.30	5.40	6.80	7.90
y	7.50	16.10	38.90	67.00	146.60	266.20
$f(x)$	6.96	16.61	37.56	68.33	146.33	266.20
$y - f(x)$	0.54	−0.51	1.34	−1.33	0.267	0.00

$$S = \sum \left[y_i - f(x_i)\right]^2 = 4.186$$

$$\sigma = \sqrt{\frac{S}{6 - 2}} = 1.023$$

Observe that the residuals and standard deviation are smaller than in Part (1), indicating a better fit, as expected.

It can be shown that fitting y_i directly (which involves the solution of a transcendental equation) results in $f(x) = 3.614e^{0.5442}$. The corresponding standard deviation is $\sigma = 1.022$, which is very close to the result in Part (2).

EXAMPLE 3.12

Write a program that fits a polynomial of arbitrary degree m to the data points shown in the following table. Use the program to determine m that best fits this data in the least-squares sense.

x	−0.04	0.93	1.95	2.90	3.83	5.00
y	−8.66	−6.44	−4.36	−3.27	−0.88	0.87
x	5.98	7.05	8.21	9.08	10.09	
y	3.31	4.63	6.19	7.40	8.85	

Solution. The program shown next prompts for m. Execution is terminated by entering an invalid character (e.g., the "return" character).

```
#!/usr/bin/python
## example3_12
import numpy as np
from polyFit import *

xData = np.array([-0.04,0.93,1.95,2.90,3.83,5.0,        \
                  5.98,7.05,8.21,9.08,10.09])
yData = np.array([-8.66,-6.44,-4.36,-3.27,-0.88,0.87, \
                  3.31,4.63,6.19,7.4,8.85])
while True:
    try:
        m = eval(input("\nDegree of polynomial ==> "))
        coeff = polyFit(xData,yData,m)
        print("Coefficients are:\n",coeff)
        print("Std. deviation =",stdDev(coeff,xData,yData))
    except SyntaxError: break
input("Finished. Press return to exit")
```

The results are

```
Degree of polynomial ==> 1
Coefficients are:
 [-7.94533287  1.72860425]
Std. deviation = 0.5112788367370911

Degree of polynomial ==> 2
Coefficients are:
 [-8.57005662  2.15121691 -0.04197119]
Std. deviation = 0.3109920728551074

Degree of polynomial ==> 3
Coefficients are:
```

```
[ -8.46603423e+00 1.98104441e+00 2.88447008e-03 -2.98524686e-03]
Std. deviation = 0.31948179156753187

Degree of polynomial ==>
Finished. Press return to exit
```

Because the quadratic $f(x) = -8.5700 + 2.1512x - 0.041971x^2$ produces the smallest standard deviation, it can be considered as the "best" fit to the data. But be warned—the standard deviation is not a reliable measure of the goodness-of-fit. It is always a good idea to plot the data points and $f(x)$ before making a final determination. The plot of our data indicates that the quadratic (solid line) is indeed a reasonable choice for the fitting function.

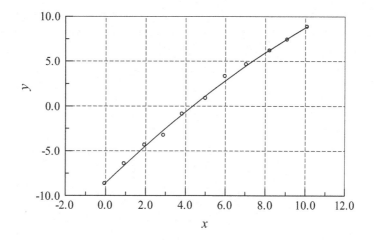

PROBLEM SET 3.2

Instructions: Plot the data points and the fitting function whenever appropriate.

1. Show that the straight line obtained by least-squares fit of unweighted data always passes through the point $(\bar{x}, \bar{y})$.
2. Use linear regression to find the line that fits the data

x	-1.0	-0.5	0	0.5	1.0
y	-1.00	-0.55	0.00	0.45	1.00

   and determine the standard deviation.
3. Three tensile tests were carried out on an aluminum bar. In each test the strain was measured at the same values of stress. The results were

Stress (MPa)	34.5	69.0	103.5	138.0
Strain (Test 1)	0.46	0.95	1.48	1.93
Strain (Test 2)	0.34	1.02	1.51	2.09
Strain (Test 3)	0.73	1.10	1.62	2.12

where the units of strain are mm/m. Use linear regression to estimate the modulus of elasticity of the bar (modulus of elasticity = stress/strain).

4. Solve Prob. 3 assuming that the third test was performed on an inferior machine, so that its results carry only half the weight of the other two tests.

5. ■ The following table shows the annual atmospheric CO_2 concentration (in parts per million) in Antarctica. Fit a straight line to the data and determine the average increase of the concentration per year.

Year	1994	1995	1996	1997	1998	1999	2000	2001
ppm	356.8	358.2	360.3	361.8	364.0	365.7	366.7	368.2

Year	2002	2003	2004	2005	2006	2007	2008	2009
ppm	370.5	372.2	374.9	376.7	378.7	381.0	382.9	384.7

6. ■ The following table displays the mass M and average fuel consumption ϕ of motor vehicles manufactured by Ford and Honda in 2008. Fit a straight line $\phi = a + bM$ to the data and compute the standard deviation.

Model	M (kg)	ϕ (km/liter)
Focus	1198	11.90
Crown Victoria	1715	6.80
Expedition	2530	5.53
Explorer	2014	6.38
F-150	2136	5.53
Fusion	1492	8.50
Taurus	1652	7.65
Fit	1168	13.60
Accord	1492	9.78
CR-V	1602	8.93
Civic	1192	11.90
Ridgeline	2045	6.38

7. ■ The relative density ρ of air was measured at various altitudes h. The results were

h (km)	0	1.525	3.050	4.575	6.10	7.625	9.150
ρ	1	0.8617	0.7385	0.6292	0.5328	0.4481	0.3741

Use a quadratic least-squares fit to determine the relative air density at $h = 10.5$ km. (This problem was solved by interpolation in Prob. 20, Problem Set 3.1.)

8. ■ The kinematic viscosity μ_k of water varies with temperature T as shown in the following table. Determine the cubic that best fits the data, and use it to compute μ_k at $T = 10°, 30°, 60°$, and $90°$C. (This problem was solved in Prob. 19, Problem Set 3.1 by interpolation.)

T (°C)	0	21.1	37.8	54.4	71.1	87.8	100
μ_k (10^{-3} m^2/s)	1.79	1.13	0.696	0.519	0.338	0.321	0.296

9. ■ Fit a straight line and a quadratic to the data in the following table.

x	1.0	2.5	3.5	4.0	1.1	1.8	2.2	3.7
y	6.008	15.722	27.130	33.772	5.257	9.549	11.098	28.828

Which is a better fit?

10. ■ The following table displays thermal efficiencies of some early steam engines[4]. Use linear regression to predict the thermal efficiency in the year 2000.

Year	Efficiency (%)	Type
1718	0.5	Newcomen
1767	0.8	Smeaton
1774	1.4	Smeaton
1775	2.7	Watt
1792	4.5	Watt
1816	7.5	Woolf compound
1828	12.0	Improved Cornish
1834	17.0	Improved Cornish
1878	17.2	Corliss compound
1906	23.0	Triple expansion

11. The following table shows the variation of relative thermal conductivity k of sodium with temperature T. Find the quadratic that fits the data in the least-squares sense.

T (°C)	79	190	357	524	690
k	1.00	0.932	0.839	0.759	0.693

12. Let $f(x) = ax^b$ be the least-squares fit of the data (x_i, y_i), $i = 0, 1, \ldots, n$, and let $F(x) = \ln a + b \ln x$ be the least-squares fit of $(\ln x_i, \ln y_i)$—see Table 3.4. Prove that $R_i \approx r_i / y_i$, where the residuals are $r_i = y_i - f(x_i)$ and $R_i = \ln y_i - F(x_i)$. Assume that $r_i << y_i$.

13. Determine a and b for which $f(x) = a \sin(\pi x/2) + b \cos(\pi x/2)$ fits the following data in the least-squares sense.

x	−0.5	−0.19	0.02	0.20	0.35	0.50
y	−3.558	−2.874	−1.995	−1.040	−0.068	0.677

14. Determine a and b so that $f(x) = ax^b$ fits the following data in the least-squares sense.

x	0.5	1.0	1.5	2.0	2.5
y	0.49	1.60	3.36	6.44	10.16

---

[4] Source: Singer, C., Holmyard, E.J., Hall, A.R., and Williams, T.H., *A History of Technology*, Oxford Press, 1958.

15. Fit the function $f(x) = axe^{bx}$ to the following data and compute the standard deviation.

x	0.5	1.0	1.5	2.0	2.5
y	0.541	0.398	0.232	0.106	0.052

16. ■ The intensity of radiation of a radioactive substance was measured at half-year intervals. The results were

t (years)	0	0.5	1	1.5	2	2.5
γ	1.000	0.994	0.990	0.985	0.979	0.977
t (years)	3	3.5	4	4.5	5	5.5
γ	0.972	0.969	0.967	0.960	0.956	0.952

where γ is the relative intensity of radiation. Knowing that radioactivity decays exponentially with time, $\gamma(t) = ae^{-bt}$, estimate the radioactive half-life of the substance.

17. Linear regression can be extended to data that depend on two or more variables (called multiple linear regression). If the dependent variable is z and independent variables are x and y, the data to be fitted have the form

x_1	y_1	z_1
x_2	y_2	z_2
x_3	y_3	z_3
$\vdots$	$\vdots$	$\vdots$
x_n	y_n	z_n

Instead of a straight line, the fitting function now represents a plane:

$$f(x, y) = a + bx + cy$$

Show that the normal equations for the coefficients are

$$\begin{bmatrix} n & \Sigma x_i & \Sigma y_i \\ \Sigma x_i & \Sigma x_i^2 & \Sigma x_i y_i \\ \Sigma y_i & \Sigma x_i y_i & \Sigma y_i^2 \end{bmatrix} \begin{bmatrix} a \\ b \\ c \end{bmatrix} = \begin{bmatrix} \Sigma z_i \\ \Sigma x_i z_i \\ \Sigma y_i z_i \end{bmatrix}$$

18. Use multiple linear regression explained in Prob. 17 to determine the function

$$f(x, y) = a + bx + cy$$

that fits the following data:

x	y	z
0	0	1.42
0	1	1.85
1	0	0.78
2	0	0.18
2	1	0.60
2	2	1.05

4 Roots of Equations

> Find the solutions of $f(x) = 0$, where the function f is given.

4.1 Introduction

A common problem encountered in engineering analysis is as follows: Given a function $f(x)$, determine the values of x for which $f(x) = 0$. The solutions (values of x) are known as the *roots* of the equation $f(x) = 0$, or the *zeroes* of the function $f(x)$.

Before proceeding further, it might be helpful to review the concept of a *function*. The equation

$$y = f(x)$$

contains three elements: an input value x, an output value y, and the rule f for computing y. The function is said to be given if the rule f is specified. In numerical computing the rule is invariably a computer algorithm. It may be a function statement, such as

$$f(x) = \cosh(x)\cos(x) - 1$$

or a complex procedure containing hundreds or thousands of lines of code. As long as the algorithm produces an output y for each input x, it qualifies as a function.

The roots of equations may be real or complex. The complex roots are seldom computed, because they rarely have physical significance. An exception is the polynomial equation

$$a_0 + a_1 x + a_1 x^2 + \cdots + a_n x^n = 0$$

where the complex roots may be meaningful (as in the analysis of damped vibrations, for example). First we will concentrate on finding the real roots of equations. Complex zeros of polynomials are treated near the end of this chapter.

In general, an equation may have any number of (real) roots or no roots at all. For example,

$$\sin x - x = 0$$

has a single root, namely $x = 0$, whereas

$$\tan x - x = 0$$

has an infinite number of roots ($x = 0, \pm 4.493, \pm 7.725, \ldots$).

All methods of finding roots are iterative procedures that require a starting point (i.e., an estimate of the root). This estimate is crucial; a bad starting value may fail to converge, or it may converge to the "wrong" root (a root different from the one sought). There is no universal recipe for estimating the value of a root. If the equation is associated with a physical problem, then the context of the problem (physical insight) might suggest the approximate location of the root. Otherwise, you may undertake a systematic numerical search for the roots. One such search method is described in the next section. Plotting the function is another means of locating the roots, but it is a visual procedure that cannot be programmed.

It is highly advisable to go a step further and *bracket* the root (determine its lower and upper bounds) before passing the problem to a root-finding algorithm. Prior bracketing is, in fact, mandatory in the methods described in this chapter.

4.2   Incremental Search Method

The approximate locations of the roots are best determined by plotting the function. Often a very rough plot, based on a few points, is sufficient to provide reasonable starting values. Another useful tool for detecting and bracketing roots is the incremental search method. It can also be adapted for computing roots, but the effort would not be worthwhile, because other methods described in this chapter are more efficient for that task.

The basic idea behind the incremental search method is simple: If $f(x_1)$ and $f(x_2)$ have opposite signs, then there is at least one root in the interval (x_1, x_2). If the interval is small enough, it is likely to contain a single root. Thus the zeros of $f(x)$ can be detected by evaluating the function at intervals Δx and looking for a change in sign.

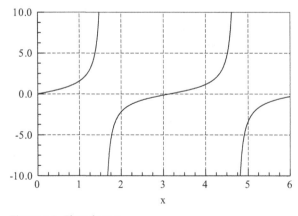

Figure 4.1. Plot of $\tan x$.

There are several potential problems with the incremental search method:

- It is possible to miss two closely spaced roots if the search increment Δx is larger than the spacing of the roots.
- A double root (two roots that coincide) will not be detected.
- Certain singularities (poles) of $f(x)$ can be mistaken for roots. For example, $f(x) = \tan x$ changes sign at $x = \pm\frac{1}{2}n\pi$, $n = 1, 3, 5, \ldots$, as shown in Figure 4.1. However, these locations are not true zeroes, because the function does not cross the x-axis.

■ rootsearch

This function searches for a zero of the user-supplied function *f(x)* in the interval (a,b) in increments of dx. It returns the bounds (x1,x2) of the root if the search was successful; x1 = x2 = None indicates that no roots were detected. After the first root (the root closest to a) has been detected, rootsearch can be called again with a replaced by x2 in order to find the next root. This can be repeated as long as rootsearch detects a root.

```
## module rootsearch
''' x1,x2 = rootsearch(f,a,b,dx).
    Searches the interval (a,b) in increments dx for
    the bounds (x1,x2) of the smallest root of f(x).
    Returns x1 = x2 = None if no roots were detected.
'''
from numpy import sign

def rootsearch(f,a,b,dx):
    x1 = a; f1 = f(a)
    x2 = a + dx; f2 = f(x2)
    while sign(f1) == sign(f2):
        if x1 >= b: return None,None
        x1 = x2; f1 = f2
        x2 = x1 + dx; f2 = f(x2)
    else:
        return x1,x2
```

EXAMPLE 4.1
A root of $x^3 - 10x^2 + 5 = 0$ lies in the interval (0, 1). Use rootsearch to compute this root with four-digit accuracy.

Solution. To obtain four-digit accuracy, we need a search increment no bigger than $\Delta x = 0.0001$. A search of the interval (0, 1) at increments Δx would thus entail 10,000 function evaluations. The following program reduces thenumber of function evalutions to 40 by closing in on the root in four stages, each stage involving 10 search intervals (and thus 10 function evaluations).

```
#!/usr/bin/python
## example4_1
from rootsearch import *

def f(x): return x**3 - 10.0*x**2 + 5.0

x1 = 0.0; x2 = 1.0
for i in range(4):
    dx = (x2 - x1)/10.0
    x1,x2 = rootsearch(f,x1,x2,dx)
x = (x1 + x2)/2.0
print('x =', '{:6.4f}'.format(x))
input("Press return to exit")
```

The output is

```
x = 0.7346
```

4.3   Method of Bisection

After a root of $f(x) = 0$ has been bracketed in the interval (x_1, x_2), several methods can be used to close in on it. The method of bisection accomplishes this by successively halving the interval until it becomes sufficiently small. This technique is also known as the *interval halving method.* Bisection is not the fastest method available for computing roots, but it is the most reliable one. Once a root has been bracketed, bisection will always close in on it.

The method of bisection uses the same principle as incremental search: If there is a root in the interval (x_1, x_2), then $f(x_1)$ and $f(x_2)$ have opposite signs. To halve the interval, we compute $f(x_3)$, where $x_3 = \frac{1}{2}(x_1 + x_2)$ is the midpoint of the interval. If $f(x_2)$ and $f(x_3)$ have opposite signs, then the root must be in (x_2, x_3), and we record this by replacing the original bound x_1 by x_3. Otherwise, the root lies in (x_1, x_3), in which case x_2 is replaced by x_3. In either case, the new interval (x_1, x_2) is half the size of the original interval. The bisection is repeated until the interval has been reduced to a small value ε, so that

$$|x_2 - x_1| \leq \varepsilon$$

It is easy to compute the number of bisections required to reach a prescribed ε. The original interval Δx is reduced to $\Delta x/2$ after one bisection, $\Delta x/2^2$ after two bisections, and after n bisections it is $\Delta x/2^n$. Setting $\Delta x/2^n = \varepsilon$ and solving for n, we get

$$n = \frac{\ln(\Delta x/\varepsilon)}{\ln 2} \tag{4.1}$$

Since n must be an integer, the ceiling of n is used (the ceiling of n is the smallest integer greater than n).

■ bisection

This function uses the method of bisection to compute the root of $f(x) = 0$ that is known to lie in the interval $(x1, x2)$. The number of bisections n required to reduce the interval to `tol` is computed from Eq. (4.1). By setting switch $= 1$, we force the routine to check whether the magnitude of $f(x)$ decreases with each interval halving. If it does not, something may be wrong (probably the "root" is not a root at all, but a pole), and root $=$ None is returned. Because this feature is not always desirable, the default value is switch $= 0$. The function error.err, which we use to terminate a program, is listed in Section 1.7.

```
## module bisection
''' root = bisection(f,x1,x2,switch=0,tol=1.0e-9).
    Finds a root of f(x) = 0 by bisection.
    The root must be bracketed in (x1,x2).
    Setting switch = 1 returns root = None if
    f(x) increases upon bisection.
'''
import math
import error
from numpy import sign

def bisection(f,x1,x2,switch=1,tol=1.0e-9):
    f1 = f(x1)
    if f1 == 0.0: return x1
    f2 = f(x2)
    if f2 == 0.0: return x2
    if sign(f1) == sign(f2):
        error.err('Root is not bracketed')
    n = int(math.ceil(math.log(abs(x2 - x1)/tol)/math.log(2.0)))

    for i in range(n):
        x3 = 0.5*(x1 + x2); f3 = f(x3)
        if (switch == 1) and (abs(f3) > abs(f1)) \
                         and (abs(f3) > abs(f2)):
            return None
```

```
        if f3 == 0.0: return x3
        if sign(f2)!= sign(f3): x1 = x3; f1 = f3
        else: x2 = x3; f2 = f3
    return (x1 + x2)/2.0
```

EXAMPLE 4.2

Use `bisection` to find the root of $x^3 - 10x^2 + 5 = 0$ that lies in the interval $(0, 1)$ to four-digit accuracy (this problem was solved with `rootsearch` in Example 4.1). How many function evaluations are involved in the procedure?

Solution. Here is the program:

```
#!/usr/bin/python
## example4_2
from bisection import *


def f(x): return x**3 - 10.0*x**2 + 5.0

x = bisection(f, 0.0, 1.0, tol = 1.0e-4)
print('x =', '{:6.4f}'.format(x))
input("Press return to exit")
```

Note that we set $\varepsilon = 0.0001$ (`tol = 1.0e-4` in `bisection`) to limit the accuracy to four significant figures. The result is

```
x = 0.7346
```

According to Eq. (4.1)

$$n = \frac{\ln\left(|\Delta x|/\varepsilon\right)}{\ln 2} = \frac{\ln(1.0/0.0001)}{\ln 2} = 13.29$$

Therefore, the number of function evaluations in the `for` loop of `bisection` is $\lceil 13.29 \rceil = 14$. There are an additional 2 evaluations at the beginning of the subroutine, making a total of 16 function evaluations.

EXAMPLE 4.3

Find *all* the zeros of $f(x) = x - \tan x$ in the interval $(0, 20)$ by the method of bisection. Use the functions `rootsearch` and `bisection`.

Solution. Note that $\tan x$ is singular and changes sign at $x = \pi/2, 3\pi/2, \ldots$. To prevent `bisection` from mistaking these points for roots, we set `switch = 1`. The closeness of roots to the singularities is another potential problem that can be alleviated by using small Δx in rootsearch. Choosing $\Delta x = 0.01$, we arrive at the following program:

```
#!/usr/bin/python
## example4_3
import math
from rootsearch import *
```

```
from bisection import *

def f(x): return x - math.tan(x)

a,b,dx = (0.0, 20.0, 0.01)
print("The roots are:")
while True:
    x1,x2 = rootsearch(f,a,b,dx)
    if x1 != None:
        a = x2
        root = bisection(f,x1,x2,1)
        if root != None: print(root)
    else:
        print("\nDone")
        break
input("Press return to exit")
```

The output from the program is

```
The roots are:
0.0
4.493409458100745
7.725251837074637
10.904121659695917
14.06619391292308
17.220755272209537
Done
```

4.4 Methods Based on Linear Interpolation

Secant and False Position Methods

The secant and the false position methods are closely related. Both methods require two starting estimates of the root, say, x_1 and x_2. The function $f(x)$ is assumed to be approximately linear near the root, so that the improved value x_3 of the root can be estimated by linear interpolation between x_1 and x_2.

Referring to Figure 4.2, the similar triangles (shaded in the figure) yield the relationship

$$\frac{f_2}{x_3 - x_2} = \frac{f_1 - f_2}{x_2 - x_1}$$

where we used the notation $f_i = f(x_i)$. Thus the improved estimate of the root is

$$x_3 = x_2 - f_2 \frac{x_2 - x_1}{f_2 - f_1} \tag{4.2}$$

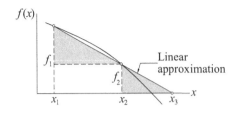

Figure 4.2. Linear interpolation.

The false position method (also known as *regula falsi*) requires x_1 and x_2 to bracket the root. After the improved root is computed from Eq. (4.2), either x_1 or x_2 is replaced by x_3. If f_3 has the same sign as f_1, we let $x_1 \leftarrow x_3$; otherwise we choose $x_2 \leftarrow x_3$. In this manner, the root is always bracketed in (x_1, x_2). The procedure is then repeated until convergence is obtained.

The secant method differs from the false position method in two ways: It does not require prior bracketing of the root, and it discards the oldest prior estimate of the root (i.e., after x_3 is computed, we let $x_1 \leftarrow x_2, x_2 \leftarrow x_3$).

The convergence of the secant method can be shown to be superlinear, with the error behaving as $E_{k+1} = cE_k^{1.618\ldots}$ (the exponent $1.618\ldots$ is the "golden ratio"). The precise order of convergence for the false position method is impossible to calculate. Generally, it is somewhat better than linear, but not by much. However, because the false position method always brackets the root, it is more reliable. We do not delve further into these methods, because both of them are inferior to Ridder's method as far as the order of convergence is concerned.

Ridder's Method

Ridder's method is a clever modification of the false position method. Assuming that the root is bracketed in (x_1, x_2), we first compute $f_3 = f(x_3)$, where x_3 is the midpoint of the bracket, as indicated in Figure 4.3(a). Next we the introduce the function

$$g(x) = f(x)e^{(x-x_1)Q} \tag{a}$$

where the constant Q is determined by requiring the points (x_1, g_1), (x_2, g_2) and (x_3, g_3) to lie on a straight line, as shown in Figure 4.3(b). As before, the notation we use is $g_i = g(x_i)$. The improved value of the root is then obtained by linear interpolation of $g(x)$ rather than $f(x)$.

Let us now look at the details. From Eq. (a) we obtain

$$g_1 = f_1 \qquad g_2 = f_2 e^{2hQ} \qquad g_3 = f_3 e^{hQ} \tag{b}$$

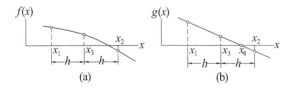

Figure 4.3. Mapping used in Ridder's method.

where $h = (x_2 - x_1)/2$. The requirement that the three points in Figure 4.3b lie on a straight line is $g_3 = (g_1 + g_2)/2$, or

$$f_3 e^{hQ} = \frac{1}{2}(f_1 + f_2 e^{2hQ})$$

which is a quadratic equation in e^{hQ}. The solution is

$$e^{hQ} = \frac{f_3 \pm \sqrt{f_3^2 - f_1 f_2}}{f_2} \tag{c}$$

Linear interpolation based on points (x_1, g_1) and (x_3, g_3) now yields for the improved root

$$x_4 = x_3 - g_3 \frac{x_3 - x_1}{g_3 - g_1} = x_3 - f_3 e^{hQ} \frac{x_3 - x_1}{f_3 e^{hQ} - f_1}$$

where in the last step we used Eqs. (b). As the final step, we substitute for e^{hQ} from Eq. (c), and obtain after some algebra

$$x_4 = x_3 \pm (x_3 - x_1) \frac{f_3}{\sqrt{f_3^2 - f_1 f_2}} \tag{4.3}$$

It can be shown that the correct result is obtained by choosing the plus sign if $f_1 - f_2 > 0$, and the minus sign if $f_1 - f_2 < 0$. After the computation of x_4, new brackets are determined for the root and Eq. (4.3) is applied again. The procedure is repeated until the difference between two successive values of x_4 becomes negligible.

Ridder's iterative formula in Eq. (4.3) has a very useful property: If x_1 and x_2 straddle the root, then x_4 is always within the interval (x_1, x_2). In other words, once the root is bracketed, it stays bracketed, making the method very reliable. The downside is that each iteration requires two function evaluations. There are competitive methods that get by with only one function evaluation per iteration (e.g., Brent's method), but they are more complex and require elaborate bookkeeping.

Ridder's method can be shown to converge quadratically, making it faster than either the secant or the false position method. It is the method to use if the derivative of $f(x)$ is impossible or difficult to compute.

■ ridder

The following is the source code for Ridder's method:

```
## module ridder
''' root = ridder(f,a,b,tol=1.0e-9).
    Finds a root of f(x) = 0 with Ridder's method.
    The root must be bracketed in (a,b).
'''
import error
import math
from numpy import sign
```

```
def ridder(f,a,b,tol=1.0e-9):
    fa = f(a)
    if fa == 0.0: return a
    fb = f(b)
    if fb == 0.0: return b
    if sign(f2)!= sign(f3): x1 = x3; f1 = f3
    for i in range(30):
      # Compute the improved root x from Ridder's formula
        c = 0.5*(a + b); fc = f(c)
        s = math.sqrt(fc**2 - fa*fb)
        if s == 0.0: return None
        dx = (c - a)*fc/s
        if (fa - fb) < 0.0: dx = -dx
        x = c + dx; fx = f(x)
      # Test for convergence
        if i > 0:
            if abs(x - xOld) < tol*max(abs(x),1.0): return x
        xOld = x
      # Re-bracket the root as tightly as possible
        if sign(fc) == sign(fx):
            if sign(fa)!= sign(fx): b = x; fb = fx
            else: a = x; fa = fx
        else:
            a = c; b = x; fa = fc; fb = fx
    return None
    print('Too many iterations')
```

EXAMPLE 4.4
Determine the root of $f(x) = x^3 - 10x^2 + 5 = 0$ that lies in $(0.6, 0.8)$ with Ridder's method.

Solution. The starting points are

$$x_1 = 0.6 \qquad f_1 = 0.6^3 - 10(0.6)^2 + 5 = 1.6160$$

$$x_2 = 0.8 \qquad f_2 = 0.8^3 - 10(0.8)^2 + 5 = -0.8880$$

First Iteration. Bisection yields the point

$$x_3 = 0.7 \qquad f_3 = 0.7^3 - 10(0.7)^2 + 5 = 0.4430$$

The improved estimate of the root can now be computed with Ridder's formula:

$$s = \sqrt{f_3^2 - f_1 f_2} = \sqrt{0.4330^2 - 1.6160(-0.8880)} = 1.2738$$

$$x_4 = x_3 \pm (x_3 - x_1)\frac{f_3}{s}$$

Because $f_1 > f_2$ we must use the plus sign. Therefore,

$$x_4 = 0.7 + (0.7 - 0.6)\frac{0.4430}{1.2738} = 0.7348$$

$$f_4 = 0.7348^3 - 10(0.7348)^2 + 5 = -0.0026$$

As the root clearly lies in the interval (x_3, x_4), we let

$$x_1 \leftarrow x_3 = 0.7 \qquad f_1 \leftarrow f_3 = 0.4430$$
$$x_2 \leftarrow x_4 = 0.7348 \qquad f_2 \leftarrow f_4 = -0.0026$$

which are the starting points for the next iteration.

Second Iteration

$$x_3 = 0.5(x_1 + x_2) = 0.5(0.7 + 0.7348) = 0.717\,4$$

$$f_3 = 0.717\,4^3 - 10(0.717\,4)^2 + 5 = 0.2226$$

$$s = \sqrt{f_3^2 - f_1 f_2} = \sqrt{0.2226^2 - 0.4430(-0.0026)} = 0.2252$$

$$x_4 = x_3 \pm (x_3 - x_1)\frac{f_3}{s}$$

Since $f_1 > f_2$, we again use the plus sign, so that

$$x_4 = 0.717\,4 + (0.717\,4 - 0.7)\frac{0.2226}{0.2252} = 0.7346$$

$$f_4 = 0.7346^3 - 10(0.7346)^2 + 5 = 0.0000$$

Thus the root is $x = 0.7346$, accurate to at least four decimal places.

EXAMPLE 4.5
Compute the zero of the function

$$f(x) = \frac{1}{(x - 0.3)^2 + 0.01} - \frac{1}{(x - 0.8)^2 + 0.04}$$

Solution. We obtain the approximate location of the root by plotting the function.

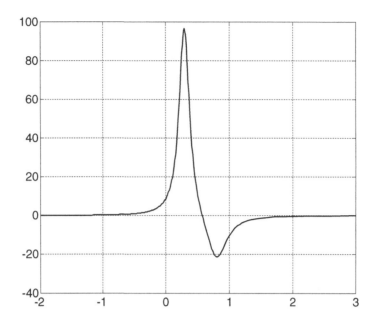

It is evident that the root of $f(x) = 0$ lies between $x = 0$ and 1. We can extract this root with the following program:

```python
#!/usr/bin/python
## example4_5
from ridder import *

def f(x):
    a = (x - 0.3)**2 + 0.01
    b = (x - 0.8)**2 + 0.04
    return 1.0/a - 1.0/b

print("root =",ridder(f,0.0,1.0))
input("Press return to exit")
```

The result is

```
root = 0.5800000000000001
```

4.5   Newton-Raphson Method

The Newton-Raphson algorithm is the best known method of finding roots for a good reason: It is simple and fast. The only drawback of the method is that it uses the derivative $f'(x)$ of the function as well as the function $f(x)$ itself. Therefore, the Newton-Raphson method is usable only in problems where $f'(x)$ can be readily computed.

The Newton-Raphson formula can be derived from the Taylor series expansion of $f(x)$ about x:

$$f(x_{i+1}) = f(x_i) + f'(x_i)(x_{i+1} - x_i) + O(x_{i+1} - x_i)^2 \tag{a}$$

where $O(z)$ is to be read as "of the order of z"—see Appendix A1. If x_{i+1} is a root of $f(x) = 0$, Eq. (a) becomes

$$0 = f(x_i) + f'(x_i)(x_{i+1} - x_i) + O(x_{i+1} - x_i)^2 \tag{b}$$

Assuming that x_i is a close to x_{i+1}, we can drop the last term in Eq. (b) and solve for x_{i+1}. The result is the Newton-Raphson formula:

$$x_{i+1} = x_i - \frac{f(x_i)}{f'(x_i)} \tag{4.3}$$

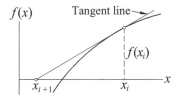

Figure 4.4. Graphical interpretation of the Newton-Raphson formula.

The graphical depiction of the Newton-Raphson formula is shown in Figure 4.4. The formula approximates $f(x)$ by the straight line that is tangent to the curve at x_i. Thus x_{i+1} is at the intersection of the x-axis and the tangent line.

The algorithm for the Newton-Raphson method is simple: It repeatedly applies Eq. (4.3), starting with an initial value x_0, until the convergence criterion

$$|x_{i+1} - x_i| < \varepsilon$$

is reached, ε being the error tolerance. Only the latest value of x has to be stored. Here is the algorithm:

> Let x be an estimate of the root of $f(x) = 0$.
> Do until $|\Delta x| < \varepsilon$:
>   Compute $\Delta x = -f(x)/f'(x)$.
>   Let $x \leftarrow x + \Delta x$.

The truncation error E in the Newton-Raphson formula can be shown to behave as

$$E_{i+1} = -\frac{f''(x)}{2f'(x)} E_i^2$$

where x is the root. This indicates that the method converges *quadratically* (the error is the square of the error in the previous step). Consequently, the number of significant figures is roughly doubled in every iteration.

Although the Newton-Raphson method converges fast near the root, its global convergence characteristics are poor. The reason is that the tangent line is not always

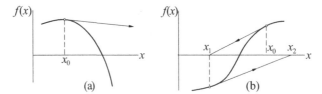

Figure 4.5. Examples where the Newton-Raphson method diverges.

an acceptable approximation of the function, as illustrated in the two examples in Figure 4.5. However, the method can be made nearly fail-safe by combining it with bisection.

■ newtonRaphson

The following *safe version* of the Newton-Raphson method assumes that the root to be computed is initially bracketed in (a,b). The midpoint of the bracket is used as the initial guess of the root. The brackets are updated after each iteration. If a Newton-Raphson iteration does not stay within the brackets, it is disregarded and replaced with bisection. Because newtonRaphson uses the function f(x) as well as its derivative, function routines for both (denoted by f and df in the listing) must be provided by the user.

```
## module newtonRaphson
''' root = newtonRaphson(f,df,a,b,tol=1.0e-9).
    Finds a root of f(x) = 0 by combining the Newton-Raphson
    method with bisection. The root must be bracketed in (a,b).
    Calls user-supplied functions f(x) and its derivative df(x).
'''
def newtonRaphson(f,df,a,b,tol=1.0e-9):
    import error
    from numpy import sign

    fa = f(a)
    if fa == 0.0: return a
    fb = f(b)
    if fb == 0.0: return b
    if sign(fa) == sign(fb): error.err('Root is not bracketed')
    x = 0.5*(a + b)
    for i in range(30):
        fx = f(x)
        if fx == 0.0: return x
      # Tighten the brackets on the root
        if sign(fa) != sign(fx): b = x
        else: a = x
```

```
    # Try a Newton-Raphson step
    dfx = df(x)
    # If division by zero, push x out of bounds
    try: dx = -fx/dfx
    except ZeroDivisionError: dx = b - a
    x = x + dx
    # If the result is outside the brackets, use bisection
    if (b - x)*(x - a) < 0.0:
        dx = 0.5*(b - a)
        x = a + dx
    # Check for convergence
    if abs(dx) < tol*max(abs(b),1.0): return x
print('Too many iterations in Newton-Raphson')
```

EXAMPLE 4.6

Use the Newton-Raphson method to obtain successive approximations of $\sqrt{2}$ as the ratio of two integers.

Solution. The problem is equivalent to finding the root of $f(x) = x^2 - 2 = 0$. Here the Newton-Raphson formula is

$$x \leftarrow x - \frac{f(x)}{f'(x)} = x - \frac{x^2 - 2}{2x} = \frac{x^2 + 2}{2x}$$

Starting with $x = 1$, successive iterations yield

$$x \leftarrow \frac{(1)^2 + 2}{2(1)} = \frac{3}{2}$$

$$x \leftarrow \frac{(3/2)^2 + 2}{2(3/2)} = \frac{17}{12}$$

$$x \leftarrow \frac{(17/12)^2 + 2}{2(17/12)} = \frac{577}{408}$$

$$\vdots$$

Note that $x = 577/408 = 1.1414216$ is already very close to $\sqrt{2} = 1.1414214$.

The results are dependent on the starting value of x. For example, $x = 2$ would produce a different sequence of ratios.

EXAMPLE 4.7

Find the smallest positive zero of

$$f(x) = x^4 - 6.4x^3 + 6.45x^2 + 20.538x - 31.752$$

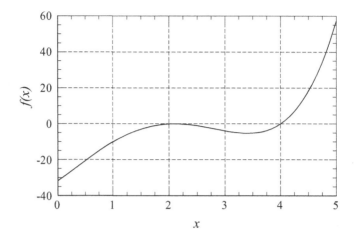

Solution. Inspecting the plot of the function, we suspect that the smallest positive zero is a double root at about $x = 2$. Bisection and Ridder's method would not work here, because they depend on the function changing its sign at the root. The same argument applies to the function `newtonRaphson`. Yet there is no reason why the unrefined version of the Newton-Raphson method should not succeed. We used the following program, which prints the number of iterations in addition to the root:

```
#!/usr/bin/python
## example4_8

def f(x): return x**4 - 6.4*x**3 + 6.45*x**2 + 20.538*x - 31.752
def df(x): return 4.0*x**3 - 19.2*x**2 + 12.9*x + 20.538

def newtonRaphson(x,tol=1.0e-9):
    for i in range(30):
        dx = -f(x)/df(x)
        x = x + dx
        if abs(dx) < tol: return x,i
    print 'Too many iterations\n'

root,numIter = newtonRaphson(2.0)
print 'Root =',root
print 'Number of iterations =',numIter
raw_input(''Press return to exit'')
```

The output is

```
Root = 2.0999999786199406
Number of iterations = 22
```

The true value of the root is $x = 2.1$. It can be shown that near a multiple root the convergence of the Newton-Raphson method is linear, rather than quadratic, which explains the large number of iterations. Convergence to a multiple root can be speeded up by replacing the Newton-Raphson formula in Eq. (4.3) with

$$x_{i+1} = x_i - m\frac{f(x_i)}{f'(x_i)}$$

where m is the multiplicity of the root ($m = 2$ in this problem). After making the change in the program, we obtained the result in only five iterations.

4.6   Systems of Equations

Introduction

Up to this point, we confined our attention to solving the single equation $f(x) = 0$. Let us now consider the n-dimensional version of the same problem, namely,

$$\mathbf{f}(\mathbf{x}) = \mathbf{0}$$

or, using scalar notation

$$f_1(x_1, x_2, \ldots, x_n) = 0$$
$$f_2(x_1, x_2, \ldots, x_n) = 0 \qquad\qquad (4.4)$$
$$\vdots$$
$$f_n(x_1, x_2, \ldots, x_n) = 0$$

Solving n simultaneous, nonlinear equations is a much more formidable task than finding the root of a single equation. The trouble is there is no a reliable method for bracketing the solution vector $\mathbf{x}$. Therefore, we cannot always provide the solution algorithm with a good starting value of $\mathbf{x}$, unless such a value is suggested by the physics of the problem.

The simplest and the most effective means of computing $\mathbf{x}$ is the Newton-Raphson method. It works well with simultaneous equations, provided that it is supplied with a good starting point. There are other methods that have better global convergence characteristics, but all of then are variants of the Newton-Raphson method.

Newton-Raphson Method

To derive the Newton-Raphson method for a system of equations, we start with the Taylor series expansion of $f_i(\mathbf{x})$ about the point $\mathbf{x}$:

$$f_i(\mathbf{x} + \Delta\mathbf{x}) = f_i(\mathbf{x}) + \sum_{j=1}^{n} \frac{\partial f_i}{\partial x_j}\Delta x_j + O(\Delta x^2) \qquad\qquad (4.5a)$$

Dropping terms of order Δx^2, we can write Eq. (4.5a) as

$$\mathbf{f}(\mathbf{x} + \Delta \mathbf{x}) = \mathbf{f}(\mathbf{x}) + \mathbf{J}(\mathbf{x}) \, \Delta \mathbf{x} \qquad (4.5b)$$

where $\mathbf{J}(\mathbf{x})$ is the *Jacobian matrix* (of size $n \times n$) made up of the partial derivatives

$$J_{ij} = \frac{\partial f_i}{\partial x_j} \qquad (4.6)$$

Note that Eq. (4.5b) is a linear approximation (vector $\Delta \mathbf{x}$ being the variable) of the vector-valued function $\mathbf{f}$ in the vicinity of point $\mathbf{x}$.

Let us now assume that $\mathbf{x}$ is the current approximation of the solution of $\mathbf{f}(\mathbf{x}) = \mathbf{0}$, and let $\mathbf{x} + \Delta \mathbf{x}$ be the improved solution. To find the correction $\Delta \mathbf{x}$, we set $\mathbf{f}(\mathbf{x} + \Delta \mathbf{x}) = \mathbf{0}$ in Eq. (4.5b). The result is a set of linear equations for $\Delta \mathbf{x}$:

$$\mathbf{J}(\mathbf{x}) \, \Delta \mathbf{x} = -\mathbf{f}(\mathbf{x}) \qquad (4.7)$$

Because analytical derivation of each $\partial f_i / \partial x_j$ can be difficult or impractical, it is preferable to let the computer calculate them from the finite difference approximation

$$\frac{\partial f_i}{\partial x_j} \approx \frac{f_i(\mathbf{x} + \mathbf{e}_j h) - f_i(\mathbf{x})}{h} \qquad (4.8)$$

where h is a small increment of x_j and $\mathbf{e}_j$ represents a unit vector in the direction of x_j. This formula can be obtained from Eq. (4.5a) after dropping the terms of order Δx^2 and setting $\Delta \mathbf{x} = \mathbf{e}_j h$. We get away with the approximation in Eq. (4.8) because the Newton-Raphson method is rather insensitive to errors in $\mathbf{J}(\mathbf{x})$. By using this approximation, we also avoid the tedium of typing the expressions for $\partial f_i / \partial x_j$ into the computer code.

The following steps constitute the Newton-Raphson method for simultaneous, nonlinear equations:

> Estimate the solution vector $\mathbf{x}$.
> Do until $|\Delta \mathbf{x}| < \varepsilon$:
>   Compute the matrix $\mathbf{J}(\mathbf{x})$ from Eq. (4.8).
>   Solve $\mathbf{J}(\mathbf{x}) \, \Delta \mathbf{x} = -\mathbf{f}(\mathbf{x})$ for $\Delta \mathbf{x}$.
>   Let $\mathbf{x} \leftarrow \mathbf{x} + \Delta \mathbf{x}$.

where ε is the error tolerance. As in the one-dimensional case, success of the Newton-Raphson procedure depends entirely on the initial estimate of $\mathbf{x}$. If a good starting point is used, convergence to the solution is very rapid. Otherwise, the results are unpredictable.

■ newtonRaphson2

This function is an implementation of the Newton-Raphson method. The nested function `jacobian` computes the Jacobian matrix from the finite difference approximation in Eq. (4.8). The simultaneous equations in Eq. (4.7) are solved by Gauss elimination with row pivoting using the function gaussPivot listed in Section 2.5. The function subroutine f that returns the array **f(x)** must be supplied by the user.

```
## module newtonRaphson2
''' soln = newtonRaphson2(f,x,tol=1.0e-9).
    Solves the simultaneous equations f(x) = 0 by
    the Newton-Raphson method using {x} as the initial
    guess. Note that {f} and {x} are vectors.
'''
import numpy as np
from gaussPivot import *
import math
def newtonRaphson2(f,x,tol=1.0e-9):

    def jacobian(f,x):
        h = 1.0e-4
        n = len(x)
        jac = np.zeros((n,n))
        f0 = f(x)
        for i in range(n):
            temp = x[i]
            x[i] = temp + h
            f1 = f(x)
            x[i] = temp
            jac[:,i] = (f1 - f0)/h
        return jac,f0

    for i in range(30):
        jac,f0 = jacobian(f,x)
        if math.sqrt(np.dot(f0,f0)/len(x)) < tol: return x
        dx = gaussPivot(jac,-f0)
        x = x + dx
        if math.sqrt(np.dot(dx,dx)) < tol*max(max(abs(x)),1.0):
            return x
    print(Too many iterations')
```

Note that the Jacobian matrix **J(x)** is recomputed in each iterative loop. Since each calculation of **J(x)** involves $n + 1$ evaluations of **f(x)** (n is the number of equations), the expense of computation can be high depending on the size of n and the complexity of **f(x)**. It is often possible to save computer time by neglecting the changes in the Jacobian matrix between iterations, thus computing **J(x)** only once. This approach will work, provided that the initial **x** is sufficiently close to the solution.

EXAMPLE 4.8

Determine the points of intersection between the circle $x^2 + y^2 = 3$ and the hyperbola $xy = 1$.

Solution. The equations to be solved are

$$f_1(x, y) = x^2 + y^2 - 3 = 0 \tag{a}$$

$$f_2(x, y) = xy - 1 = 0 \tag{b}$$

The Jacobian matrix is

$$\mathbf{J}(x, y) = \begin{bmatrix} \partial f_1/\partial x & \partial f_1/\partial y \\ \partial f_2/\partial x & \partial f_2/\partial y \end{bmatrix} = \begin{bmatrix} 2x & 2y \\ y & x \end{bmatrix}$$

Thus the linear equations $\mathbf{J}(\mathbf{x})\Delta\mathbf{x} = -\mathbf{f}(\mathbf{x})$ associated with the Newton-Raphson method are

$$\begin{bmatrix} 2x & 2y \\ y & x \end{bmatrix} \begin{bmatrix} \Delta x \\ \Delta y \end{bmatrix} = \begin{bmatrix} -x^2 - y^2 + 3 \\ -xy + 1 \end{bmatrix} \tag{c}$$

By plotting the circle and the hyperbola, we see that there are four points of intersection. It is sufficient, however, to find only one of these points, because the others can be deduced from symmetry. From the plot we also get a rough estimate of the coordinates of an intersection point, $x = 0.5$, $y = 1.5$, which we use as the starting values.

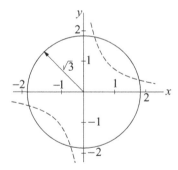

The computations then proceed as follows.

First Iteration. Substituting $x = 0.5$, $y = 1.5$ in Eq. (c), we get

$$\begin{bmatrix} 1.0 & 3.0 \\ 1.5 & 0.5 \end{bmatrix} \begin{bmatrix} \Delta x \\ \Delta y \end{bmatrix} = \begin{bmatrix} 0.50 \\ 0.25 \end{bmatrix}$$

the solution of which is $\Delta x = \Delta y = 0.125$. Therefore, the improved coordinates of the intersection point are

$$x = 0.5 + 0.125 = 0.625 \qquad y = 1.5 + 0.125 = 1.625$$

Second Iteration. Repeating the procedure using the latest values of x and y, we obtain

$$\begin{bmatrix} 1.250 & 3.250 \\ 1.625 & 0.625 \end{bmatrix} \begin{bmatrix} \Delta x \\ \Delta y \end{bmatrix} = \begin{bmatrix} -0.031250 \\ -0.015625 \end{bmatrix}$$

which yields $\Delta x = \Delta y = -0.00694$. Thus

$$x = 0.625 - 0.006\,94 = 0.618\,06 \qquad y = 1.625 - 0.006\,94 = 1.618\,06$$

Third Iteration. Substitution of the latest x and y into Eq. (c) yields

$$\begin{bmatrix} 1.236\,12 & 3.23612 \\ 1.618\,06 & 0.61806 \end{bmatrix} \begin{bmatrix} \Delta x \\ \Delta y \end{bmatrix} = \begin{bmatrix} -0.000\,116 \\ -0.000\,058 \end{bmatrix}$$

The solution is $\Delta x = \Delta y = -0.00003$, so that

$$x = 0.618\,06 - 0.000\,03 = 0.618\,03$$

$$y = 1.618\,06 - 0.000\,03 = 1.618\,03$$

Subsequent iterations would not change the results within five significant figures. Therefore, the coordinates of the four intersection points are

$$\pm(0.618\,03, 1.618\,03) \text{ and } \pm (1.618\,03, 0.618\,03)$$

Alternate Solution. If there are only a few equations, it may be possible to eliminate all but one of the unknowns. Then we would be left with a single equation that can be solved by the methods described in Sections 4.2–4.5. In this problem, we obtain from Eq. (b)

$$y = \frac{1}{x}$$

which upon substitution into Eq. (a) yields $x^2 + 1/x^2 - 3 = 0$, or

$$x^4 - 3x^2 + 1 = 0$$

The solutions of this biquadratic equation are $x = \pm 0.618\,03$ and $\pm 1.618\,03$, which agree with the results obtained by the Newton-Raphson method.

EXAMPLE 4.9

Find a solution of

$$\sin x + y^2 + \ln z - 7 = 0$$

$$3x + 2^y - z^3 + 1 = 0$$

$$x + y + z - 5 = 0$$

using newtonRaphson2. Start with the point $(1, 1, 1)$.

Solution. Letting $x_1 = x$, $x_2 = y$ and $x_3 = z$, we obtain the following program:

```
#!/usr/bin/python
## example4_10
```

```
import numpy as np
import math
from newtonRaphson2 import *

def f(x):
    f = np.zeros(len(x))
    f[0] = math.sin(x[0]) + x[1]**2 + math.log(x[2]) - 7.0
    f[1] = 3.0*x[0] + 2.0**x[1] - x[2]**3 + 1.0
    f[2] = x[0] + x[1] + x[2] - 5.0
    return f

x = np.array([1.0, 1.0, 1.0])
print(newtonRaphson2(f,x))
input("\nPress return to exit")
```

The output is

```
[ 0.59905376  2.3959314   2.00501484]
```

PROBLEM SET 4.1

1. Use the Newton-Raphson method and a four-function calculator ($+ - \times \div$ operations only) to compute $\sqrt[3]{75}$ with four significant figure accuracy.

2. Find the smallest positive (real) root of $x^3 - 3.23x^2 - 5.54x + 9.84 = 0$ by the method of bisection.

3. The smallest positive, nonzero root of $\cosh x \cos x - 1 = 0$ lies in the interval (4, 5). Compute this root by Ridder's method.

4. Solve Problem 3 by the Newton-Raphson method.

5. A root of the equation $\tan x - \tanh x = 0$ lies in (7.0, 7.4). Find this root with three decimal place accuracy by the method of bisection.

6. Determine the two roots of $\sin x + 3\cos x - 2 = 0$ that lie in the interval (−2, 2). Use the Newton-Raphson method.

7. Solve Prob. 6 using the secant formula in Eq. (4.2).

8. Draw a plot of $f(x) = \cosh x \cos x - 1$ in the range $0 \le x \le 10$. (a) Verify from the plot that the smallest positive, nonzero root of $f(x) = 0$ lies in the interval (4, 5). (b) Show graphically that the Newton-Raphson formula would not converge to this root if it is started with $x = 4$.

9. The equation $x^3 - 1.2x^2 - 8.19x + 13.23 = 0$ has a double root close to $x = 2$. Determine this root with the Newton-Raphson method within four decimal places.

10. ■ Write a program that computes all the roots of $f(x) = 0$ in a given interval with Ridder's method. Use the functions `rootsearch` and `ridder`. You may use the program in Example 4.3 as a model. Test the program by finding the roots of $x \sin x + 3\cos x - x = 0$ in (−6, 6).

11. ■ Repeat Prob. 10 with the Newton-Raphson method.

12. ■ Determine all real roots of $x^4 + 0.9x^3 - 2.3x^2 + 3.6x - 25.2 = 0$.

13. ■ Compute all positive real roots of $x^4 + 2x^3 - 7x^2 + 3 = 0$.

14. ■ Find all positive, nonzero roots of $\sin x - 0.1x = 0$.

15. ■ The natural frequencies of a uniform cantilever beam are related to the roots β_i of the frequency equation $f(\beta) = \cosh \beta \cos \beta + 1 = 0$, where

$$\beta_i^4 = (2\pi f_i)^2 \frac{mL^3}{EI}$$

$f_i = i$th natural frequency (cps)

$m = $ mass of the beam

$L = $ length of the beam

$E = $ modulus of elasticity

$I = $ moment of inertia of the cross section

Determine the lowest two frequencies of a steel beam 0.9 m. long, with a rectangular cross section 25 mm wide and 2.5 mm in. high. The mass density of steel is 7850 kg/m^3 and $E = 200$ GPa.

16. ■

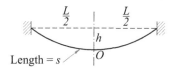

Length = s

A cable is suspended as shown in the figure. Its length s and the sag h are related to the span L by

$$s = \frac{2}{\lambda} \sinh \frac{\lambda L}{2} \qquad h = \frac{1}{\lambda} \left(\cosh \frac{\lambda L}{2} - 1 \right)$$

where

$\lambda = w_0 / T_0$
$w_0 = $ weight of cable per unit length
$T_0 = $ cable tension at O

Compute s for $L = 160$ m and $h = 15$ m.

17. ■

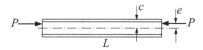

The aluminum W 310×202 (wide flange) column is subjected to an eccentric axial load P as shown. The maximum compressive stress in the column is given by the so-called *secant formula*:

$$\sigma_{\max} = \bar{\sigma} \left[1 + \frac{ec}{r^2} \sec \left(\frac{L}{2r} \sqrt{\frac{\bar{\sigma}}{E}} \right) \right]$$

where

$$\bar{\sigma} = P/A = \text{average stress}$$

$$A = 25\,800 \text{ mm}^2 = \text{cross-sectional area of the column}$$

$$e = 85 \text{ mm} = \text{eccentricity of the load}$$

$$c = 170 \text{ mm} = \text{half depth of the column}$$

$$r = 142 \text{ mm} = \text{radius of gyration of the cross section}$$

$$L = 7\,100 \text{ mm} = \text{length of the column}$$

$$E = 71 \times 10^9 \text{ Pa} = \text{modulus of elasticity}$$

Determine the maximum load P that the column can carry if the maximum stress is not to exceed 120×10^6 Pa.

18. ■

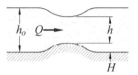

Bernoulli's equation for fluid flow in an open channel with a small bump is

$$\frac{Q^2}{2gb^2h_0^2} + h_0 = \frac{Q^2}{2gb^2h^2} + h + H$$

where

$$Q = 1.2 \text{ m}^3/\text{s} = \text{volume rate of flow}$$

$$g = 9.81 \text{ m/s}^2 = \text{gravitational acceleration}$$

$$b = 1.8 \text{ m} = \text{width of channel}$$

$$h_0 = 0.6 \text{ m} = \text{upstream water level}$$

$$H = 0.075 \text{ m} = \text{height of bump}$$

$$h = \text{ water level above the bump}$$

Determine h.

19. ■ The speed v of a Saturn V rocket in vertical flight near the surface of earth can be approximated by

$$v = u \ln \frac{M_0}{M_0 - \dot{m}t} - gt$$

where

$$u = 2\,510 \text{ m/s} = \text{velocity of exhaust relative to the rocket}$$

$$M_0 = 2.8 \times 10^6 \text{ kg} = \text{mass of rocket at liftoff}$$

$$\dot{m} = 13.3 \times 10^3 \text{ kg/s} = \text{rate of fuel consumption}$$

$$g = 9.81 \text{ m/s}^2 = \text{gravitational acceleration}$$

$$t = \text{time measured from liftoff}$$

Determine the time when the rocket reaches the speed of sound (335 m/s).

20. ■

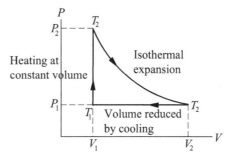

The figure shows the thermodynamic cycle of an engine. The efficiency of this engine for monatomic gas is

$$\eta = \frac{\ln(T_2/T_1) - (1 - T_1/T_2)}{\ln(T_2/T_1) + (1 - T_1/T_2)/(\gamma - 1)}$$

where T is the absolute temperature and $\gamma = 5/3$. Find T_2/T_1 that results in 30% efficiency ($\eta = 0.3$).

21. ■

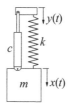

Consider the forced vibration of the spring-mass-dashpot system shown. When the harmonic displacement $y(t) = Y\sin\omega t$ is imposed on the support, the response of the mass is the displacement $x(t) = X\sin(\omega t - \beta)$, where

$$\frac{X}{Y} = \sqrt{(1 + Z\cos\phi)^2 + (Z\sin\phi)^2} \qquad \tan\beta = \frac{Z\sin\phi}{1 + Z\cos\phi}$$

In these two equations we used this notation:

$$Z = \frac{(\omega/p)^2}{\sqrt{\left[1 - (\omega/p)^2\right]^2 + (2\zeta\omega/p)^2}} \qquad \tan\phi = \frac{2\zeta\omega/p}{1 - (\omega/p)^2}$$

$$p = \sqrt{\frac{k}{m}} = \text{natural frequency of the system}$$

$$\zeta = \frac{c}{2mp} = \text{damping factor}$$

If $m = 0.2$ kg, $k = 2\,880$ N/m, and $\omega = 96$ rad/s, determine the smallest c (the coefficient of damping) for which X/Y does not exceed 1.5.

22. ■

The cylindrical oil tank of radius r and length L is filled to depth h. The resulting volume of oil in the tank is

$$V = r^2 L \left[\phi - \left(1 - \frac{h}{r} \right) \sin \phi \right]$$

where

$$\phi = \cos^{-1} \left(1 - \frac{h}{r} \right)$$

If the tank is 3/4 full, determine h/r.

23. ■ Determine the coordinates of the two points where the circles $(x - 2)^2 + y^2 = 4$ and $x^2 + (y - 3)^2 = 4$ intersect. Start by estimating the locations of the points from a sketch of the circles, and then use the Newton-Raphson method to compute the coordinates.

24. ■ The equations

$$\sin x + 3 \cos x - 2 = 0$$

$$\cos x - \sin y + 0.2 = 0$$

have a solution in the vicinity of the point $(1, 1)$. Use the Newton-Raphson method to refine the solution.

25. ■ Use any method to find *all* real solutions of the simultaneous equations

$$\tan x - y = 1$$

$$\cos x - 3 \sin y = 0$$

in the region $0 \leq x \leq 1.5$.

26. ■ The equation of a circle is

$$(x - a)^2 + (y - b)^2 = R^2$$

where R is the radius and (a, b) are the coordinates of the center. If the coordinates of three points on the circle are

x	8.21	0.34	5.96
y	0.00	6.62	−1.12

determine R, a, and b.

27. ■

The trajectory of a satellite orbiting the earth is

$$R = \frac{C}{1 + e\sin(\theta + \alpha)}$$

where (R, θ) are the polar coordinates of the satellite, and C, e, and α are constants (e is known as the eccentricity of the orbit). If the satellite was observed at the following three positions

θ	−30°	0°	30°
R (km)	6870	6728	6615

determine the smallest R of the trajectory and the corresponding value of θ.

28. ■

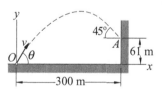

A projectile is launched at O with the velocity v at the angle θ to the horizontal. The parametric equations of the trajectory are

$$x = (v\cos\theta)t$$

$$y = -\frac{1}{2}gt^2 + (v\sin\theta)t$$

where t is the time measured from the instant of launch, and $g = 9.81$ m/s^2 represents the gravitational acceleration. If the projectile is to hit the target A at the 45° angle shown in the figure, determine v, θ, and the time of flight.

29. ■

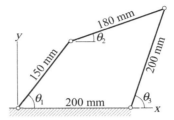

The three angles of the four-bar linkage are related by

$$150 \cos \theta_1 + 180 \cos \theta_2 - 200 \cos \theta_3 = 200$$

$$150 \sin \theta_1 + 180 \sin \theta_2 - 200 \sin \theta_3 = 0$$

Determine θ_1 and θ_2 when $\theta_3 = 75°$. Note that there are two solutions.

30. ■

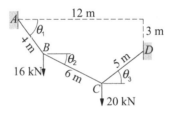

The 15-m cable is suspended from A and D and carries concentrated loads at B and C. The vertical equilibrium equations of joints B and C are

$$T(- \tan \theta_2 + \tan \theta_1) = 16$$

$$T(\tan \theta_3 + \tan \theta_2) = 20$$

where T is the horizontal component of the cable force (it is the same in all segments of the cable). In addition, there are two geometric constraints imposed by the positions of the supports:

$$-4 \sin \theta_1 - 6 \sin \theta_2 + 5 \sin \theta_2 = -3$$

$$4 \cos \theta_1 + 6 \cos \theta_2 + 5 \cos \theta_3 = 12$$

Determine the angles θ_1, θ_2, and θ_3.

31. ■ Given the data points

x	0	0.25	0.50	0.75	1.0	1.25	1.5
y	0	-1.2233	-2.2685	-2.8420	-2.2130	2.5478	55.507

determine the nonzero root of $y(x) = 0$. *Hint*: Use rational function interpolation to compute y.

*4.7   Zeros of Polynomials

Introduction

A polynomial of degree n has the form

$$P_n(x) = a_0 + a_1 x + a_2 x^2 + \cdots + a_n x^n \tag{4.9}$$

where the coefficients a_i may be real or complex. We focus here on polynomials with real coefficients, but the algorithms presented in this section also work with complex coefficients.

The polynomial equation $P_n(x) = 0$ has exactly n roots, which may be real or complex. If the coefficients are real, the complex roots always occur in conjugate pairs $(x_r + ix_i, \ x_r - ix_i)$, where x_r and x_i are the real and imaginary parts, respectively. For real coefficients, the number of real roots can be estimated from the *rule of Descartes*:

- The number of positive, real roots equals the number of sign changes in the expression for $P_n(x)$, or less by an even number.
- The number of negative, real roots is equal to the number of sign changes in $P_n(-x)$, or less by an even number.

As an example, consider $P_3(x) = x^3 - 2x^2 - 8x + 27$. Since the sign changes twice, $P_3(x) = 0$ has either two or zero positive real roots. In contrast, $P_3(-x) = -x^3 - 2x^2 + 8x + 27$ contains a single sign change; hence $P_3(x)$ possesses one negative real zero.

The real zeros of polynomials with real coefficients can always be computed by one of the methods already described. Yet if complex roots are to be computed, it is best to use a method that specializes in polynomials. Here we present a method due to Laguerre, which is reliable and simple to implement. Before proceeding to Laguerre's method, we must first develop two numerical tools that are needed in any method capable of determining the zeros of a polynomial. The first tool is an efficient algorithm for evaluating a polynomial and its derivatives. The second algorithm we need is for the *deflation* of a polynomial; that is, for dividing the $P_n(x)$ by $x - r$, where r is a root of $P_n(x) = 0$.

Evaluation of Polynomials

It is tempting to evaluate the polynomial in Eq. (4.9) from left to right by the following algorithm (we assume that the coefficients are stored in the array **a**):

```
p = 0.0
for i in range(n+1):
    p = p + a[i]*x**i
```

Since x^k is evaluated as $x \times x \times \cdots \times x$ ($k - 1$ multiplications), we deduce that the number of multiplications in this algorithm is

$$1 + 2 + 3 + \cdots + n - 1 = \frac{1}{2}n(n - 1)$$

If n is large, the number of multiplications can be reduced considerably if we evaluate the polynomial from right to left. For an example, take

$$P_4(x) = a_0 + a_1 x + a_2 x^2 + a_3 x^3 + a_4 x^4$$

After rewriting the polynomial as

$$P_4(x) = a_0 + x\{a_1 + x[a_2 + x(a_3 + xa_4)]\}$$

the preferred computational sequence becomes obvious:

$$P_0(x) = a_4$$
$$P_1(x) = a_3 + x P_0(x)$$
$$P_2(x) = a_2 + x P_1(x)$$
$$P_3(x) = a_1 + x P_2(x)$$
$$P_4(x) = a_0 + x P_3(x)$$

For a polynomial of degree n, the procedure can be summarized as

$$P_0(x) = a_n$$
$$P_i(x) = a_{n-i} + x P_{i-1}(x), \quad i = 1, 2, \ldots, n \tag{4.10}$$

leading to the algorithm

```
p = a[n]
for i in range(1,n+1):
    p = a[n-i] + p*x
```

The last algorithm involves only n multiplications, making it more efficient for $n > 3$. Yet computational economy is not the primary reason why this algorithm should be used. Because the result of each multiplication is rounded off, the procedure with the least number of multiplications invariably accumulates the smallest roundoff error.

Some root-finding algorithms, including Laguerre's method, also require evaluation of the first and second derivatives of $P_n(x)$. From Eq. (4.10) we obtain by differentiation

$$P_0'(x) = 0 \quad P_i'(x) = P_{i-1}(x) + x P_{i-1}'(x), \quad i = 1, 2, \ldots, n \tag{4.11a}$$

$$P_0''(x) = 0 \quad P_i''(x) = 2 P_{i-1}'(x) + x P_{i-1}''(x), \quad i = 1, 2, \ldots, n \tag{4.11b}$$

■ evalPoly

Here is the function that evaluates a polynomial and its derivatives:

```
## module evalPoly
'''  p,dp,ddp = evalPoly(a,x).
     Evaluates the polynomial
     p = a[0] + a[1]*x + a[2]*x^2 +...+ a[n]*x^n
     with its derivatives dp = p' and ddp = p''
     at x.
'''

def evalPoly(a,x):
    n = len(a) - 1
    p = a[n]
    dp = 0.0 + 0.0j
    ddp = 0.0 + 0.0j
    for i in range(1,n+1):
        ddp = ddp*x + 2.0*dp
        dp = dp*x + p
        p = p*x + a[n-i]
    return p,dp,ddp
```

Deflation of Polynomials

After a root r of $P_n(x) = 0$ has been computed, it is desirable to factor the polynomial as follows:

$$P_n(x) = (x - r)P_{n-1}(x) \tag{4.12}$$

This procedure, known as deflation or *synthetic division*, involves nothing more than computing the coefficients of $P_{n-1}(x)$. Because the remaining zeros of $P_n(x)$ are also the zeros of $P_{n-1}(x)$, the root-finding procedure can now be applied to $P_{n-1}(x)$ rather than $P_n(x)$. Deflation thus makes it progressively easier to find successive roots, because the degree of the polynomial is reduced every time a root is found. Moreover, by eliminating the roots that have already been found, the chance of computing the same root more than once is eliminated.

If we let

$$P_{n-1}(x) = b_0 + b_1 x + b_2 x^2 + \cdots + b_{n-1} x^{n-1}$$

then Eq. (4.12) becomes

$$a_0 + a_1 x + a_2 x^2 + \cdots + a_{n-1} x^{n-1} + a_n x^n$$
$$= (x - r)(b_0 + b_1 x + b_2 x^2 + \cdots + b_{n-1} x^{n-1})$$

Equating the coefficients of like powers of x, we obtain

$$b_{n-1} = a_n \qquad b_{n-2} = a_{n-1} + r b_{n-1} \qquad \cdots \qquad b_0 = a_1 + r b_1 \tag{4.13}$$

which leads to *Horner's deflation algorithm:*

```
b[n-1] = a[n]
for i in range(n-2,-1,-1):
    b[i] = a[i+1] + r*b[i+1]
```

Laguerre's Method

Laquerre's formulas are not easily derived for a general polynomial $P_n(x)$. However, the derivation is greatly simplified if we consider the special case where the polynomial has a zero at $x = r$ and an $(n - 1)$ zeros at $x = q$. Hence the polynomial can be written as

$$P_n(x) = (x - r)(x - q)^{n-1} \tag{a}$$

Our problem is now this: Given the polynomial in Eq. (a) in the form

$$P_n(x) = a_0 + a_1 x + a_2 x^2 + \cdots + a_n x^n$$

determine r (note that q is also unknown). It turns out that the result, which is exact for the special case considered here, works well as an iterative formula with any polynomial.

Differentiating Eq. (a) with respect to x, we get

$$P_n'(x) = (x - q)^{n-1} + (n - 1)(x - r)(x - q)^{n-2}$$

$$= P_n(x) \left(\frac{1}{x - r} + \frac{n - 1}{x - q} \right)$$

Thus

$$\frac{P_n'(x)}{P_n(x)} = \frac{1}{x - r} + \frac{n - 1}{x - q} \tag{b}$$

which upon differentiation yields

$$\frac{P_n''(x)}{P_n(x)} - \left[\frac{P_n'(x)}{P_n(x)} \right]^2 = -\frac{1}{(x - r)^2} - \frac{n - 1}{(x - q)^2} \tag{c}$$

It is convenient to introduce the notation

$$G(x) = \frac{P_n'(x)}{P_n(x)} \qquad H(x) = G^2(x) - \frac{P_n''(x)}{P_n(x)} \tag{4.14}$$

so that Eqs. (b) and (c) become

$$G(x) = \frac{1}{x - r} + \frac{n - 1}{x - q} \tag{4.15a}$$

$$H(x) = \frac{1}{(x - r)^2} + \frac{n - 1}{(x - q)^2} \tag{4.15b}$$

If we solve Eq. (4.15a) for $x - q$ and substitute the result into Eq. (4.15b), we obtain a quadratic equation for $x - r$. The solution of this equation is *Laguerre's formula*:

$$x - r = \frac{n}{G(x) \pm \sqrt{(n-1)\left[nH(x) - G^2(x)\right]}} \tag{4.16}$$

The procedure for finding a zero of a polynomial by Laguerre's formula is as follows:

> Let x be a guess for the root of $P_n(x) = 0$ (any value will do).
> Do until $|P_n(x)| < \varepsilon$ or $|x - r| < \varepsilon$ (ε is the error tolerance):
> > Evaluate $P_n(x)$, $P'_n(x)$ and $P''_n(x)$ using `evalPoly`.
> > Compute $G(x)$ and $H(x)$ from Eqs. (4.14).
> > Determine the improved root r from Eq. (4.16) choosing the sign
> > > that results in the *larger magnitude of the denominator*.
> > Let $x \leftarrow r$.

One nice property of Laguerre's method is that it converges to a root, with very few exceptions, from any starting value of x.

■ polyRoots

The function `polyRoots` in this module computes all the roots of $P_n(x) = 0$, where the polynomial $P_n(x)$ is defined by its coefficient array $\mathbf{a} = [a_0, a_1, \ldots, a_n]$. After the first root is computed by the nested function `laguerre`, the polynomial is deflated using `deflPoly`, and the next zero is computed by applying `laguerre` to the deflated polynomial. This process is repeated until all n roots have been found. If a computed root has a very small imaginary part, it is more than likely that it represents roundoff error. Therefore, `polyRoots` replaces a tiny imaginary part by zero.

```
## module polyRoots
''' roots = polyRoots(a).
    Uses Laguerre's method to compute all the roots of
    a[0] + a[1]*x + a[2]*x^2 +...+ a[n]*x^n = 0.
    The roots are returned in the array 'roots',
'''
from evalPoly import *
import numpy as np
import cmath
from random import random

def polyRoots(a,tol=1.0e-12):

    def laguerre(a,tol):
        x = random()    # Starting value (random number)
        n = len(a) - 1
```

```
        for i in range(30):
            p,dp,ddp = evalPoly(a,x)
            if abs(p) < tol: return x
            g = dp/p
            h = g*g - ddp/p
            f = cmath.sqrt((n - 1)*(n*h - g*g))
            if abs(g + f) > abs(g - f): dx = n/(g + f)
            else: dx = n/(g - f)
            x = x - dx
            if abs(dx) < tol: return x
        print('Too many iterations')

    def deflPoly(a,root):   # Deflates a polynomial
        n = len(a)-1
        b = [(0.0 + 0.0j)]*n
        b[n-1] = a[n]
        for i in range(n-2,-1,-1):
            b[i] = a[i+1] + root*b[i+1]
        return b

    n = len(a) - 1
    roots = np.zeros((n),dtype=complex)
    for i in range(n):
        x = laguerre(a,tol)
        if abs(x.imag) < tol: x = x.real
        roots[i] = x
        a = deflPoly(a,x)
    return roots
```

Because the roots are computed with finite accuracy, each deflation introduces small errors in the coefficients of the deflated polynomial. The accumulated roundoff error increases with the degree of the polynomial and can become severe if the polynomial is ill-conditioned (small changes in the coefficients produce large changes in the roots). Hence the results should be viewed with caution when dealing with polynomials of high degree.

The errors caused by deflation can be reduced by recomputing each root using the original undeflated polynomial. The roots obtained previously in conjunction with deflation are employed as the starting values.

EXAMPLE 4.10
A zero of the polynomial $P_4(x) = 3x^4 - 10x^3 - 48x^2 - 2x + 12$ is $x = 6$. Deflate the polynomial with Horner's algorithm; that is, find $P_3(x)$ so that $(x - 6)P_3(x) = P_4(x)$.

Solution. With $r = 6$ and $n = 4$, Eqs. (4.13) become

$$b_3 = a_4 = 3$$
$$b_2 = a_3 + 6b_3 = -10 + 6(3) = 8$$
$$b_1 = a_2 + 6b_2 = -48 + 6(8) = 0$$
$$b_0 = a_1 + 6b_1 = -2 + 6(0) = -2$$

Therefore,

$$P_3(x) = 3x^3 + 8x^2 - 2$$

EXAMPLE 4.11

A root of the equation $P_3(x) = x^3 - 4.0x^2 - 4.48x + 26.1$ is approximately $x = 3 - i$. Find a more accurate value of this root by one application of Laguerre's iterative formula.

Solution. Use the given estimate of the root as the starting value. Thus

$$x = 3 - i \qquad x^2 = 8 - 6i \qquad x^3 = 18 - 26i$$

Substituting these values in $P_3(x)$ and its derivatives, we get

$$P_3(x) = x^3 - 4.0x^2 - 4.48x + 26.1$$
$$= (18 - 26i) - 4.0(8 - 6i) - 4.48(3 - i) + 26.1 = -1.34 + 2.48i$$
$$P_3'(x) = 3.0x^2 - 8.0x - 4.48$$
$$= 3.0(8 - 6i) - 8.0(3 - i) - 4.48 = -4.48 - 10.0i$$
$$P_3''(x) = 6.0x - 8.0 = 6.0(3 - i) - 8.0 = 10.0 - 6.0i$$

Equations (4.14) then yield

$$G(x) = \frac{P_3'(x)}{P_3(x)} = \frac{-4.48 - 10.0i}{-1.34 + 2.48i} = -2.36557 + 3.08462i$$

$$H(x) = G^2(x) - \frac{P_3''(x)}{P_3(x)} = (-2.36557 + 3.08462i)^2 - \frac{10.0 - 6.0i}{-1.34 + 2.48i}$$
$$= 0.35995 - 12.48452i$$

The term under the square root sign of the denominator in Eq. (4.16) becomes

$$F(x) = \sqrt{(n - 1)\left[n\,H(x) - G^2(x)\right]}$$
$$= \sqrt{2\left[3(0.35995 - 12.48452i) - (-2.36557 + 3.08462i)^2\right]}$$
$$= \sqrt{5.67822 - 45.71946i} = 5.08670 - 4.49402i$$

Now we must find which sign in Eq. (4.16) produces the larger magnitude of the denominator:

$$|G(x) + F(x)| = |(-2.36557 + 3.08462i) + (5.08670 - 4.49402i)|$$

$$= |2.72113 - 1.40940i| = 3.06448$$

$$|G(x) - F(x)| = |(-2.36557 + 3.08462i) - (5.08670 - 4.49402i)|$$

$$= |-7.45227 + 7.57864i| = 10.62884$$

Using the minus sign, Eq. (4.16) yields the following improved approximation for the root:

$$r = x - \frac{n}{G(x) - F(x)} = (3 - i) - \frac{3}{-7.45227 + 7.57864i}$$

$$= 3.19790 - 0.79875i$$

Thanks to the good starting value, this approximation is already quite close to the exact value $r = 3.20 - 0.80i$.

EXAMPLE 4.12

Use `polyRoots` to compute *all* the roots of $x^4 - 5x^3 - 9x^2 + 155x - 250 = 0$.

Solution. The program

```
#!/usr/bin/python
## example4_12
from polyRoots import *
import numpy as np

c = np.array([-250.0,155.0,-9.0,-5.0,1.0])
print('Roots are:\n',polyRoots(c))
input('Press return to exit')
```

produced the output

```
Roots are:
 [ 2.+0.j  4.-3.j  4.+3.j -5.+0.j]
```

PROBLEM SET 4.2

Problems 1–5 A zero $x = r$ of $P_n(x)$ is given. Verify that r is indeed a zero, and then deflate the polynomial; that is, find $P_{n-1}(x)$ so that $P_n(x) = (x - r)P_{n-1}(x)$.

1. $P_3(x) = 3x^3 + 7x^2 - 36x + 20, r = -5$.
2. $P_4(x) = x^4 - 3x^2 + 3x - 1, r = 1$.
3. $P_5(x) = x^5 - 30x^4 + 361x^3 - 2178x^2 + 6588x - 7992, r = 6$.
4. $P_4(x) = x^4 - 5x^3 - 2x^2 - 20x - 24, r = 2i$.
5. $P_3(x) = 3x^3 - 19x^2 + 45x - 13, r = 3 - 2i$.

Problems 6–9 A zero $x = r$ of $P_n(x)$ is given. Determine all the other zeros of $P_n(x)$ by using a calculator. You should need no tools other than deflation and the quadratic formula.

6. $P_3(x) = x^3 + 1.8x^2 - 9.01x - 13.398$, $r = -3.3$.
7. $P_3(x) = x^3 - 6.64x^2 + 16.84x - 8.32$, $r = 0.64$.
8. $P_3(x) = 2x^3 - 13x^2 + 32x - 13$, $r = 3 - 2i$.
9. $P_4(x) = x^4 - 3x^2 + 10x^2 - 6x - 20$, $r = 1 + 3i$.

Problems 10–15 Find all the zeros of the given $P_n(x)$.

10. ■ $P_4(x) = x^4 + 2.1x^3 - 2.52x^2 + 2.1x - 3.52$.
11. ■ $P_5(x) = x^5 - 156x^4 - 5x^3 + 780x^2 + 4x - 624$.
12. ■ $P_6(x) = x^6 + 4x^5 - 8x^4 - 34x^3 + 57x^2 + 130x - 150$.
13. ■ $P_7(x) = 8x^7 + 28x^6 + 34x^5 - 13x^4 - 124x^3 + 19x^2 + 220x - 100$.
14. ■ $P_3(x) = 2x^3 - 6(1 + i)x^2 + x - 6(1 - i)$
15. ■ $P_4(x) = x^4 + (5 + i)x^3 - (8 - 5i)x^2 + (30 - 14i)x - 84$.
16. ■

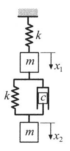

The two blocks of mass m each are connected by springs and a dashpot. The stiffness of each spring is k, and c is the coefficient of damping of the dashpot. When the system is displaced and released, the displacement of each block during the ensuing motion has the form

$$x_k(t) = A_k e^{\omega_r t} \cos(\omega_i t + \phi_k), \ k = 1, 2$$

where A_k and ϕ_k are constants, and $\omega = \omega_r \pm i\omega_i$ are the roots of

$$\omega^4 + 2\frac{c}{m}\omega^3 + 3\frac{k}{m}\omega^2 + \frac{c}{m}\frac{k}{m}\omega + \left(\frac{k}{m}\right)^2 = 0$$

Determine the two possible combinations of ω_r and ω_i if $c/m = 12$ s^{-1} and $k/m = 1\,500$ s^{-2}.

17.

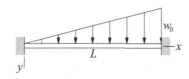

The lateral deflection of the beam shown is

$$y = \frac{w_0}{120\,EI}(x^5 - 3L^2 x^3 + 2L^3 x^2)$$

where w_0 is the maximum load intensity and EI represents the bending rigidity. Determine the value of x/L where the maximum deflection occurs.

4.8 Other Methods

The most prominent root-finding algorithm omitted from this chapter is *Brent's method*, which combines bisection and quadratic interpolation. It is potentially more efficient than Ridder's method, requiring only one function evaluation per iteration (as compared to two evaluations in Ridder's method), but this advantage is somewhat negated by its required elaborate bookkeeping.

There are many methods for finding zeroes of polynomials. Of these, the *Jenkins-Traub algorithm*[1] deserves special mention because of its robustness and widespread use in packaged software.

The zeros of a polynomial can also be obtained by calculating the eigenvalues of the $n \times n$ "companion matrix"

$$\mathbf{A} = \begin{bmatrix} -a_{n-1}/a_n & -a_2/a_n & \cdots & -a_1/a_n & -a_0/a_n \\ 1 & 0 & \cdots & 0 & 0 \\ 0 & 1 & \cdots & 0 & 0 \\ \vdots & \vdots & \ddots & \vdots & \vdots \\ 0 & 0 & \cdots & 1 & 0 \end{bmatrix}$$

where a_i are the coefficients of the polynomial. The characteristic equation (see Section 9.1) of this matrix is

$$x^n + \frac{a_{n-1}}{a_n}x^{n-1} + \frac{a_{n-2}}{a_n}x^{n-2} + \cdots + \frac{a_1}{a_n}x + \frac{a_0}{a_n} = 0$$

which is equivalent to $P_n(x) = 0$. Thus the eigenvalues of $\mathbf{A}$ are the zeroes of $P_n(x)$. The eigenvalue method is robust, but considerably slower than Laguerre's method. However, it is worthy of consideration if a good program for eigenvalue problems is available.

---

[1] Jenkins, M. and Traub, J., *SIAM Journal on Numerical Analysis*, Vol. 7 (1970), p. 545.

5 Numerical Differentiation

$$\boxed{\text{Given the function } f(x), \text{ compute } d^n f/dx^n \text{ at given } x}$$

5.1 Introduction

Numerical differentiation deals with the following problem: We are given the function $y = f(x)$ and wish to obtain one of its derivatives at the point $x = x_k$. The term "given" means that we either have an algorithm for computing the function, or we possess a set of discrete data points (x_i, y_i), $i = 0, 1, \ldots, n$. In either case, we have access to a finite number of (x, y) data pairs from which to compute the derivative. If you suspect by now that numerical differentiation is related to interpolation, you are correct—one means of finding the derivative is to approximate the function locally by a polynomial and then to differentiate it. An equally effective tool is the Taylor series expansion of $f(x)$ about the point x_k, which has the advantage of providing us with information about the error involved in the approximation.

Numerical differentiation is not a particularly accurate process. It suffers from a conflict between roundoff errors (caused by limited machine precision) and errors inherent in interpolation. For this reason, a derivative of a function can never be computed with the same precision as the function itself.

5.2 Finite Difference Approximations

The derivation of the finite difference approximations for the derivatives of $f(x)$ is based on forward and backward Taylor series expansions of $f(x)$ about x, such as

$$f(x + h) = f(x) + hf'(x) + \frac{h^2}{2!}f''(x) + \frac{h^3}{3!}f'''(x) + \frac{h^4}{4!}f^{(4)}(x) + \cdots \qquad (a)$$

$$f(x - h) = f(x) - hf'(x) + \frac{h^2}{2!}f''(x) - \frac{h^3}{3!}f'''(x) + \frac{h^4}{4!}f^{(4)}(x) - \cdots \qquad (b)$$

$$f(x + 2h) = f(x) + 2hf'(x) + \frac{(2h)^2}{2!}f''(x) + \frac{(2h)^3}{3!}f'''(x) + \frac{(2h)^4}{4!}f^{(4)}(x) + \cdots \quad \text{(c)}$$

$$f(x - 2h) = f(x) - 2hf'(x) + \frac{(2h)^2}{2!}f''(x) - \frac{(2h)^3}{3!}f'''(x) + \frac{(2h)^4}{4!}f^{(4)}(x) - \cdots \quad \text{(d)}$$

We also record the sums and differences of the series:

$$f(x + h) + f(x - h) = 2f(x) + h^2 f''(x) + \frac{h^4}{12}f^{(4)}(x) + \cdots \quad \text{(e)}$$

$$f(x + h) - f(x - h) = 2hf'(x) + \frac{h^3}{3}f'''(x) + \ldots \quad \text{(f)}$$

$$f(x + 2h) + f(x - 2h) = 2f(x) + 4h^2 f''(x) + \frac{4h^4}{3}f^{(4)}(x) + \cdots \quad \text{(g)}$$

$$f(x + 2h) - f(x - 2h) = 4hf'(x) + \frac{8h^3}{3}f'''(x) + \cdots \quad \text{(h)}$$

Note that the sums contain only even derivatives, whereas the differences retain just the odd derivatives. Equations (a)–(h) can be viewed as simultaneous equations that can be solved for various derivatives of $f(x)$. The number of equations involved and the number of terms kept in each equation depend on the order of the derivative and the desired degree of accuracy.

First Central Difference Approximations

The solution of Eq. (f) for $f'(x)$ is

$$f'(x) = \frac{f(x + h) - f(x - h)}{2h} - \frac{h^2}{6}f'''(x) - \cdots$$

or

$$f'(x) = \frac{f(x + h) - f(x - h)}{2h} + \mathcal{O}(h^2) \quad (5.1)$$

which is called the *first central difference approximation* for $f'(x)$. The term $\mathcal{O}(h^2)$ reminds us that the truncation error behaves as h^2.

Similarly, Eq. (e) yields the first central difference approximation for $f''(x)$:

$$f''(x) = \frac{f(x + h) - 2f(x) + f(x - h)}{h^2} + \frac{h^2}{12}f^{(4)}(x) + \ldots$$

or

$$f''(x) = \frac{f(x + h) - 2f(x) + f(x - h)}{h^2} + \mathcal{O}(h^2) \quad (5.2)$$

Central difference approximations for other derivatives can be obtained from Eqs. (a)–(h) in the same manner. For example, eliminating $f'(x)$ from Eqs. (f) and (h) and solving for $f'''(x)$ yields

$$f'''(x) = \frac{f(x+2h) - 2f(x+h) + 2f(x-h) - f(x-2h)}{2h^3} + \mathcal{O}(h^2) \qquad (5.3)$$

The approximation

$$f^{(4)}(x) = \frac{f(x+2h) - 4f(x+h) + 6f(x) - 4f(x-h) + f(x-2h)}{h^4} + \mathcal{O}(h^2) \qquad (5.4)$$

is available from Eq. (e) and (g) after eliminating $f''(x)$. Table 5.1 summarizes the results.

	$f(x-2h)$	$f(x-h)$	$f(x)$	$f(x+h)$	$f(x+2h)$
$2hf'(x)$		-1	0	1	
$h^2 f''(x)$		1	-2	1	
$2h^3 f'''(x)$	-1	2	0	-2	1
$h^4 f^{(4)}(x)$	1	-4	6	-4	1

Table 5.1. Coefficients of Central Finite Difference Approximations of $\mathcal{O}(h^2)$

First Noncentral Finite Difference Approximations

Central finite difference approximations are not always usable. For example, consider the situation where the function is given at the n discrete points $x_0, x_1, \ldots, x_n$. Because central differences use values of the function on each side of x, we would be unable to compute the derivatives at x_0 and x_n. Clearly, there is a need for finite difference expressions that require evaluations of the function only on one side of x. These expressions are called *forward* and *backward* finite difference approximations.

Noncentral finite differences can also be obtained from Eqs. (a)–(h). Solving Eq. (a) for $f'(x)$ we get

$$f'(x) = \frac{f(x+h) - f(x)}{h} - \frac{h}{2}f''(x) - \frac{h^2}{6}f'''(x) - \frac{h^3}{4!}f^{(4)}(x) - \cdots$$

Keeping only the first term on the right-hand side leads to the *first forward difference approximation:*

$$f'(x) = \frac{f(x+h) - f(x)}{h} + \mathcal{O}(h) \qquad (5.5)$$

Similarly, Eq. (b) yields the *first backward difference approximation:*

$$f'(x) = \frac{f(x) - f(x-h)}{h} + \mathcal{O}(h) \qquad (5.6)$$

Note that the truncation error is now $\mathcal{O}(h)$, which is not as good as $\mathcal{O}(h^2)$ in central difference approximations.

We can derive the approximations for higher derivatives in the same manner. For example, Eqs. (a) and (c) yield

$$f''(x) = \frac{f(x+2h) - 2f(x+h) + f(x)}{h^2} + \mathcal{O}(h) \tag{5.7}$$

The third and fourth derivatives can be derived in a similar fashion. The results are shown in Tables 5.2a and 5.2b.

	$f(x)$	$f(x+h)$	$f(x+2h)$	$f(x+3h)$	$f(x+4h)$
$hf'(x)$	-1	1			
$h^2 f''(x)$	1	-2	1		
$h^3 f'''(x)$	-1	3	-3	1	
$h^4 f^{(4)}(x)$	1	-4	6	-4	1

Table 5.2a. Coefficients of Forward Finite Difference Approximations of $\mathcal{O}(h)$

	$f(x-4h)$	$f(x-3h)$	$f(x-2h)$	$f(x-h)$	$f(x)$
$hf'(x)$				-1	1
$h^2 f''(x)$			1	-2	1
$h^3 f'''(x)$		-1	3	-3	1
$h^4 f^{(4)}(x)$	1	-4	6	-4	1

Table 5.2b. Coefficients of Backward Finite Difference Approximations of $\mathcal{O}(h)$

Second Noncentral Finite Difference Approximations

Finite difference approximations of $\mathcal{O}(h)$ are not popular because of reasons that are explained shortly. The common practice is to use expressions of $\mathcal{O}(h^2)$. To obtain noncentral difference formulas of this order, we have to retain more term in the Taylor series. As an illustration, we derive the expression for $f'(x)$. We start with Eqs. (a) and (c), which are

$$f(x+h) = f(x) + hf'(x) + \frac{h^2}{2}f''(x) + \frac{h^3}{6}f'''(x) + \frac{h^4}{24}f^{(4)}(x) + \cdots$$

$$f(x+2h) = f(x) + 2hf'(x) + 2h^2 f''(x) + \frac{4h^3}{3}f'''(x) + \frac{2h^4}{3}f^{(4)}(x) + \cdots$$

We eliminate $f''(x)$ by multiplying the first equation by 4 and subtracting it from the second equation. The result is

$$f(x+2h) - 4f(x+h) = -3f(x) - 2hf'(x) + \frac{h^4}{2}f^{(4)}(x) + \cdots$$

Therefore,

$$f'(x) = \frac{-f(x+2h) + 4f(x+h) - 3f(x)}{2h} + \frac{h^2}{4} f^{(4)}(x) + \cdots$$

or

$$f'(x) = \frac{-f(x+2h) + 4f(x+h) - 3f(x)}{2h} + \mathcal{O}(h^2) \tag{5.8}$$

Equation (5.8) is called the *second forward finite difference approximation*.

Derivation of finite difference approximations for higher derivatives involves additional Taylor series. Thus the forward difference approximation for $f''(x)$ uses series for $f(x+h)$, $f(x+2h)$, and $f(x+3h)$; the approximation for $f'''(x)$ involves Taylor expansions for $f(x+h)$, $f(x+2h)$, $f(x+3h)$, $f(x+4h)$, and so on. As you can see, the computations for high-order derivatives can become rather tedious. The results for both the forward and backward finite differences are summarized in Tables 5.3a and 5.3b.

	$f(x)$	$f(x+h)$	$f(x+2h)$	$f(x+3h)$	$f(x+4h)$	$f(x+5h)$
$2hf'(x)$	-3	4	-1			
$h^2 f''(x)$	2	-5	4	-1		
$2h^3 f'''(x)$	-5	18	-24	14	-3	
$h^4 f^{(4)}(x)$	3	-14	26	-24	11	-2

Table 5.3a. Coefficients of Forward Finite Difference Approximations of $\mathcal{O}(h^2)$

	$f(x-5h)$	$f(x-4h)$	$f(x-3h)$	$f(x-2h)$	$f(x-h)$	$f(x)$
$2hf'(x)$				1	-4	3
$h^2 f''(x)$			-1	4	-5	2
$2h^3 f'''(x)$		3	-14	24	-18	5
$h^4 f^{(4)}(x)$	-2	11	-24	26	-14	3

Table 5.3b. Coefficients of Backward Finite Difference Approximations of $\mathcal{O}(h^2)$

Errors in Finite Difference Approximations

Observe that in all finite difference expressions the sum of the coefficients is zero. The effect on the *roundoff error* can be profound. If h is very small, the values of $f(x)$, $f(x \pm h)$, $f(x \pm 2h)$, and so on, will be approximately equal. When they are multiplied by the coefficients and added, several significant figures can be lost. Yet we cannot make h too large, because then the *truncation error* would become excessive. This unfortunate situation has no remedy, but we can obtain some relief by taking the following precautions:

- Use double-precision arithmetic.
- Employ finite difference formulas that are accurate to at least $\mathcal{O}(h^2)$.

To illustrate the errors, let us compute the second derivative of $f(x) = e^{-x}$ at $x = 1$ from the central difference formula, Eq. (5.2). We carry out the calculations with six- and eight-digit precision, using different values of h. The results, shown in Table 5.4, should be compared with $f''(1) = e^{-1} = 0.367\,879\,44$.

h	Six-Digit Precision	Eight-Digit Precision
0.64	0.380 610	0.380 609 11
0.32	0.371 035	0.371 029 39
0.16	0.368 711	0.368 664 84
0.08	0.368 281	0.368 076 56
0.04	0.368 75	0.367 831 25
0.02	0.37	0.3679
0.01	0.38	0.3679
0.005	0.40	0.3676
0.0025	0.48	0.3680
0.00125	1.28	0.3712

Table 5.4. $(e^{-x})''$ at $x = 1$ from Central Finite Difference Approximation

In the six-digit computations, the optimal value of h is 0.08, yielding a result accurate to three significant figures. Hence three significant figures are lost due to a combination of truncation and roundoff errors. Above optimal h, the dominant error is caused by truncation; below it, the roundoff error becomes pronounced. The best result obtained with the eight-digit computation is accurate to four significant figures. Because the extra precision decreases the roundoff error, the optimal h is smaller (about 0.02) than in the six-figure calculations.

5.3 Richardson Extrapolation

Richardson extrapolation is a simple method for boosting the accuracy of certain numerical procedures, including finite difference approximations (we also use it later in other applications).

Suppose that we have an approximate means of computing some quantity G. Moreover, assume that the result depends on a parameter h. Denoting the approximation by $g(h)$, we have $G = g(h) + E(h)$, where $E(h)$ represents the error. Richardson extrapolation can remove the error, provided that it has the form $E(h) = ch^p$, c and p being constants. We start by computing $g(h)$ with some value of h, say $h = h_1$. In that case we have

$$G = g(h_1) + ch_1^p \qquad \text{(i)}$$

Then we repeat the calculation with $h = h_2$, so that

$$G = g(h_2) + ch_2^p \qquad \text{(j)}$$

Eliminating c and solving for G, Eqs. (i) and (j) yield

$$G = \frac{(h_1/h_2)^p g(h_2) - g(h_1)}{(h_1/h_2)^p - 1} \tag{5.8}$$

which is the *Richardson extrapolation formula.* It is common practice to use $h_2 = h_1/2$, in which case Eq. (5.8) becomes

$$G = \frac{2^p g(h_1/2) - g(h_1)}{2^p - 1} \tag{5.9}$$

Let us illustrate Richardson extrapolation by applying it to the finite difference approximation of $(e^{-x})''$ at $x = 1$. We work with six-digit precision and use the results in Table 5.4. Because the extrapolation works only on the truncation error, we must confine h to values that produce negligible roundoff. In Table 5.4 we have

$$g(0.64) = 0.380\,610 \qquad g(0.32) = 0.371\,035$$

The truncation error in central difference approximation is $E(h) = \mathcal{O}(h^2) = c_1 h^2 + c_2 h^4 + c_3 h^6 + \ldots$. Therefore, we can eliminate the first (dominant) error term if we substitute $p = 2$ and $h_1 = 0.64$ in Eq. (5.9). The result is

$$G = \frac{2^2 g(0.32) - g(0.64)}{2^2 - 1} = \frac{4(0.371\,035) - 0.380\,610}{3} = 0.367\,843$$

which is an approximation of $(e^{-x})''$ with the error $\mathcal{O}(h^4)$. Note that it is as accurate as the best result obtained with eight-digit computations in Table 5.4.

EXAMPLE 5.1
Given the evenly spaced data points

x	0	0.1	0.2	0.3	0.4
$f(x)$	0.0000	0.0819	0.1341	0.1646	0.1797

compute $f'(x)$ and $f''(x)$ at $x = 0$ and 0.2 using finite difference approximations of $\mathcal{O}(h^2)$.

Solution. We will use finite difference approximations of $\mathcal{O}(h^2)$. From the forward difference tables Table 5.3a we get

$$f'(0) = \frac{-3f(0) + 4f(0.1) - f(0.2)}{2(0.1)} = \frac{-3(0) + 4(0.0819) - 0.1341}{0.2} = 0.967$$

$$f''(0) = \frac{2f(0) - 5f(0.1) + 4f(0.2) - f(0.3)}{(0.1)^2}$$

$$= \frac{2(0) - 5(0.0819) + 4(0.1341) - 0.1646}{(0.1)^2} = -3.77$$

The central difference approximations in Table 5.1 yield

$$f'(0.2) = \frac{-f(0.1) + f(0.3)}{2(0.1)} = \frac{-0.0819 + 0.1646}{0.2} = 0.4135$$

$$f''(0.2) = \frac{f(0.1) - 2f(0.2) + f(0.3)}{(0.1)^2} = \frac{0.0819 - 2(0.1341) + 0.1646}{(0.1)^2} = -2.17$$

EXAMPLE 5.2
Use the data in Example 5.1 to compute $f'(0)$ as accurately as you can.

Solution. One solution is to apply Richardson extrapolation to finite difference approximations. We start with two forward difference approximations of $\mathcal{O}(h^2)$ for $f'(0)$: one using $h = 0.2$ and the other one using $h = 0.1$. Referring to the formulas of $O(h^2)$ in Table 5.3a, we get

$$g(0.2) = \frac{-3f(0) + 4f(0.2) - f(0.4)}{2(0.2)} = \frac{3(0) + 4(0.1341) - 0.1797}{0.4} = 0.8918$$

$$g(0.1) = \frac{-3f(0) + 4f(0.1) - f(0.2)}{2(0.1)} = \frac{-3(0) + 4(0.0819) - 0.1341}{0.2} = 0.9675$$

Recall that the error in both approximations is of the form $E(h) = c_1 h^2 + c_2 h^4 + c_3 h^6 + \dots$. We can now use Richardson extrapolation, Eq. (5.9), to eliminate the dominant error term. With $p = 2$ we obtain

$$f'(0) \approx G = \frac{2^2 g(0.1) - g(0.2)}{2^2 - 1} = \frac{4(0.9675) - 0.8918}{3} = 0.9927$$

which is a finite difference approximation of $O(h^4)$.

EXAMPLE 5.3

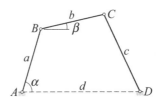

The linkage shown has the dimensions $a = 100$ mm, $b = 120$ mm, $c = 150$ mm, and $d = 180$ mm. It can be shown by geometry that the relationship between the angles α and β is

α (deg)	0	5	10	15	20	25	30
β (rad)	1.6595	1.5434	1.4186	1.2925	1.1712	1.0585	0.9561

If link AB rotates with the constant angular velocity of 25 rad/s, use finite difference approximations of $\mathcal{O}(h^2)$ to tabulate the angular velocity $d\beta/dt$ of link BC against α.

Solution. The angular speed of BC is

$$\frac{d\beta}{dt} = \frac{d\beta}{d\alpha}\frac{d\alpha}{dt} = 25\frac{d\beta}{d\alpha}\ \text{rad/s}$$

where $d\beta/d\alpha$ can be computed from finite difference approximations using the data in the table. Forward and backward differences of $\mathcal{O}(h^2)$ are used at the endpoints, and central differences elsewhere. Note that the increment of α is

$$h = (5\,\text{deg})\left(\frac{\pi}{180}\text{rad / deg}\right) = 0.087\,266\,\text{rad}$$

The computations yield

$$\dot\beta(0°) = 25\frac{-3\beta(0°) + 4\beta(5°) - \beta(10°)}{2h} = 25\frac{-3(1.6595) + 4(1.5434) - 1.4186}{2\,(0.087\,266)}$$

$$= -32.01\,\text{rad/s}.$$

$$\dot\beta(5°) = 25\frac{\beta(10°) - \beta(0°)}{2h} = 25\frac{1.4186 - 1.6595}{2(0.087\,266)} = -34.51\,\text{rad/s}$$

etc.

The complete set of results is

α (deg)	0	5	10	15	20	25	30
$\dot\beta$ (rad/s)	−32.01	−34.51	−35.94	−35.44	−33.52	−30.81	−27.86

5.4 Derivatives by Interpolation

If $f(x)$ is given as a set of discrete data points, interpolation can be a very effective means of computing its derivatives. The idea is to approximate the derivative of $f(x)$ by the derivative of the interpolant. This method is particularly useful if the data points are located at uneven intervals of x, when the finite difference approximations listed in Section 5.2 are not applicable[1].

Polynomial Interpolant

The idea here is simple: Fit the polynomial of degree n

$$P_{n-1}(x) = a_0 + a_1 x + a_2 x^2 + \cdots + a_n x^n$$

through $n + 1$ data points and then evaluate its derivatives at the given x. As pointed out in Section 3.2, it is generally advisable to limit the degree of the polynomial to less than six in order to avoid spurious oscillations of the interpolant. Because these oscillations are magnified with each differentiation, their effect can devastating.

---

[1] It is possible to derive finite difference approximations for unevenly spaced data, but they would not be as accurate as the formulas derived in Section 5.2.

In view of this limitation, the interpolation is usually a local one, involving no more than a few nearest-neighbor data points.

For evenly spaced data points, polynomial interpolation and finite difference approximations produce identical results. In fact, the finite difference formulas are equivalent to polynomial interpolation.

Several methods of polynomial interpolation were introduced in Section 3.2. Unfortunately, none are suited for the computation of derivatives of the interpolant. The method that we need is one that determines the coefficients $a_0, a_1, \ldots, a_n$ of the polynomial. There is only one such method discussed in Chapter 3: the *least-squares fit*. Although this method is designed mainly for smoothing of data, it will carry out interpolation if we use $m = n$ in Eq. (3.22)—recall that m is the degree of the interpolating polynomial and $n + 1$ represents the number of data points to be fitted. If the data contain noise, then the least-squares fit should be used in the smoothing mode; that is, with $m < n$. After the coefficients of the polynomial have been found, the polynomial and its first two derivatives can be evaluated efficiently by the function `evalPoly` listed in Section 4.7.

Cubic Spline Interpolant

Because of its stiffness, a cubic spline is a good global interpolant; moreover, it is easy to differentiate. The first step is to determine the second derivatives k_i of the spline at the knots by solving Eqs. (3.11). This can be done with the function `curvatures` in the module `cubicSpline` listed in Section 3.3. The first and second derivatives are then be computed from

$$f'_{i,i+1}(x) = \frac{k_i}{6}\left[\frac{3(x - x_{i+1})^2}{x_i - x_{i+1}} - (x_i - x_{i+1})\right]$$

$$- \frac{k_{i+1}}{6}\left[\frac{3(x - x_i)^2}{x_i - x_{i+1}} - (x_i - x_{i+1})\right] + \frac{y_i - y_{i+1}}{x_i - x_{i+1}} \tag{5.10}$$

$$f''_{i,i+1}(x) = k_i \frac{x - x_{i+1}}{x_i - x_{i+1}} - k_{i+1}\frac{x - x_i}{x_i - x_{i+1}} \tag{5.11}$$

which are obtained by differentiation of Eq. (3.10).

EXAMPLE 5.4

Given the data

x	1.5	1.9	2.1	2.4	2.6	3.1
$f(x)$	1.0628	1.3961	1.5432	1.7349	1.8423	2.0397

compute $f'(2)$ and $f''(2)$ using (1) polynomial interpolation over three nearest-neighbor points, and (2) a natural cubic spline interpolant spanning all the data points.

Solution of Part (1). The interpolant is $P_2(x) = a_0 + a_1 x + a_2 x^2$ passing through the points at $x = 1.9, 2.1$, and 2.4. The normal equations, Eqs. (3.22), of the least-squares fit are

$$\begin{bmatrix} n & \sum x_i & \sum x_i^2 \\ \sum x_i & \sum x_i^2 & \sum x_i^3 \\ \sum x_i^2 & \sum x_i^3 & \sum x_i^4 \end{bmatrix} \begin{bmatrix} a_0 \\ a_1 \\ a_2 \end{bmatrix} = \begin{bmatrix} \sum y_i \\ \sum y_i x_i \\ \sum y_i x_i^2 \end{bmatrix}$$

After substituting the data, we get

$$\begin{bmatrix} 3 & 6.4 & 13.78 \\ 6.4 & 13.78 & 29.944 \\ 13.78 & 29.944 & 65.6578 \end{bmatrix} \begin{bmatrix} a_0 \\ a_1 \\ a_2 \end{bmatrix} = \begin{bmatrix} 4.6742 \\ 10.0571 \\ 21.8385 \end{bmatrix}$$

which yields $\mathbf{a} = \begin{bmatrix} -0.7714 & 1.5075 & -0.1930 \end{bmatrix}^T$.

The derivatives of the interpolant are $P_2'(x) = a_1 + 2a_2 x$ and $P_2''(x) = 2a_2$. Therefore,

$$f'(2) \approx P_2'(2) = 1.5075 + 2(-0.1930)(2) = 0.7355$$

$$f''(2) \approx P_2''(2) = 2(-0.1930) = -0.3860$$

Solution of Part (2). We must first determine the second derivatives k_i of the spline at its knots, after which the derivatives of $f(x)$ can be computed from Eqs. (5.10) and (5.11). The first part can be carried out by the following small program:

```
#!/usr/bin/python
## example5_4
from cubicSpline import curvatures
from LUdecomp3 import *
import numpy as np

xData = np.array([1.5,1.9,2.1,2.4,2.6,3.1])
yData = np.array([1.0628,1.3961,1.5432,1.7349,1.8423, 2.0397])
print(curvatures(xData,yData))
input("Press return to exit")
```

The output of the program, consisting of k_0 to k_5, is

```
[ 0.    -0.4258431 -0.37744139 -0.38796663 -0.55400477   0.   ]
```

Because $x = 2$ lies between knots 1 and 2, we must use Eqs. (5.10) and (5.11) with $i = 1$. This yields

$$f'(2) \approx f'_{1,2}(2) = \frac{k_1}{6}\left[\frac{3(x - x_2)^2}{x_1 - x_2} - (x_1 - x_2)\right]$$

$$- \frac{k_2}{6}\left[\frac{3(x - x_1)^2}{x_1 - x_2} - (x_1 - x_2)\right] + \frac{y_1 - y_2}{x_1 - x_2}$$

$$= \frac{(-0.4258)}{6}\left[\frac{3(2 - 2.1)^2}{(-0.2)} - (-0.2)\right]$$

$$- \frac{(-0.3774)}{6}\left[\frac{3(2 - 1.9)^2}{(-0.2)} - (-0.2)\right] + \frac{1.3961 - 1.5432}{(-0.2)}$$

$$= 0.7351$$

$$f''(2) \approx f''_{1,2}(2) = k_1\frac{x - x_2}{x_1 - x_2} - k_2\frac{x - x_1}{x_1 - x_2}$$

$$= (-0.4258)\frac{2 - 2.1}{(-0.2)} - (-0.3774)\frac{2 - 1.9}{(-0.2)} = -0.4016$$

Note that the solutions for $f'(2)$ in parts (1) and (2) differ only in the fourth significant figure, but the values of $f''(2)$ are much farther apart. This is not unexpected, considering the general rule: The higher the order of the derivative, the lower the precision with which it can be computed. It is impossible to tell which of the two results is better without knowing the expression for $f(x)$. In this particular problem, the data points fall on the curve $f(x) = x^2 e^{-x/2}$, so that the "true" values of the derivatives are $f'(2) = 0.7358$ and $f''(2) = -0.3679$.

EXAMPLE 5.5
Determine $f'(0)$ and $f'(1)$ from the following noisy data:

x	0	0.2	0.4	0.6
$f(x)$	1.9934	2.1465	2.2129	2.1790

x	0.8	1.0	1.2	1.4
$f(x)$	2.0683	1.9448	1.7655	1.5891

Solution. We used the program listed in Example 3.10 to find the best polynomial fit (in the least-squares sense) to the data. The program was run three times with the following results:

```
Degree of polynomial ==> 2
Coefficients are:
[ 2.0261875   0.64703869 -0.70239583]
Std. deviation = 0.0360968935809

Degree of polynomial ==> 3
Coefficients are:
[ 1.99215    1.09276786 -1.55333333  0.40520833]
```

```
Std. deviation = 0.0082604082973

Degree of polynomial ==> 4
Coefficients are:
[ 1.99185568  1.10282373 -1.59056108  0.44812973 -0.01532907]
Std. deviation = 0.00951925073521
```

Based on standard deviation, the cubic seems to be the best candidate for the interpolant. Before accepting the result, we compare the plots of the data points and the interpolant—see the following figure. The fit does appear to be satisfactory.

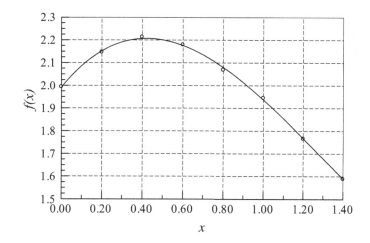

Approximating $f(x)$ by the interpolant, we have

$$f(x) \approx a_0 + a_1 x + a_2 x^2 + a_3 x^3$$

so that

$$f'(x) \approx a_1 + 2a_2 x + 3a_3 x^2$$

Therefore,

$$f'(0) \approx a_1 = 1.093$$

$$f'(1) = a_1 + 2a_2 + 3a_3 = 1.093 + 2(-1.553) + 3(0.405) = -0.798$$

In general, derivatives obtained from noisy data are at best rough approximations. In this problem, the data represent $f(x) = (x+2)/\cosh x$ with added random noise. Thus $f'(x) = \left[1 - (x+2)\tanh x\right]/\cosh x$, so that the "correct" derivatives are $f'(0) = 1.000$ and $f'(1) = -0.833$.

PROBLEM SET 5.1

1. Given the values of $f(x)$ at the points x, $x - h_1$, and $x + h_2$, where $h_1 \neq h_2$, determine the finite difference approximation for $f''(x)$. What is the order of the truncation error?

2. Given the first backward finite difference approximations for $f'(x)$ and $f''(x)$, derive the first backward finite difference approximation for $f'''(x)$ using the operation $f'''(x) = [f''(x)]'$.

3. Derive the central difference approximation for $f''(x)$ accurate to $\mathcal{O}(h^4)$ by applying Richardson extrapolation to the central difference approximation of $\mathcal{O}(h^2)$.

4. Derive the second forward finite difference approximation for $f'''(x)$ from the Taylor series.

5. Derive the first central difference approximation for $f^{(4)}(x)$ from the Taylor series.

6. Use finite difference approximations of $O(h^2)$ to compute $f'(2.36)$ and $f''(2.36)$ from the following data:

x	2.36	2.37	2.38	2.39
$f(x)$	0.85866	0.86289	0.86710	0.87129

7. Estimate $f'(1)$ and $f''(1)$ from the following data:

x	0.97	1.00	1.05
$f(x)$	0.85040	0.84147	0.82612

8. Given the data

x	0.84	0.92	1.00	1.08	1.16
$f(x)$	0.431711	0.398519	0.367879	0.339596	0.313486

calculate $f''(1)$ as accurately as you can.

9. Use the data in the table to compute $f'(0.2)$ as accurately: as possible:

x	0	0.1	0.2	0.3	0.4
$f(x)$	0.000 000	0.078 348	0.138 910	0.192 916	0.244 981

10. Using five significant figures in the computations, determine $d(\sin x)/dx$ at $x = 0.8$ from (a) the first forward difference approximation and (b) the first central difference approximation. In each case, use h that gives the most accurate result (this requires experimentation).

11. ∎ Use polynomial interpolation to compute f' and f'' at $x = 0$, using the data

x	−2.2	−0.3	0.8	1.9
$f(x)$	15.180	10.962	1.920	−2.040

Given that $f(x) = x^3 - 0.3x^2 - 8.56x + 8.448$, gauge the accuracy of the result.

12. ∎

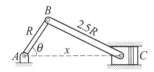

The crank AB of length $R = 90$ mm is rotating at a constant angular speed of $d\theta/dt = 5\,000$ rev/min. The position of the piston C can be shown to vary with the angle θ as

$$x = R\left(\cos\theta + \sqrt{2.5^2 - \sin^2\theta}\right)$$

Write a program that plots the acceleration of the piston at $\theta = 0°, 5°, 10°, \ldots,$ $180°$. Use numerical differentiation to compute the acceleration.

13. ∎

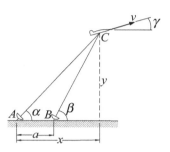

The radar stations A and B, separated by the distance $a = 500$ m, track the plane C by recording the angles α and β at one-second intervals. If three successive readings are

t (s)	9	10	11
α	54.80°	54.06°	53.34°
β	65.59°	64.59°	63.62°

calculate the speed v of the plane and the climb angle γ at $t = 10$ s. The coordinates of the plane can be shown to be

$$x = a\frac{\tan\beta}{\tan\beta - \tan\alpha} \qquad y = a\frac{\tan\alpha\tan\beta}{\tan\beta - \tan\alpha}$$

14. ∎

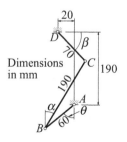

Dimensions in mm

Geometric analysis of the linkage shown resulted in the following table relating the angles θ and β:

θ (deg)	0	30	60	90	120	150
β (deg)	59.96	56.42	44.10	25.72	−0.27	−34.29

Assuming that member AB of the linkage rotates with the constant angular velocity $d\theta/dt = 1$ rad/s, compute $d\beta/dt$ in rad/s at the tabulated values of θ. Use cubic spline interpolation.

15. ■ The relationship between stress σ and strain ε of some biological materials in uniaxial tension is

$$\frac{d\sigma}{d\varepsilon} = a + b\sigma$$

where a and b are constants ($d\sigma/d\varepsilon$ is called *tangent modulus*). The following table gives the results of a tension test on such a material:

Strain ε	Stress σ (MPa)
0	0
0.05	0.252
0.10	0.531
0.15	0.840
0.20	1.184
0.25	1.558
0.30	1.975
0.35	2.444
0.40	2.943
0.45	3.500
0.50	4.115

Determine the parameters a and b by linear regression.

6 Numerical Integration

Compute $\int_a^b f(x)\,dx$, where $f(x)$ is a given function.

6.1 Introduction

Numerical integration, also known as *quadrature*, is intrinsically a much more accurate procedure than numerical differentiation. Quadrature approximates the definite integral

$$\int_a^b f(x)\,dx$$

by the sum

$$I = \sum_{i=0}^{n} A_i f(x_i)$$

where the *nodal abscissas* x_i and *weights* A_i depend on the particular rule used for the quadrature. All rules of quadrature are derived from polynomial interpolation of the integrand. Therefore, they work best if $f(x)$ can be approximated by a polynomial.

Methods of numerical integration can be divided into two groups: Newton-Cotes formulas and Gaussian quadrature. Newton-Cotes formulas are characterized by equally spaced abscissas and include well-known methods such as the trapezoidal rule and Simpson's rule. They are most useful if $f(x)$ has already been computed at equal intervals or can be computed at low cost. Because Newton-Cotes formulas are based on local interpolation, they require only a piecewise fit to a polynomial.

In Gaussian quadrature the locations of the abscissas are chosen to yield the best possible accuracy. Because Gaussian quadrature requires fewer evaluations of the integrand for a given level of precision, it is popular in cases where $f(x)$ is expensive to evaluate. Another advantage of Gaussian quadrature is its ability to handle integrable singularities, enabling us to evaluate expressions such as

$$\int_0^1 \frac{g(x)}{\sqrt{1-x^2}}\,dx$$

provided that $g(x)$ is a well-behaved function.

6.2　Newton-Cotes Formulas

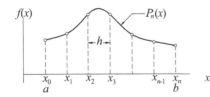

Figure 6.1. Polynomial approximation of $f(x)$.

Consider the definite integral

$$\int_a^b f(x)\,dx \tag{6.1}$$

We divide the range of integration (a, b) into n equal intervals of length $h = (b - a)/n$, as shown in Figure 6.1, and denote the abscissas of the resulting nodes by $x_0, x_1, \ldots, x_n$. Next we approximate $f(x)$ by a polynomial of degree n that intersects all the nodes. Lagrange's form of this polynomial, Eq. (3.1a), is

$$P_n(x) = \sum_{i=0}^n f(x_i)\ell_i(x)$$

where $\ell_i(x)$ are the cardinal functions defined in Eq. (3.1b). Therefore, an approximation to the integral in Eq. (6.1) is

$$I = \int_a^b P_n(x)dx = \sum_{i=0}^n \left[f(x_i) \int_a^b \ell_i(x)dx \right] = \sum_{i=0}^n A_i f(x_i) \tag{6.2a}$$

where

$$A_i = \int_a^b \ell_i(x)dx, \quad i = 0, 1, \ldots, n \tag{6.2b}$$

Equations (6.2) are the *Newton-Cotes formulas*. Classical examples of these formulas are the *trapezoidal rule* ($n = 1$), *Simpson's rule* ($n = 2$), and *3/8 Simpson's rule* ($n = 3$). The most important of these formulas is the trapezoidal rule. It can be combined with Richardson extrapolation into an efficient algorithm known as *Romberg integration*, which makes the other classical rules somewhat redundant.

Trapezoidal Rule

If $n = 1$ (one panel), as illustrated in Figure 6.2, we have $\ell_0 = (x - x_1)/(x_0 - x_1) = -(x - b)/h$. Therefore,

$$A_0 = \frac{1}{h} \int_a^b (x - b)\,dx = \frac{1}{2h}(b - a)^2 = \frac{h}{2}$$

Also $\ell_1 = (x - x_0)/(x_1 - x_0) = (x - a)/h$, so that

$$A_1 = \frac{1}{h} \int_a^b (x - a)\,dx = \frac{1}{2h}(b - a)^2 = \frac{h}{2}$$

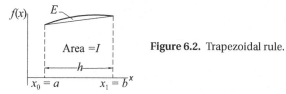

Figure 6.2. Trapezoidal rule.

Substitution in Eq. (6.2a) yields

$$I = \left[f(a) + f(b)\right]\frac{h}{2} \tag{6.3}$$

which is known as the *trapezoidal rule*. It represents the area of the trapezoid in Figure 6.2.

The error in the trapezoidal rule

$$E = \int_a^b f(x)dx - I$$

is the area of the region between $f(x)$ and the straight-line interpolant, as indicated in Figure 6.2. It can be obtained by integrating the interpolation error in Eq. (3.3):

$$E = \frac{1}{2!}\int_a^b (x - x_0)(x - x_1)f''(\xi)dx = \frac{1}{2}f''(\xi)\int_a^b (x - a)(x - b)dx$$

$$= -\frac{1}{12}(b - a)^3 f''(\xi) = -\frac{h^3}{12}f''(\xi) \tag{6.4}$$

Composite Trapezoidal Rule

In practice the trapezoidal rule is applied in a piecewise fashion. Figure 6.3 shows the region (a, b) divided into n panels, each of width h. The function $f(x)$ to be integrated is approximated by a straight line in each panel. From the trapezoidal rule we obtain for the approximate area of a typical (ith) panel,

$$I_i = \left[f(x_i) + f(x_{i+1})\right]\frac{h}{2}$$

Hence total area, representing $\int_a^b f(x)\,dx$, is

$$I = \sum_{i=0}^{n-1} I_i = \left[f(x_0) + 2f(x_1) + 2f(x_2) + \ldots + 2f(x_{n-1}) + f(x_n)\right]\frac{h}{2} \tag{6.5}$$

which is the *composite trapezoidal rule*.

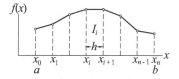

Figure 6.3. Composite trapezoidal rule.

The truncation error in the area of a panel is–see Eq. (6.4)—

$$E_i = -\frac{h^3}{12} f''(\xi_i)$$

where ξ_i lies in (x_i, x_{i+1}). Hence the truncation error in Eq. (6.5) is

$$E = \sum_{i=0}^{n-1} E_i = -\frac{h^3}{12} \sum_{i=0}^{n-1} f''(\xi_i) \tag{a}$$

But

$$\sum_{i=0}^{n-1} f''(\xi_i) = n\bar{f}''$$

where $\bar{f}''$ is the arithmetic mean of the second derivatives. If $f''(x)$ is continuous, there must be a point ξ in (a, b) at which $f''(\xi) = \bar{f}''$, enabling us to write

$$\sum_{i=0}^{n-1} f''(\xi_i) = nf''(\xi) = \frac{b-a}{h} f''(\xi)$$

Therefore, Eq. (a) becomes

$$E = -\frac{(b-a)h^2}{12} f''(\xi) \tag{6.6}$$

It would be incorrect to conclude from Eq. (6.6) that $E = ch^2$ (c being a constant), because $f''(\xi)$ is not entirely independent of h. A deeper analysis of the error[1] shows that if $f(x)$ and its derivatives are finite in (a, b), then

$$E = c_1 h^2 + c_2 h^4 + c_3 h^6 + \ldots \tag{6.7}$$

Recursive Trapezoidal Rule

Let I_k be the integral evaluated with the composite trapezoidal rule using 2^{k-1} panels. Note that if k is increased by one, the number of panels is doubled. Using the notation

$$H = b - a$$

Eq. (6.5) yields the following results for $k = 1, 2$, and 3.
    $k = 1$ (one panel):

$$I_1 = [f(a) + f(b)] \frac{H}{2} \tag{6.8}$$

    $k = 2$ (two panels):

$$I_2 = \left[f(a) + 2f\left(a + \frac{H}{2}\right) + f(b)\right] \frac{H}{4} = \frac{1}{2} I_1 + f\left(a + \frac{H}{2}\right) \frac{H}{2}$$

---

[1] The analysis requires familiarity with the *Euler-Maclaurin summation formula*, which is covered in advanced texts.

$k = 3$ (four panels):

$$I_3 = \left[f(a) + 2f\left(a + \frac{H}{4}\right) + 2f\left(a + \frac{H}{2}\right) + 2f\left(a + \frac{3H}{4}\right) + f(b) \right] \frac{H}{8}$$

$$= \frac{1}{2} I_2 + \left[f\left(a + \frac{H}{4}\right) + f\left(a + \frac{3H}{4}\right) \right] \frac{H}{4}$$

We can now see that for arbitrary $k > 1$ we have

$$I_k = \frac{1}{2} I_{k-1} + \frac{H}{2^{k-1}} \sum_{i=1}^{2^{k-2}} f\left[a + \frac{(2i-1)H}{2^{k-1}} \right], \quad k = 2, 3, \ldots \tag{6.9a}$$

which is the *recursive trapezoidal rule*. Observe that the summation contains only the new nodes that were created when the number of panels was doubled. Therefore, the computation of the sequence $I_1, I_2, I_3, \ldots, I_k$ from Eqs. (6.8) and (6.9) involves the same amount of algebra as the calculation of I_k directly from Eq. (6.5). The advantage of using the recursive trapezoidal rule is that it allows us to monitor convergence and terminate the process when the difference between I_{k-1} and I_k becomes sufficiently small. A form of Eq. (6.9a) that is easier to remember is

$$I(h) = \frac{1}{2} I(2h) + h \sum f(x_{\text{new}}) \tag{6.9b}$$

where $h = H/n$ is the width of each panel.

■ **trapezoid**

The function trapezoid computes I_k (Inew), given I_{k-1} (Iold) using Eqs. (6.8) and (6.9). We can compute $\int_a^b f(x)\, dx$ by calling trapezoid with $k = 1, 2, \ldots$ until the desired precision is attained.

```
## module trapezoid
''' Inew = trapezoid(f,a,b,Iold,k).
    Recursive trapezoidal rule:
    old = Integral of f(x) from x = a to b computed by
    trapezoidal rule with 2^(k-1) panels.
    Inew = Same integral computed with 2^k panels.
'''
def trapezoid(f,a,b,Iold,k):
    if k == 1:Inew = (f(a) + f(b))*(b - a)/2.0
    else:
        n = 2**(k -2 )        # Number of new points
        h = (b - a)/n         # Spacing of new points
        x = a + h/2.0
        sum = 0.0
        for i in range(n):
            sum = sum + f(x)
            x = x + h
        Inew = (Iold + h*sum)/2.0
    return Inew
```

Simpson's Rules

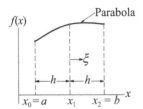

Figure 6.4. Simpson's 1/3 rule.

Simpson's 1/3 rule can be obtained from Newton-Cotes formulas with $n = 2$; that is, by passing a parabolic interpolant through three adjacent nodes, as shown in Figure 6.4. The area under the parabola, which represents an approximation of $\int_a^b f(x)\,dx$, is (see derivation in Example 6.1)

$$I = \left[f(a) + 4f\left(\frac{a+b}{2}\right) + f(b) \right] \frac{h}{3} \tag{a}$$

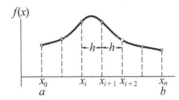

Figure 6.5. Composite Simpson's 1/3 rule.

To obtain the *composite Simpson's 1/3 rule*, the integration range (a, b) is divided into n panels (n even) of width $h = (b - a)/n$ each, as indicated in Figure 6.5. Applying Eq. (a) to two adjacent panels, we have

$$\int_{x_i}^{x_{i+2}} f(x)\,dx \approx \left[f(x_i) + 4f(x_{i+1}) + f(x_{i+2}) \right] \frac{h}{3} \tag{b}$$

Substituting Eq. (b) into

$$\int_a^b f(x)\,dx = \int_{x_0}^{x_m} f(x)\,dx = \sum_{i=0,2,\dots}^{n} \left[\int_{x_i}^{x_{i+2}} f(x)\,dx \right]$$

yields

$$\int_a^b f(x)\,dx \approx I = [f(x_0) + 4f(x_1) + 2f(x_2) + 4f(x_3) + \dots \tag{6.10}$$

$$\dots + 2f(x_{n-2}) + 4f(x_{n-1}) + f(x_n)] \frac{h}{3}$$

The composite Simpson's 1/3 rule in Eq. (6.10) is perhaps the best known method of numerical integration. However, its reputation is somewhat undeserved, because the trapezoidal rule is more robust and Romberg integration is more efficient.

The error in the composite Simpson's rule is

$$E = \frac{(b-a)h^4}{180} f^{(4)}(\xi) \tag{6.11}$$

from which we conclude that Eq. (6.10) is exact if $f(x)$ is a polynomial of degree three or less.

Simpson's 1/3 rule requires the number of panels n to be even. If this condition is not satisfied, we can integrate over the first (or last) three panels with *Simpson's 3/8 rule,*

$$I = \left[f(x_0) + 3f(x_1) + 3f(x_2) + f(x_3) \right] \frac{3h}{8} \qquad (6.12)$$

and use Simpson's 1/3 rule for the remaining panels. The error in Eq. (6.12) is of the same order as in Eq. (6.10).

EXAMPLE 6.1
Derive Simpson's 1/3 rule from Newton-Cotes formulas.

Solution. Referring to Figure 6.4, Simpson's 1/3 rule uses three nodes located at $x_0 = a$, $x_1 = (a + b)/2$, and $x_2 = b$. The spacing of the nodes is $h = (b - a)/2$. The cardinal functions of Lagrange's three-point interpolation are—see Section 3.2—

$$\ell_0(x) = \frac{(x - x_1)(x - x_2)}{(x_0 - x_1)(x_0 - x_2)} \qquad \ell_1(x) = \frac{(x - x_0)(x - x_2)}{(x_1 - x_0)(x_1 - x_2)}$$

$$\ell_2(x) = \frac{(x - x_0)(x - x_1)}{(x_2 - x_0)(x_2 - x_1)}$$

The integration of these functions is easier if we introduce the variable ξ with origin at x_1. Then the coordinates of the nodes are $\xi_0 = -h$, $\xi_1 = 0$ and $\xi_2 = h$ and Eq. (6.2b) becomes $A_i = \int_a^b \ell_i(x) = \int_{-h}^h \ell_i(\xi) d\xi$. Therefore,

$$A_0 = \int_{-h}^h \frac{(\xi - 0)(\xi - h)}{(-h)(-2h)} d\xi = \frac{1}{2h^2} \int_{-h}^h (\xi^2 - h\xi) d\xi = \frac{h}{3}$$

$$A_1 = \int_{-h}^h \frac{(\xi + h)(\xi - h)}{(h)(-h)} d\xi = -\frac{1}{h^2} \int_{-h}^h (\xi^2 - h^2) d\xi = \frac{4h}{3}$$

$$A_2 = \int_{-h}^h \frac{(\xi + h)(\xi - 0)}{(2h)(h)} d\xi = \frac{1}{2h^2} \int_{-h}^h (\xi^2 + h\xi) d\xi = \frac{h}{3}$$

Equation (6.2a) then yields

$$I = \sum_{i=0}^2 A_i f(x_i) = \left[f(a) + 4f\left(\frac{a + b}{2}\right) + f(b) \right] \frac{h}{3}$$

which is Simpson's 1/3 rule.

EXAMPLE 6.2
Evaluate the bounds on $\int_0^\pi \sin(x) \, dx$ with the composite trapezoidal rule using (1) 8 panels; and (2) 16 panels.

Solution of Part (1). With eight panels there are nine nodes spaced at $h = \pi/8$. The abscissas of the nodes are $x_i = i\pi/8$, $i = 0, 1, \ldots, 8$. From Eq. (6.5) we get

$$I = \left[\sin 0 + 2 \sum_{i=1}^{7} \sin \frac{i\pi}{8} + \sin \pi \right] \frac{\pi}{16} = 1.97423$$

The error is given by Eq. (6.6):

$$E = -\frac{(b-a)h^2}{12} f''(\xi) = -\frac{(\pi - 0)(\pi/8)^2}{12} (-\sin \xi) = \frac{\pi^3}{768} \sin \xi$$

where $0 < \xi < \pi$. Because we do not know the value of ξ, we cannot evaluate E, but we can determine its bounds:

$$E_{\min} = \frac{\pi^3}{768} \sin(0) = 0 \qquad E_{\max} = \frac{\pi^3}{768} \sin \frac{\pi}{2} = 0.040\,37$$

Therefore, $I + E_{\min} < \int_0^\pi \sin(x)\, dx < I + E_{\max}$, or

$$1.974\,23 < \int_0^\pi \sin(x)\, dx < 2.014\,60$$

The exact integral is, of course, 2.

Solution of Part (2). The new nodes created by the doubling of panels are located at the midpoints of the old panels. Their abscissas are

$$x_j = \pi/16 + j\pi/8 = (1 + 2j)\pi/16, \quad j = 0, 1, \ldots, 7$$

Using the recursive trapezoidal rule in Eq. (6.9b), we get

$$I = \frac{1.974\,23}{2} + \frac{\pi}{16} \sum_{j=0}^{7} \sin \frac{(1 + 2j)\pi}{16} = 1.993\,58$$

and the bounds on the error become (note that E is quartered when h is halved) $E_{\min} = 0$, $E_{\max} = 0.040\,37/4 = 0.010\,09$. Hence

$$1.993\,58 < \int_0^\pi \sin(x)\, dx < 2.003\,67$$

EXAMPLE 6.3
Estimate $\int_0^{2.5} f(x)\, dx$ from the following data:

x	0	0.5	1.0	1.5	2.0	2.5
$f(x)$	1.5000	2.0000	2.0000	1.6364	1.2500	0.9565

Solution. We use Simpson's rules because they are more accurate than the trapezoidal rule. Because the number of panels is odd, we compute the integral over the first three panels by Simpson's 3/8 rule, and use the 1/3 rule for the last two panels:

$$I = \left[f(0) + 3f(0.5) + 3f(1.0) + f(1.5) \right] \frac{3(0.5)}{8}$$

$$+ \left[f(1.5) + 4f(2.0) + f(2.5) \right] \frac{0.5}{3}$$

$$= 2.8381 + 1.2655 = 4.1036$$

EXAMPLE 6.4

Use the recursive trapezoidal rule to evaluate $\int_0^\pi \sqrt{x}\cos x\, dx$ to six decimal places. How many panels are needed to achieve this result?

Solution

```
#!/usr/bin/python
## example6_4
import math
from trapezoid import *

def f(x): return math.sqrt(x)*math.cos(x)

Iold = 0.0
for k in range(1,21):
    Inew = trapezoid(f,0.0,math.pi,Iold,k)
    if (k > 1) and (abs(Inew - Iold)) < 1.0e-6: break
    Iold = Inew
print("Integral =",Inew)
print("nPanels =",2**(k-1))
input("\nPress return to exit")
```

The output from the program is

```
Integral = -0.8948316648532865
nPanels = 32768
```

Hence $\int_0^\pi \sqrt{x}\cos x\, dx = -0.894\,832$ requiring $32\,768$ panels. The slow convergence is the result of all the derivatives of $f(x)$ being singular at $x = 0$. Consequently, the error does not behave as shown in Eq. (6.7): $E = c_1 h^2 + c_2 h^4 + \cdots$, but is unpredictable. Difficulties of this nature can often be remedied by a change in variable. In this case, we introduce $t = \sqrt{x}$, so that $dt = dx/(2\sqrt{x}) = dx/(2t)$, or $dx = 2t\, dt$. Thus

$$\int_0^\pi \sqrt{x}\cos x\, dx = \int_0^{\sqrt{\pi}} 2t^2 \cos t^2 dt$$

Evaluation of the integral on the right-hand-side was completed with 4096 panels.

6.3   Romberg Integration

Romberg integration combines the trapezoidal rule with Richardson extrapolation (see Section 5.3). Let us first introduce the notation

$$R_{i,1} = I_i$$

where, as before, I_i represents the approximate value of $\int_a^b f(x)dx$ computed by the recursive trapezoidal rule using 2^{i-1} panels. Recall that the error in this approximation is $E = c_1 h^2 + c_2 h^4 + \ldots$, where

$$h = \frac{b-a}{2^{i-1}}$$

is the width of a panel.

Romberg integration starts with the computation of $R_{1,1} = I_1$ (one panel) and $R_{2,1} = I_2$ (two panels) from the trapezoidal rule. The leading error term $c_1 h^2$ is then eliminated by Richardson extrapolation. Using $p = 2$ (the exponent in the leading error term) in Eq. (5.9) and denoting the result by $R_{2,2}$, we obtain

$$R_{2,2} = \frac{2^2 R_{2,1} - R_{1,1}}{2^2 - 1} = \frac{4}{3} R_{2,1} - \frac{1}{3} R_{1,1} \tag{a}$$

It is convenient to store the results in an array of the form

$$\begin{bmatrix} R_{1,1} \\ R_{2,1} & R_{2,2} \end{bmatrix}$$

The next step is to calculate $R_{3,1} = I_3$ (four panels) and repeat Richardson extrapolation with $R_{2,1}$ and $R_{3,1}$, storing the result as $R_{3,2}$:

$$R_{3,2} = \frac{4}{3} R_{3,1} - \frac{1}{3} R_{2,1} \tag{b}$$

The elements of array **R** calculated so far are

$$\begin{bmatrix} R_{1,1} \\ R_{2,1} & R_{2,2} \\ R_{3,1} & R_{3,2,} \end{bmatrix}$$

Both elements of the second column have an error of the form $c_2 h^4$, which can also be eliminated with Richardson extrapolation. Using $p = 4$ in Eq. (5.9), we get

$$R_{3,3} = \frac{2^4 R_{3,2} - R_{2,2}}{2^4 - 1} = \frac{16}{15} R_{3,2} - \frac{1}{15} R_{2,2} \tag{c}$$

This result has an error of $\mathcal{O}(h^6)$. The array has now expanded to

$$\begin{bmatrix} R_{1,1} \\ R_{2,1} & R_{2,2} \\ R_{3,1} & R_{3,2} & R_{3,3} \end{bmatrix}$$

After another round of calculations we get

$$\begin{bmatrix} R_{1,1} \\ R_{2,1} & R_{2,2} \\ R_{3,1} & R_{3.2} & R_{3,3} \\ R_{4,1} & R_{4,2} & R_{4,3} & R_{4,4} \end{bmatrix}$$

where the error in $R_{4,4}$ is $\mathcal{O}(h^8)$. Note that the most accurate estimate of the integral is always the last diagonal term of the array. This process is continued until the difference between two successive diagonal terms becomes sufficiently small. The general extrapolation formula used in this scheme is

$$R_{i,j} = \frac{4^{j-1}R_{i,j-1} - R_{i-1,j-1}}{4^{j-1} - 1}, \quad i > 1, \quad j = 2, 3, \ldots, i \tag{6.13a}$$

A pictorial representation of Eq. (6.13a) is

$$\boxed{R_{i-1,j-1}}$$

$$\searrow$$

$$\alpha \tag{6.13b}$$

$$\searrow$$

$$\boxed{R_{i,j-1}} \quad \rightarrow \quad \beta \quad \rightarrow \quad \boxed{R_{i,j}}$$

where the multipliers α and β depend on j in the following manner:

j	2	3	4	5	6
α	$-1/3$	$-1/15$	$-1/63$	$-1/255$	$-1/1023$
β	$4/3$	$16/15$	$64/63$	$256/255$	$1024/1023$

$$\tag{6.13c}$$

The triangular array is convenient for hand computations, but computer implementation of the Romberg algorithm can be carried out within a one-dimensional array $\mathbf{R}'$. After the first extrapolation—see Eq. (a)— $R_{1,1}$ is never used again, so that it can be replaced with $R_{2,2}$. As a result, we have the array

$$\begin{bmatrix} R_1' = R_{2,2} \\ R_2' = R_{2,1} \end{bmatrix}$$

In the second extrapolation round, defined by Eqs. (b) and (c), $R_{3,2}$ overwrites $R_{2,1}$, and $R_{3,3}$ replaces $R_{2,2}$, so that the array contains

$$\begin{bmatrix} R_1' = R_{3,3} \\ R_2' = R_{3,2} \\ R_3' = R_{3,1} \end{bmatrix}$$

and so on. In this manner, R_1' always contains the best current result. The extrapolation formula for the kth round is

$$R_j' = \frac{4^{k-j}R_{j+1}' - R_j'}{4^{k-j} - 1}, \quad j = k - 1, k - 2, \ldots, 1 \tag{6.14}$$

■ **romberg**

The algorithm for Romberg integration is implemented in the function `romberg`. It returns the integral and the number of panels used. Richardson's extrapolation is carried out by the subfunction `richardson`.

```
## module romberg
''' I,nPanels = romberg(f,a,b,tol=1.0e-6).
    Romberg integration of f(x) from x = a to b.
    Returns the integral and the number of panels used.
'''
```

```
import numpy as np
from trapezoid import *

def romberg(f,a,b,tol=1.0e-6):

    def richardson(r,k):
        for j in range(k-1,0,-1):
            const = 4.0**(k-j)
            r[j] = (const*r[j+1] - r[j])/(const - 1.0)
        return r

    r = np.zeros(21)
    r[1] = trapezoid(f,a,b,0.0,1)
    r_old = r[1]
    for k in range(2,21):
        r[k] = trapezoid(f,a,b,r[k-1],k)
        r = richardson(r,k)
        if abs(r[1]-r_old) < tol*max(abs(r[1]),1.0):
            return r[1],2**(k-1)
        r_old = r[1]
    print("Romberg quadrature did not converge")
```

EXAMPLE 6.5
Show that $R_{k,2}$ in Romberg integration is identical to composite Simpson's 1/3 rule in Eq. (6.10) with 2^{k-1} panels.

Solution. Recall that in Romberg integration $R_{k,1} = I_k$ denoted the approximate integral obtained by the composite trapezoidal rule with $n = 2^{k-1}$ panels. Denoting the abscissas of the nodes by $x_0, x_1, \ldots, x_n$, we have from the composite trapezoidal rule in Eq. (6.5)

$$R_{k,1} = I_k = \left[f(x_0) + 2 \sum_{i=1}^{n-1} f(x_i) + f(x_n), \right] \frac{h}{2}$$

When we halve the number of panels (panel width $2h$), only the even-numbered abscissas enter the composite trapezoidal rule, yielding

$$R_{k-1,1} = I_{k-1} = \left[f(x_0) + 2 \sum_{i=2,4,\ldots}^{n-2} f(x_i) + f(x_n) \right] h$$

Applying Richardson extrapolation yields

$$R_{k,2} = \frac{4}{3} R_{k,1} - \frac{1}{3} R_{k-1,1}$$

$$= \left[\frac{1}{3} f(x_0) + \frac{4}{3} \sum_{i=1,3,\ldots}^{n-1} f(x_i) + \frac{2}{3} \sum_{i=2,4,\ldots}^{n-2} f(x_i) + \frac{1}{3} f(x_n) \right] h$$

which agrees with Eq. (6.10).

EXAMPLE 6.6

Use Romberg integration to evaluate $\int_0^\pi f(x)\,dx$, where $f(x) = \sin x$. Work with four decimal places.

Solution. From the recursive trapezoidal rule in Eq. (6.9b) we get

$$R_{1,1} = I(\pi) = \frac{\pi}{2}\left[f(0) + f(\pi)\right] = 0$$

$$R_{2,1} = I(\pi/2) = \frac{1}{2}I(\pi) + \frac{\pi}{2}f(\pi/2) = 1.5708$$

$$R_{3,1} = I(\pi/4) = \frac{1}{2}I(\pi/2) + \frac{\pi}{4}\left[f(\pi/4) + f(3\pi/4)\right] = 1.8961$$

$$R_{4,1} = I(\pi/8) = \frac{1}{2}I(\pi/4) + \frac{\pi}{8}\left[f(\pi/8) + f(3\pi/8) + f(5\pi/8) + f(7\pi/8)\right]$$

$$= 1.9742$$

Using the extrapolation formulas in Eqs. (6.13), we can now construct the following table:

$$\begin{bmatrix} R_{1,1} & & & \\ R_{2,1} & R_{2,2} & & \\ R_{3,1} & R_{3,2} & R_{3,3} & \\ R_{4,1} & R_{4,2} & R_{4,3} & R_{4,4} \end{bmatrix} = \begin{bmatrix} 0 & & & \\ 1.5708 & 2.0944 & & \\ 1.8961 & 2.0046 & 1.9986 & \\ 1.9742 & 2.0003 & 2.0000 & 2.0000 \end{bmatrix}$$

It appears that the procedure has converged. Therefore, $\int_0^\pi \sin x\,dx = R_{4,4} = 2.0000$, which is, of course, the correct result.

EXAMPLE 6.7

Use Romberg integration to evaluate $\int_0^{\sqrt{\pi}} 2x^2 \cos x^2\,dx$ and compare the results with Example 6.4.

Solution

```
#!usr/bin/python
## example6_7
import math
from romberg import *

def f(x): return 2.0*(x**2)*math.cos(x**2)

I,n = romberg(f,0,math.sqrt(math.pi))
print("Integral =",I)
print("numEvals =",n)
input("\nPress return to exit")
```

The results of running the program are

```
Integral = -0.894831469504
nPanels = 64
```

It is clear that Romberg integration is considerably more efficient than the trapezoidal rule—it required only 64 panels as compared to 4096 panels for the trapezoidal rule in Example 6.4.

PROBLEM SET 6.1

1. Use the recursive trapezoidal rule to evaluate $\int_0^{\pi/4} \ln(1 + \tan x)dx$. Explain the results.

2. The following table shows the power P supplied to the driving wheels of a car as a function of the speed v. If the mass of the car is $m = 2000$ kg, determine the time Δt it takes for the car to accelerate from 1 m/s to 6 m/s. Use the trapezoidal rule for integration. *Hint*: $\Delta t = m \int_{1s}^{6s} (v/P)\, dv$, which can be derived from Newton's law $F = m(dv/dt)$ and the definition of power $P = Fv$.

v (m/s)	0	1.0	1.8	2.4	3.5	4.4	5.1	6.0
P (kW)	0	4.7	12.2	19.0	31.8	40.1	43.8	43.2

3. Evaluate $\int_{-1}^{1} \cos(2\cos^{-1} x)dx$ with Simpson's 1/3 rule using two, four, and six panels. Explain the results.

4. Determine $\int_1^\infty (1 + x^4)^{-1}dx$ with the trapezoidal rule using five panels and compare the result with the "exact" integral 0.243 75. *Hint*: Use the transformation $x^3 = 1/t$.

5.

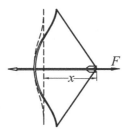

The following table gives the pull F of the bow as a function of the draw x. If the bow is drawn 0.5 m, determine the speed of the 0.075-kg arrow when it leaves the bow. *Hint*: The kinetic energy of the arrow equals the work done in drawing the bow; that is, $mv^2/2 = \int_0^{0.5m} F\, dx$.

x (m)	0.00	0.05	0.10	0.15	0.20	0.25
F (N)	0	37	71	104	134	161

x (m)	0.30	0.35	0.40	0.45	0.50	
F (N)	185	207	225	239	250	

6. Evaluate $\int_0^2 (x^5 + 3x^3 - 2)\, dx$ by Romberg integration.

7. Estimate $\int_0^\pi f(x)\,dx$ as accurately as possible, where $f(x)$ is defined by the following data:

x	0	$\pi/4$	$\pi/2$	$3\pi/4$	π
$f(x)$	1.0000	0.3431	0.2500	0.3431	1.0000

8. Evaluate

$$\int_0^1 \frac{\sin x}{\sqrt{x}}\,dx$$

   with Romberg integration. *Hint*: Use transformation of the variable to eliminate the singularity at $x = 0$.

9. Newton-Cotes formulas for evaluating $\int_a^b f(x)\,dx$ were based on polynomial approximations of $f(x)$. Show that if $y = f(x)$ is approximated by a natural cubic spline with evenly spaced knots at $x_0, x_1, \ldots, x_n$, the quadrature formula becomes

$$I = \frac{h}{2}\left(y_0 + 2y_1 + 2y_2 + \cdots + 2y_{n-1} + y_n\right)$$

$$-\frac{h^3}{24}\left(k_0 + 2k_1 + k_2 + \cdots + 2k_{n-1} + k_n\right)$$

   where h is the distance between the knots and $k_i = y_i''$. Note that the first part is the composite trapezoidal rule; the second part may be viewed as a "correction" for curvature.

10. ■ Evaluate

$$\int_0^{\pi/4} \frac{dx}{\sqrt{\sin x}}$$

   with Romberg integration. *Hint*: Use the transformation $\sin x = t^2$.

11. ■ The period of a simple pendulum of length L is $\tau = 4\sqrt{L/g}h(\theta_0)$, where g is the gravitational acceleration, θ_0 represents the angular amplitude, and

$$h(\theta_0) = \int_0^{\pi/2} \frac{d\theta}{\sqrt{1 - \sin^2(\theta_0/2)\sin^2\theta}}$$

   Compute $h(15°)$, $h(30°)$, and $h(45°)$, and compare these values with $h(0) = \pi/2$ (the approximation used for small amplitudes).

12. ■

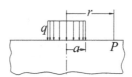

The figure shows an elastic half-space that carries uniform loading of intensity q over a circular area of radius a. The vertical displacement of the surface at point P can be shown to be

$$w(r) = w_0 \int_0^{\pi/2} \frac{\cos^2 \theta}{\sqrt{(r/a)^2 - \sin^2 \theta}} d\theta \qquad r \geq a$$

where w_0 is the displacement at $r = a$. Use numerical integration to determine w/w_0 at $r = 2a$.

13. ■

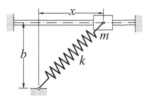

The mass m is attached to a spring of free length b and stiffness k. The coefficient of friction between the mass and the horizontal rod is μ. The acceleration of the mass can be shown to be (you may wish to prove this) $\ddot{x} = -f(x)$, where

$$f(x) = \mu g + \frac{k}{m}(\mu b + x)\left(1 - \frac{b}{\sqrt{b^2 + x^2}}\right)$$

If the mass is released from rest at $x = b$, its speed at $x = 0$ is given by

$$v_0 = \sqrt{2 \int_0^b f(x)dx}$$

Compute v_0 by numerical integration using the data $m = 0.8$ kg, $b = 0.4$ m, $\mu = 0.3$, $k = 80$ N/m, and $g = 9.81$ m/s^2.

14. ■ Debye's formula for the heat capacity C_V as a solid is $C_V = 9Nkg(u)$, where

$$g(u) = u^3 \int_0^{1/u} \frac{x^4 e^x}{(e^x - 1)^2} dx$$

The terms in this equation are

$$N = \text{number of particles in the solid}$$

$$k = \text{Boltzmann constant}$$

$$u = T/\Theta_D$$

$$T = \text{absolute temperature}$$

$$\Theta_D = \text{Debye temperature}$$

Compute $g(u)$ from $u = 0$ to 1.0 in intervals of 0.05 and plot the results.

15. ■ A power spike in an electric circuit results in the current

$$i(t) = i_0 e^{-t/t_0} \sin(2t/t_0)$$

across a resistor. The energy E dissipated by the resistor is

$$E = \int_0^\infty R\left[i(t)\right]^2 dt$$

Find E using the data $i_0 = 100$ A, $R = 0.5\ \Omega$, and $t_0 = 0.01$ s.

16. ■ An alternating electric current is described by

$$i(t) = i_0\left(\sin\frac{\pi t}{t_0} - \beta \sin\frac{2\pi t}{t_0}\right)$$

where $i_0 = 1$ A, $t_0 = 0.05$ s, and $\beta = 0.2$. Compute the root-mean-square current, defined as

$$i_{\mathrm{rms}} = \sqrt{\frac{1}{t_0}\int_0^{t_0} i^2(t)\, dt}$$

17. (a) Derive the composite trapezoidal rule for unevenly spaced data. (b) Consider the stress-strain diagram obtained from a uniaxial tension test.

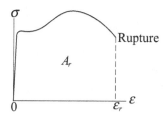

The area under the diagram is

$$A_r = \int_{\varepsilon=0}^{\varepsilon_r} \sigma\, d\varepsilon$$

where ε_r is the strain at rupture. This area represents the work that must be performed on a unit volume of the test specimen to cause rupture; it is called the *modulus of toughness*. Use the result of Part (a) to estimate the modulus of toughness for nickel steel from the following test data:

σ (MPa)	ε
586	0.001
662	0.025
765	0.045
841	0.068
814	0.089
122	0.122
150	0.150

Note that the spacing of data is uneven.

18. ■ Write a function that computes the sine integral

$$\text{Si}(x) = \int_0^x t^{-1} \sin t \, dt$$

for any given value of x. Test the program by computing $\text{Si}(1.0)$, and compare the result with the tabulated value 0.946 08.

6.4   Gaussian Integration

Gaussian Integration Formulas

Newton-Cotes formulas for approximating $\int_a^b f(x)dx$ work best if $f(x)$ is a smooth function, such as a polynomial. This is also true for Gaussian quadrature. However, Gaussian formulas are also good at estimating integrals of the form

$$\int_a^b w(x)f(x)dx \tag{6.15}$$

where $w(x)$, called the *weighting function*, can contain singularities, as long as they are integrable. An example of such integral is $\int_0^1 (1 + x^2) \ln x \, dx$. Sometimes infinite limits, as in $\int_0^\infty e^{-x} \sin x \, dx$, can also be accommodated.

Gaussian integration formulas have the same form as Newton-Cotes rules

$$I = \sum_{i=0}^n A_i f(x_i) \tag{6.16}$$

where, as before, I represents the approximation to the integral in Eq. (6.15). The difference lies in the way that the weights A_i and nodal abscissas x_i are determined. In Newton-Cotes integration the nodes were evenly spaced in (a, b) (i.e., their locations were predetermined). In Gaussian quadrature the nodes and weights are chosen so that Eq. (6.16) yields the exact integral if $f(x)$ is a polynomial of degree $2n + 1$ or less; that is,

$$\int_a^b w(x) P_m(x)dx = \sum_{i=0}^n A_i P_m(x_i), \quad m \le 2n + 1 \tag{6.17}$$

One way of determining the weights and abscissas is to substitute $P_0(x) = 1$, $P_1(x) = x, \ldots, P_{2n+1}(x) = x^{2n+1}$ in Eq. (6.17) and solve the resulting $2n + 2$ equations

$$\int_a^b w(x)x^j dx = \sum_{i=0}^n A_i x_i^j, \quad j = 0, 1, \ldots, 2n + 1$$

for the unknowns A_i and x_i.

As an illustration, let $w(x) = e^{-x}$, $a = 0$, $b = \infty$, and $n = 1$. The four equations determining x_0, x_1, A_0, and A_1 are as follows:

$$\int_0^\infty e^{-x} dx = A_0 + A_1$$

$$\int_0^1 e^{-x} x \, dx = A_0 x_0 + A_1 x_1$$

$$\int_0^1 e^{-x} x^2 dx = A_0 x_0^2 + A_1 x_1^2$$

$$\int_0^1 e^{-x} x^3 dx = A_0 x_0^3 + A_1 x_1^3$$

After evaluating the integrals, we get these results:

$$A_0 + A_1 = 1$$
$$A_0 x_0 + A_1 x_1 = 1$$
$$A_0 x_0^2 + A_1 x_1^2 = 2$$
$$A_0 x_0^3 + A_1 x_1^3 = 6$$

The solution is

$$x_0 = 2 - \sqrt{2} \qquad A_0 = \frac{\sqrt{2}+1}{2\sqrt{2}}$$

$$x_1 = 2 + \sqrt{2} \qquad A_1 = \frac{\sqrt{2}-1}{2\sqrt{2}}$$

so that the integration formula becomes

$$\int_0^\infty e^{-x} f(x) dx \approx \frac{1}{2\sqrt{2}} \left[(\sqrt{2}+1) f\left(2 - \sqrt{2}\right) + (\sqrt{2}-1) f\left(2 + \sqrt{2}\right) \right]$$

Because of the nonlinearity of the equations, this approach does not work well for large n. Practical methods of finding x_i and A_i require some knowledge of orthogonal polynomials and their relationship to Gaussian quadrature. There are, however, several "classical" Gaussian integration formulas for which the abscissas and weights have been computed with great precision and then tabulated. These formulas can be used without knowing the theory behind them, because all one needs for Gaussian integration are the values of x_i and A_i. If you do not intend to venture outside the classical formulas, you can skip the next two topics.

*Orthogonal Polynomials

Orthogonal polynomials are employed in many areas of mathematics and numerical analysis. They have been studied thoroughly, and many of their properties are known. What follows is a very small compendium of a large topic.

The polynomials $\varphi_n(x)$, $n = 0, 1, 2, \ldots$ (n is the degree of the polynomial) are said to form an *orthogonal set* in the interval (a, b) with respect to the *weighting function* $w(x)$ if

$$\int_a^b w(x)\varphi_m(x)\varphi_n(x)dx = 0, \quad m \neq n \tag{6.18}$$

The set is determined, except for a constant factor, by the choice of the weighting function and the limits of integration. That is, each set of orthogonal polynomials is associated with certain $w(x)$, a, and b. The constant factor is specified by standardization. Some of the classical orthogonal polynomials, named after well-known mathematicians, are listed in Table 6.1. The last column in the table shows the standardization used.

Name	Symbol	a	b	$w(x)$	$\int_a^b w(x)\left[\varphi_n(x)\right]^2 dx$
Legendre	$p_n(x)$	-1	1	1	$2/(2n+1)$
Chebyshev	$T_n(x)$	-1	1	$(1-x^2)^{-1/2}$	$\pi/2 \quad (n>0)$
Laguerre	$L_n(x)$	0	∞	e^{-x}	1
Hermite	$H_n(x)$	$-\infty$	∞	e^{-x^2}	$\sqrt{\pi}2^n n!$

Table 6.1. Classical orthogonal polynomials

Orthogonal polynomials obey recurrence relations of the form

$$a_n\varphi_{n+1}(x) = (b_n + c_n x)\varphi_n(x) - d_n\varphi_{n-1}(x) \tag{6.19}$$

If the first two polynomials of the set are known, the other members of the set can be computed from Eq. (6.19). The coefficients in the recurrence formula, together with $\varphi_0(x)$ and $\varphi_1(x)$ are given in Table 6.2.

Name	$\varphi_0(x)$	$\varphi_1(x)$	a_n	b_n	c_n	d_n
Legendre	1	x	$n+1$	0	$2n+1$	n
Chebyshev	1	x	1	0	2	1
Laguerre	1	$1-x$	$n+1$	$2n+1$	-1	n
Hermite	1	$2x$	1	0	2	2

Table 6.2. Recurrence coefficients

The classical orthogonal polynomials are also obtainable from these formulas

$$p_n(x) = \frac{(-1)^n}{2^n n!} \frac{d^n}{dx^n}\left[(1-x^2)^n\right]$$

$$T_n(x) = \cos(n\cos^{-1} x), \quad n > 0$$

$$L_n(x) = \frac{e^x}{n!} \frac{d^n}{dx^n}\left(x^n e^{-x}\right) \tag{6.20}$$

$$H_n(x) = (-1)^n e^{x^2} \frac{d^n}{dx^n}(e^{-x^2})$$

and their derivatives can be calculated from

$$(1 - x^2) p_n'(x) = n\left[-xp_n(x) + p_{n-1}(x)\right]$$
$$(1 - x^2) T_n'(x) = n\left[-x T_n(x) + n T_{n-1}(x)\right]$$
$$x L_n'(x) = n\left[L_n(x) - L_{n-1}(x)\right] \tag{6.21}$$
$$H_n'(x) = 2n H_{n-1}(x)$$

Other properties of orthogonal polynomials that have relevance to Gaussian integration are as follows:

- $\varphi_n(x)$ has n real, distinct zeros in the interval (a, b).
- The zeros of $\varphi_n(x)$ lie between the zeros of $\varphi_{n+1}(x)$.
- Any polynomial $P_n(x)$ of degree n can be expressed in the form

$$P_n(x) = \sum_{i=0}^{n} c_i \varphi_i(x) \tag{6.22}$$

- It follows from Eq. (6.22) and the orthogonality property in Eq. (6.18) that

$$\int_a^b w(x) P_n(x) \varphi_{n+m}(x) dx = 0, \quad m \geq 0 \tag{6.23}$$

*Determination of Nodal Abscissas and Weights

Theorem The nodal abscissas $x_0, x_1, \ldots, x_n$ are the zeroes of the polynomial $\varphi_{n+1}(x)$ that belongs to the orthogonal set defined in Eq. (6.18).

Proof We start the proof by letting $f(x) = P_{2n+1}(x)$ be a polynomial of degree $2n + 1$. Because the Gaussian integration with $n + 1$ nodes is exact for this polynomial, we have

$$\int_a^b w(x) P_{2n+1}(x) dx = \sum_{i=0}^{n} A_i P_{2n+1}(x_i) \tag{a}$$

A polynomial of degree $2n + 1$ can always be written in the form

$$P_{2n+1}(x) = Q_n(x) + R_n(x) \varphi_{n+1}(x) \tag{b}$$

where $Q_n(x)$, $R_n(x)$, and $\varphi_{n+1}(x)$ are polynomials of the degree indicated by the subscripts.[2] Therefore,

$$\int_a^b w(x) P_{2n+1}(x) dx = \int_a^b w(x) Q_n(x) dx + \int_a^b w(x) R_n(x) \varphi_{n+1}(x) dx$$

But according to Eq. (6.23) the second integral on the right-hand-side vanishes, so that

$$\int_a^b w(x) P_{2n+1}(x) dx = \int_a^b w(x) Q_n(x) dx \tag{c}$$

[2] It can be shown that $Q_n(x)$ and $R_n(x)$ are unique for a given $P_{2n+1}(x)$ and $\varphi_{n+1}(x)$.

Because a polynomial of degree n is uniquely defined by $n + 1$ points, it is always possible to find A_i such that

$$\int_a^b w(x)\, Q_n(x)\, dx = \sum_{i=0}^{n} A_i\, Q_n(x_i) \tag{d}$$

To arrive at Eq. (a), we must choose for the nodal abscissas x_i the roots of $\varphi_{n+1}(x) = 0$. According to Eq. (b) we then have

$$P_{2n+1}(x_i) = Q_n(x_i), \quad i = 0, 1, \ldots, n \tag{e}$$

which together with Eqs. (c) and (d) leads to

$$\int_a^b w(x)\, P_{2n+1}(x)\, dx = \int_a^b w(x)\, Q_n(x)\, dx = \sum_{i=0}^{n} A_i\, P_{2n+1}(x_i)$$

This completes the proof.

Theorem

$$A_i = \int_a^b w(x)\ell_i(x)\, dx, \quad i = 0, 1, \ldots, n \tag{6.24}$$

where $\ell_i(x)$ are the Lagrange's cardinal functions spanning the nodes at $x_0, x_1, \ldots x_n$. These functions were defined in Eq. (4.2).

Proof Applying Lagrange's formula, Eq. (4.1), to $Q_n(x)$ yields

$$Q_n(x) = \sum_{i=0}^{n} Q_n(x_i)\ell_i(x)$$

which upon substitution in Eq. (d) gives us

$$\sum_{i=0}^{n} \left[Q_n(x_i) \int_a^b w(x)\ell_i(x)\, dx \right] = \sum_{i=0}^{n} A_i\, Q_n(x_i)$$

or

$$\sum_{i=0}^{n} Q_n(x_i) \left[A_i - \int_a^b w(x)\ell_i(x)\, dx \right] = 0$$

This equation can be satisfied for arbitrary $Q(x)$ of degree n only if

$$A_i - \int_a^b w(x)\ell_i(x)\, dx = 0, \quad i = 0, 1, \ldots, n$$

which is equivalent to Eq. (6.24).

It is not difficult to compute the zeros x_i, $i = 0, 1, \ldots, n$ of a polynomial $\varphi_{n+1}(x)$ belonging to an orthogonal set by one of the methods discussed in Chapter 4. Once the zeros are known, the weights A_i, $i = 0, 1, \ldots, n$ could be found from Eq. (6.24).

However the following formulas (given without proof) are easier to compute:

$$\text{Gauss-Legendre} \quad A_i = \frac{2}{(1 - x_i^2)\left[p'_{n+1}(x_i)\right]^2}$$

$$\text{Gauss-Laguerre} \quad A_i = \frac{1}{x_i\left[L'_{n+1}(x_i)\right]^2} \qquad (6.25)$$

$$\text{Gauss-Hermite} \quad A_i = \frac{2^{n+2}(n+1)!\sqrt{\pi}}{\left[H'_{n+1}(x_i)\right]^2}$$

Abscissas and Weights for Classical Gaussian Quadratures

Here we list some classical Gaussian integration formulas. The tables of nodal abscissas and weights, covering $n = 1$ to 5, have been rounded off to six decimal places. These tables should be adequate for hand computation, but in programming you may need more precision or a larger number of nodes. In that case you should consult other references.[3] or use a subroutine to compute the abscissas and weights within the integration program.[4]

The truncation error in Gaussian quadrature

$$E = \int_a^b w(x)f(x)dx - \sum_{i=0}^n A_i f(x_i)$$

has the form $E = K(n)f^{(2n+2)}(c)$, where $a < c < b$ (the value of c is unknown; only its bounds are given). The expression for $K(n)$ depends on the particular quadrature being used. If the derivatives of $f(x)$ can be evaluated, the error formulas are useful in estimating the error bounds.

Gauss-Legendre Quadrature

$$\int_{-1}^1 f(\xi)d\xi \approx \sum_{i=0}^n A_i f(\xi_i) \qquad (6.26)$$

Gauss-Legendre quadrature is the most often used Gaussian integration formula. As seen in Table 6.3, the nodes are arranged symmetrically about $\xi = 0$, and the weights associated with a symmetric pair of nodes are equal. For example, for $n = 1$ we have $\xi_0 = -\xi_1$ and $A_0 = A_1$. The truncation error in Eq. (6.26) is

$$E = \frac{2^{2n+3}\left[(n+1)!\right]^4}{(2n+3)\left[(2n+2)!\right]^3}f^{(2n+2)}(c), \quad -1 < c < 1 \qquad (6.27)$$

---

[3] Abramowitz, M. and Stegun, I.A, *Handbook of Mathematical Functions*, Dover Publications, 1965; Stroud, A.H.and Secrest, D., *Gaussian Quadrature Formulas*, Prentice-Hall, 1966.

[4] Several such subroutines are listed in W.H. Press et al, *Numerical Recipes in Fortran 90*, Cambridge University Press, 1996.

$\pm\xi_i$	A_i	$\pm\xi_i$	A_i
$n = 1$		$n = 4$	
0.577 350	1.000 000	0.000 000	0.568 889
$n = 2$		0.538 469	0.478 629
0.000 000	0.888 889	0.906 180	0.236 927
0.774 597	0.555 556	$n = 5$	
$n = 3$		0.238 619	0.467 914
0.339 981	0.652 145	0.661 209	0.360 762
0.861 136	0.347 855	0.932 470	0.171 324

Table 6.3. Nodes and weights for Gauss–Legendre quadrature.

To apply the Gauss-Legendre quadrature to the integral $\int_a^b f(x)\,dx$, we must first map the integration range (a, b) into the "standard" range $(-1, 1)$. We can accomplish this by the transformation

$$x = \frac{b+a}{2} + \frac{b-a}{2}\xi \tag{6.28}$$

Now $dx = d\xi\,(b - a)/2$, and the quadrature becomes

$$\int_a^b f(x)\,dx \approx \frac{b-a}{2} \sum_{i=1}^{n} A_i f(x_i) \tag{6.29}$$

where the abscissas x_i must be computed from Eq. (6.28). The truncation error here is

$$E = \frac{(b-a)^{2n+3}\left[(n+1)!\right]^4}{(2n+3)\left[(2n+2)!\right]^3} f^{(2n+2)}(c), \quad a < c < b \tag{6.30}$$

Gauss-Chebyshev Quadrature

$$\int_{-1}^{1} \left(1 - x^2\right)^{-1/2} f(x)\,dx \approx \frac{\pi}{n+1} \sum_{i=0}^{n} f(x_i) \tag{6.31}$$

Note that all the weights are equal: $A_i = \pi/(n + 1)$. The abscissas of the nodes, which are symmetric about $x = 0$, are given by

$$x_i = \cos \frac{(2i + 1)\pi}{2n + 2} \tag{6.32}$$

The truncation error is

$$E = \frac{2\pi}{2^{2n+2}(2n + 2)!} f^{(2n+2)}(c), \quad -1 < c < 1 \tag{6.33}$$

Gauss-Laguerre Quadrature

$$\int_0^\infty e^{-x} f(x)\,dx \approx \sum_{i=0}^n A_i f(x_i) \tag{6.34}$$

x_i	A_i	x_i	A_i
$n = 1$		$n = 4$	
0.585 786	0.853 554	0.263 560	0.521 756
3.414 214	0.146 447	1.413 403	0.398 667
$n = 2$		3.596 426	$(-1)0.759\,424$
0.415 775	0.711 093	7.085 810	$(-2)0.361\,175$
2.294 280	0.278 517	12.640 801	$(-4)0.233\,670$
6.289 945	$(-1)0.103\,892$	$n = 5$	
$n = 3$		0.222 847	0.458 964
0.322 548	0.603 154	1.188 932	0.417 000
1.745 761	0.357 418	2.992 736	0.113 373
4.536 620	$(-1)0.388\,791$	5.775 144	$(-1)0.103\,992$
9.395 071	$(-3)0.539\,295$	9.837 467	$(-3)0.261\,017$
		15.982 874	$(-6)0.898\,548$

Table 6.4. Nodes and weights for Gauss–Laguerre quadrature (Multiply numbers by 10^k, where k is given in parentheses.)

$$E = \frac{[(n+1)!]^2}{(2n+2)!} f^{(2n+2)}(c), \quad 0 < c < \infty \tag{6.35}$$

Gauss-Hermite Quadrature

$$\int_{-\infty}^\infty e^{-x^2} f(x)\,dx \approx \sum_{i=0}^n A_i f(x_i) \tag{6.36}$$

$\pm x_i$	A_i	$\pm x_i$	A_i
$n = 1$		$n = 4$	
0.707 107	0.886 227	0.000 000	0.945 308
$n = 2$		0.958 572	0.393 619
0.000 000	1.181 636	2.020 183	$(-1)\,0.199\,532$
1.224 745	0.295 409	$n = 5$	
$n = 3$		0.436 077	0.724 629
0.524 648	0.804 914	1.335 849	0.157 067
1.650 680	$(-1)0.813\,128$	2.350 605	$(-2)0.453\,001$

Table 6.5. Nodes and weights for Gauss–Hermite quadrature (Multiply numbers by 10^k, where k is given in parentheses.)

The nodes are placed symmetrically abut $x = 0$.

$$E = \frac{\sqrt{\pi}(n+1)!}{2^2(2n+2)!} f^{(2n+2)}(c), \quad 0 < c < \infty \tag{6.37}$$

Gauss Quadrature with Logarithmic Singularity

$$\int_0^1 f(x)\ln(x)\,dx \approx -\sum_{i=0}^{n} A_i f(x_i) \tag{6.38}$$

x_i	A_i	x_i	A_i
$n = 1$		$n = 4$	
0.112 009	0.718 539	(-1)0.291 345	0.297 893
0.602 277	0.281 461	0.173 977	0.349 776
$n = 2$		0.411 703	0.234 488
(-1)0.638 907	0.513 405	0.677 314	(-1)0.989 305
0.368 997	0.391 980	0.894 771	(-1)0.189 116
0.766 880	(-1)0.946 154	$n = 5$	
$n = 3$		(-1)0.216 344	0.238 764
(-1)0.414 485	0.383 464	0.129 583	0.308 287
0.245 275	0.386 875	0.314 020	0.245 317
0.556 165	0.190 435	0.538 657	0.142 009
0.848 982	(-1)0.392 255	0.756 916	(-1)0.554 546
		0.922 669	(-1)0.101 690

Table 6.6. Nodes and weights for quadrature with logarithmic singularily (Multiply numbers by 10^k, where k is given in parentheses.)

$$E = \frac{k(n)}{(2n+1)!} f^{(2n+1)}(c), \quad 0 < c < 1 \tag{6.39}$$

where $k(1) = 0.00\,285$, $k(2) = 0.000\,17$, and $k(3) = 0.000\,01$.

■ gaussNodes

The function gaussNodes listed next[5] computes the nodal abscissas x_i and the corresponding weights A_i used in Gauss-Legendre quadrature over the "standard" interval $(-1, 1)$. It can be shown that the approximate values of the abscissas are

$$x_i = \cos\frac{\pi(i+0.75)}{m+0.5}$$

where $m = n + 1$ is the number of nodes, also called the *integration order*. Using these approximations as the starting values, the nodal abscissas are computed by finding

---

[5] This function is an adaptation of a routine in Press, W.H et al., *Numerical Recipes in Fortran 90*, Cambridge University Press, 1996.

the non-negative zeros of the Legendre polynomial $p_m(x)$ with Newton's method (the negative zeros are obtained from symmetry). Note that gaussNodes calls the sub-function legendre, which returns $p_m(t)$ and its derivative as the tuple (p,dp).

```
## module gaussNodes
''' x,A = gaussNodes(m,tol=10e-9)
    Returns nodal abscissas {x} and weights {A} of
    Gauss-Legendre m-point quadrature.
'''
import math
import numpy as np

def gaussNodes(m,tol=10e-9):

    def legendre(t,m):
        p0 = 1.0; p1 = t
        for k in range(1,m):
            p = ((2.0*k + 1.0)*t*p1 - k*p0)/(1.0 + k )
            p0 = p1; p1 = p
        dp = m*(p0 - t*p1)/(1.0 - t**2)
        return p,dp

    A = np.zeros(m)
    x = np.zeros(m)
    nRoots = int((m + 1)/2)          # Number of non-neg. roots
    for i in range(nRoots):
        t = math.cos(math.pi*(i + 0.75)/(m + 0.5))# Approx. root
        for j in range(30):
            p,dp = legendre(t,m)      # Newton-Raphson
            dt = -p/dp; t = t + dt    # method
            if abs(dt) < tol:
                x[i] = t; x[m-i-1] = -t
                A[i] = 2.0/(1.0 - t**2)/(dp**2) # Eq.(6.25)
                A[m-i-1] = A[i]
                break
    return x,A
```

■ gaussQuad

The function gaussQuad uses gaussNodes to evaluate $\int_a^b f(x)\,dx$ with Gauss-Legendre quadrature using m nodes. The function routine for $f(x)$ must be supplied by the user.

```
## module gaussQuad
''' I = gaussQuad(f,a,b,m).
```

Computes the integral of f(x) from x = a to b
with Gauss-Legendre quadrature using m nodes.
''',

from gaussNodes import *

def gaussQuad(f,a,b,m):
    c1 = (b + a)/2.0
    c2 = (b - a)/2.0
    x,A = gaussNodes(m)
    sum = 0.0
    for i in range(len(x)):
        sum = sum + A[i]*f(c1 + c2*x[i])
    return c2*sum

EXAMPLE 6.8

Evaluate $\int_{-1}^{1}(1 - x^2)^{3/2}dx$ as accurately as possible with Gaussian integration.

Solution. Because the integrand is smooth and free of singularities, we could use Gauss-Legendre quadrature. However, the exact integral can obtained with the Gauss-Chebyshev formula. We write

$$\int_{-1}^{1}\left(1 - x^2\right)^{3/2} dx = \int_{-1}^{1}\frac{\left(1 - x^2\right)^2}{\sqrt{1 - x^2}}\,dx$$

The numerator $f(x) = (1 - x^2)^2$ is a polynomial of degree four, so that Gauss-Chebyshev quadrature is exact with three nodes.

The abscissas of the nodes are obtained from Eq. (6.32). Substituting $n = 2$, we get

$$x_i = \cos\frac{(2i + 1)\pi}{6}, \quad i = 0, 1, 2$$

Therefore,

$$x_0 = \cos\frac{\pi}{6} = \frac{\sqrt{3}}{2}$$

$$x_1 = \cos\frac{\pi}{2} = 0$$

$$x_2 = \cos\frac{5\pi}{6} = \frac{\sqrt{3}}{2}$$

and Eq. (6.31) yields

$$\int_{-1}^{1}\left(1 - x^2\right)^{3/2} dx \approx \frac{\pi}{3}\sum_{i=0}^{2}\left(1 - x_i^2\right)^2$$

$$= \frac{\pi}{3}\left[\left(1 - \frac{3}{4}\right)^2 + (1 - 0)^2 + \left(1 - \frac{3}{4}\right)^2\right] = \frac{3\pi}{8}$$

EXAMPLE 6.9

Use Gaussian integration to evaluate $\int_0^{0.5} \cos \pi x \ln x \, dx$.

Solution. We split the integral into two parts:

$$\int_0^{0.5} \cos \pi x \ln x \, dx = \int_0^1 \cos \pi x \ln x \, dx - \int_{0.5}^1 \cos \pi x \ln x \, dx$$

The first integral on the right-hand-side, which contains a logarithmic singularity at $x = 0$, can be computed with the special Gaussian quadrature in Eq. (6.38). Choosing $n = 3$, we have

$$\int_0^1 \cos \pi x \ln x \, dx \approx -\sum_{i=0}^3 A_i \cos \pi x_i$$

The sum is evaluated in the following table:

x_i	$\cos \pi x_i$	A_i	$A_i \cos \pi x_i$
0.041 448	0.991 534	0.383 464	0.380 218
0.245 275	0.717 525	0.386 875	0.277 592
0.556 165	−0.175 533	0.190 435	−0.033 428
0.848 982	−0.889 550	0.039 225	−0.034 892
			$\Sigma = 0.589\,490$

Thus

$$\int_0^1 \cos \pi x \ln x \, dx \approx -0.589\,490$$

The second integral is free of singularities, so that it can be evaluated with Gauss-Legendre quadrature. Choosing $n = 3$, we have

$$\int_{0.5}^1 \cos \pi x \ln x \, dx \approx 0.25 \sum_{i=0}^3 A_i \cos \pi x_i \ln x_i$$

where the nodal abscissas are—see Eq. (6.28)—

$$x_i = \frac{1 + 0.5}{2} + \frac{1 - 0.5}{2} \xi_i = 0.75 + 0.25 \xi_i$$

Looking up ξ_i and A_i in Table 6.3 leads to the following computations:

ξ_i	x_i	$\cos \pi x_i \ln x_i$	A_i	$A_i \cos \pi x_i \ln x_i$
−0.861 136	0.534 716	0.068 141	0.347 855	0.023 703
−0.339 981	0.665 005	0.202 133	0.652 145	0.131 820
0.339 981	0.834 995	0.156 638	0.652 145	0.102 151
0.861 136	0.965 284	0.035 123	0.347 855	0.012 218
				$\Sigma = 0.269\,892$

from which

$$\int_{0.5}^{1} \cos \pi x \ln x \, dx \approx 0.25(0.269\,892) = 0.067\,473$$

Therefore,

$$\int_{0}^{1} \cos \pi x \ln x \, dx \approx -0.589\,490 - 0.067\,473 = -0.656\,963$$

which is correct to six decimal places.

EXAMPLE 6.10
Evaluate as accurately as possible

$$F = \int_{0}^{\infty} \frac{x+3}{\sqrt{x}} e^{-x} dx$$

Solution. In its present form, the integral is not suited to any of the Gaussian quadratures listed in this section. But using the transformation

$$x = t^2 \qquad dx = 2t \, dt$$

the integral becomes

$$F = 2\int_{0}^{\infty} (t^2 + 3)e^{-t^2} dt = \int_{-\infty}^{\infty} (t^2 + 3)e^{-t^2} dt$$

which can be evaluated exactly with the Gauss-Hermite formula using only two nodes ($n = 1$). Thus

$$F = A_0(t_0^2 + 3) + A_1(t_1^2 + 3)$$

$$= 0.886\,227\left[(0.707\,107)^2 + 3\right] + 0.886\,227\left[(-0.707\,107)^2 + 3\right]$$

$$= 6.203\,59$$

EXAMPLE 6.11
Determine how many nodes are required to evaluate

$$\int_{0}^{\pi} \left(\frac{\sin x}{x}\right)^2 dx$$

with Gauss-Legendre quadrature to six decimal places. The exact integral, rounded to six places, is 1.418 15.

Solution. The integrand is a smooth function; hence it is suited for Gauss-Legendre integration. There is an indeterminacy at $x = 0$, but it does not bother the quadrature because the integrand is never evaluated at that point. We used the following program that computes the quadrature with 2, 3, ... nodes until the desired accuracy is reached:

```
## example 6_11
import math
from gaussQuad import *
```

```
def f(x): return (math.sin(x)/x)**2

a = 0.0; b = math.pi;
Iexact = 1.41815
for m in range(2,12):
    I = gaussQuad(f,a,b,m)
    if abs(I - Iexact) < 0.00001:
        print("Number of nodes =",m)
        print("Integral =", gaussQuad(f,a,b,m))
        break
input("\nPress return to exit")
```

The program output is

```
Number of nodes = 5
Integral = 1.41815026778
```

EXAMPLE 6.12

Evaluate numerically $\int_{1.5}^{3} f(x)\, dx$, where $f(x)$ is represented by the following unevenly spaced data:

x	1.2	1.7	2.0	2.4	2.9	3.3
$f(x)$	$-0.362\,36$	$0.128\,84$	$0.416\,15$	$0.737\,39$	$0.970\,96$	$0.987\,48$

Knowing that the data points lie on the curve $f(x) = -\cos x$, evaluate the accuracy of the solution.

Solution. We approximate $f(x)$ by the polynomial $P_5(x)$ that intersects all the data points, and then evaluate $\int_{1.5}^{3} f(x)dx \approx \int_{1.5}^{3} P_5(x)dx$ with the Gauss-Legendre formula. Because the polynomial is of degree five, only three nodes ($n = 2$) are required in the quadrature.

From Eq. (6.28) and Table 6.6, we obtain for the abscissas of the nodes

$$x_0 = \frac{3+1.5}{2} + \frac{3-1.5}{2}(-0.774597) = 1.6691$$

$$x_1 = \frac{3+1.5}{2} = 2.25$$

$$x_2 = \frac{3+1.5}{2} + \frac{3-1.5}{2}(0.774597) = 2.8309$$

We now compute the values of the interpolant $P_5(x)$ at the nodes. This can be done using the functions `newtonPoly` or `neville` listed in Section 3.2. The results are

$$P_5(x_0) = 0.098\,08 \qquad P_5(x_1) = 0.628\,16 \qquad P_5(x_2) = 0.952\,16$$

From Gauss-Legendre quadrature

$$I = \int_{1.5}^{3} P_5(x)dx = \frac{3-1.5}{2} \sum_{i=0}^{2} A_i P_5(x_i)$$

we get

$$I = 0.75 \, [0.555\,556(0.098\,08) + 0.888\,889(0.628\,16) + 0.555\,556(0.952\,16)]$$

$$= 0.856\,37$$

Comparison with $-\int_{1.5}^{3} \cos x \, dx = 0.856\,38$ shows that the discrepancy is within the roundoff error.

PROBLEM SET 6.2

1. Evaluate

$$\int_{1}^{\pi} \frac{\ln(x)}{x^2 - 2x + 2} \, dx$$

   with Gauss-Legendre quadrature. Use (a) two nodes; and (b) four nodes.

2. Use Gauss-Laguerre quadrature to evaluate $\int_{0}^{\infty} (1 - x^2)^3 e^{-x} \, dx$.

3. Use Gauss-Chebyshev quadrature with six nodes to evaluate

$$\int_{0}^{\pi/2} \frac{dx}{\sqrt{\sin x}}$$

   Compare the result with the "exact" value 2.62206. *Hint*: Substitute $\sin x = t^2$.

4. The integral $\int_{0}^{\pi} \sin x \, dx$ is evaluated with Gauss-Legendre quadrature using four nodes. What are the bounds on the truncation error resulting from the quadrature?

5. How many nodes are required in Gauss-Laguerre quadrature to evaluate $\int_{0}^{\infty} e^{-x} \sin x \, dx$ to six decimal places?

6. Evaluate as accurately as possible

$$\int_{0}^{1} \frac{2x + 1}{\sqrt{x(1 - x)}} \, dx$$

   Hint: Substitute $x = (1 + t)/2$.

7. Compute $\int_{0}^{\pi} \sin x \ln x \, dx$ to four decimal places.

8. Calculate the bounds on the truncation error if $\int_{0}^{\pi} x \sin x \, dx$ is evaluated with Gauss-Legendre quadrature using three nodes. What is the actual error?

9. Evaluate $\int_{0}^{2} (\sinh x / x) \, dx$ to four decimal places.

10. Evaluate the integral

$$\int_{0}^{\infty} \frac{x \, dx}{e^x + 1}$$

    by Gauss-Legendre quadrature to six decimal places. *Hint*: Substitute $e^x = \ln(1/t)$.

11. ■ The equation of an ellipse is $x^2/a^2 + y^2/b^2 = 1$. Write a program that computes the length

$$S = 2 \int_{-a}^{a} \sqrt{1 + (dy/dx)^2} \, dx$$

    of the circumference to five decimal places for a given a and b. Test the program with $a = 2$ and $b = 1$.

12. ■ The error function, which is of importance in statistics, is defined as

$$\mathrm{erf}(x) = \frac{2}{\sqrt{\pi}} \int_0^x e^{-t^2} dt$$

Write a program that uses Gauss-Legendre quadrature to evaluate erf(x) for a given x to six decimal places. Note that erf(x) = 1.000 000 (correct to 6 decimal places) when $x > 5$. Test the program by verifying that erf(1.0) = 0.842 701.

13. ■

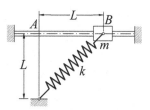

The sliding weight of mass m is attached to a spring of stiffness k that has an undeformed length L. When the mass is released from rest at B, the time it takes to reach A can be shown to be $t = C\sqrt{m/k}$, where

$$C = \int_0^1 \left[\left(\sqrt{2} - 1\right)^2 - \left(\sqrt{1 + z^2} - 1\right)^2 \right]^{-1/2} dz$$

Compute C to six decimal places. *Hint*: The integrand has singularity at $z = 1$ that behaves as $(1 - z^2)^{-1/2}$.

14. ■

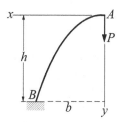

A uniform beam forms the semi-parabolic cantilever arch AB. The vertical displacement of A due to the force P can be shown to be

$$\delta_A = \frac{Pb^3}{EI} C\left(\frac{h}{b}\right)$$

where EI is the bending rigidity of the beam and

$$C\left(\frac{h}{b}\right) = \int_0^1 z^2 \sqrt{1 + \left(\frac{2h}{b} z\right)^2} \, dz$$

Write a program that computes $C(h/b)$ for any given value of h/b to four decimal places. Use the program to compute $C(0.5)$, $C(1.0)$, and $C(2.0)$.

15. ■ There is no elegant way to compute $I = \int_0^{\pi/2} \ln(\sin x)\, dx$. A "brute force" method that works is to split the integral into several parts: from $x = 0$ to 0.01,

from 0.01 to 0.2, and from $x = 0.02$ to $\pi/2$. In the first part we can use the approximation $\sin x \approx x$, which allows us to obtain the integral analytically. The other two parts can be evaluated with Gauss-Legendre quadrature. Use this method to evaluate I to six decimal places.

16. ■

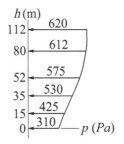

The pressure of wind was measured at various heights on a vertical wall, as shown on the diagram. Find the height of the pressure center, which is defined as

$$\bar{h} = \frac{\int_0^{112\,\text{m}} h\, p(h)\, dh}{\int_0^{112\,\text{m}} p(h)\, dh}$$

Hint: Fit a cubic polynomial to the data and then apply Gauss-Legendre quadrature.

17. ■ Write a function that computes $\int_{x_1}^{x_n} y(x)\, dx$ from a given set of data points of the form

x_1	x_2	x_3	$\cdots$	x_n
y_1	y_2	y_3	$\cdots$	y_n

The function must work for unevenly spaced x-values. Test the function with the data given in Prob. 17, Problem Set 6.1. *Hint*: Fit a cubic spline to the data points and apply Gauss-Legendre quadrature to each segment of the spline.

*6.5   Multiple Integrals

Multiple integrals, such as the area integral $\int \int_A f(x, y)\, dx\, dy$, can also be evaluated by quadrature. The computations are straightforward if the region of integration has a simple geometric shape, such as a triangle or a quadrilateral. Because of complications in specifying the limits of integration, quadrature is not a practical means of evaluating integrals over irregular regions. However, an irregular region A can always be approximated as an assembly of triangular or quadrilateral subregions $A_1, A_2, \ldots$, called *finite elements*, as illustrated in Figure 6.6. The integral over A can then be evaluated by summing the integrals over the finite elements:

$$\int \int_A f(x, y)\, dx\, dy \approx \sum_i \int \int_{A_i} f(x, y)\, dx\, dy$$

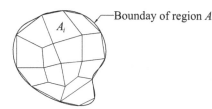

Figure 6.6. Finite element model of an irregular region.

Volume integrals can be computed in a similar manner, using tetrahedra or rectangular prisms for the finite elements.

Gauss-Legendre Quadrature over a Quadrilateral Element

Consider the double integral

$$I = \int_{-1}^{1} \int_{-1}^{1} f(\xi, \eta) \, d\eta \, d\xi$$

over the rectangular element shown in Figure 6.7(a). Evaluating each integral in turn by Gauss-Legendre quadrature using $n + 1$ integration points in each coordinate direction, we obtain

$$I = \int_{-1}^{1} \sum_{i=0}^{n} A_i f(\xi_i, \eta) \, d\eta = \sum_{j=0}^{n} A_j \left[\sum_{i=0}^{n} A_i f(\xi_i, \eta_j) \right]$$

or

$$I = \sum_{i=0}^{n} \sum_{j=0}^{n} A_i A_j f(\xi_i, \eta_j) \tag{6.40}$$

As noted previously, the number of integration points in each coordinate direction, $m = n + 1$, is called the *integration order*. Figure 6.7(a) shows the locations of the integration points used in third-order integration ($m = 3$). Because the integration limits are the "standard" limits $(-1, 1)$ of Gauss-Legendre quadrature, the weights and the coordinates of the integration points are as listed in Table 6.3.

To apply quadrature to the quadrilateral element in Figure 6.7(b), we must first map the quadrilateral into the "standard" rectangle in Figure 6.7(a). By mapping we mean a coordinate transformation $x = x(\xi, \eta)$, $y = y(\xi, \eta)$ that results in one-to-one

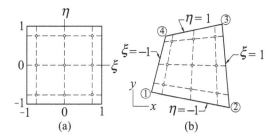

Figure 6.7. Mapping of quadrilateral into a rectangle.

correspondence between points in the quadrilateral and in the rectangle. The transformation that does the job is

$$x(\xi, \eta) = \sum_{k=1}^{4} N_k(\xi, \eta) x_k \qquad y(\xi, \eta) = \sum_{k=1}^{4} N_k(\xi, \eta) y_k \qquad (6.41)$$

where (x_k, y_k) are the coordinates of corner k of the quadrilateral and

$$N_1(\xi, \eta) = \frac{1}{4}(1 - \xi)(1 - \eta)$$

$$N_2(\xi, \eta) = \frac{1}{4}(1 + \xi)(1 - \eta) \qquad (6.42)$$

$$N_3(\xi, \eta) = \frac{1}{4}(1 + \xi)(1 + \eta)$$

$$N_4(\xi, \eta) = \frac{1}{4}(1 - \xi)(1 + \eta)$$

The functions $N_k(\xi, \eta)$, known as the *shape functions*, are bilinear (linear in each coordinate). Consequently, straight lines remain straight upon mapping. In particular, note that the sides of the quadrilateral are mapped into the lines $\xi = \pm 1$ and $\eta = \pm 1$.

Because mapping distorts areas, an infinitesimal area element $dA = dx\,dy$ of the quadrilateral is not equal to its counterpart $dA' = d\xi\,d\eta$ of the rectangle. It can be shown that the relationship between the areas is

$$dx\,dy = |J(\xi, \eta)|\ d\xi\,d\eta \qquad (6.43)$$

where

$$J(\xi, \eta) = \begin{bmatrix} \dfrac{\partial x}{\partial \xi} & \dfrac{\partial y}{\partial \xi} \\[2mm] \dfrac{\partial x}{\partial \eta} & \dfrac{\partial y}{\partial \eta} \end{bmatrix} \qquad (6.44\mathrm{a})$$

is known as the *Jacobian matrix* of the mapping. Substituting from Eqs. (6.41) and (6.42) and differentiating, the components of the Jacobian matrix are

$$J_{11} = \frac{1}{4}[-(1 - \eta)x_1 + (1 - \eta)x_2 + (1 + \eta)x_3 - (1 - \eta)x_4]$$

$$J_{12} = \frac{1}{4}[-(1 - \eta)y_1 + (1 - \eta)y_2 + (1 + \eta)y_3 - (1 - \eta)y_4] \qquad (6.44\mathrm{b})$$

$$J_{21} = \frac{1}{4}[-(1 - \xi)x_1 - (1 + \xi)x_2 + (1 + \xi)x_3 + (1 - \xi)x_4]$$

$$J_{22} = \frac{1}{4}[-(1 - \xi)y_1 - (1 + \xi)y_2 + (1 + \xi)y_3 + (1 - \xi)y_4]$$

We can now write

$$\iint_A f(x, y)\,dx\,dy = \int_{-1}^{1}\int_{-1}^{1} f[x(\xi, \eta), y(\xi, \eta)]\ |J(\xi, \eta)|\ d\xi\,d\eta \qquad (6.45)$$

Since the right-hand-side integral is taken over the "standard" rectangle, it can be evaluated using Eq. (6.40). Replacing $f(\xi, \eta)$ in Eq. (6.40) by the integrand in Eq. (6.45), we get the following formula for Gauss-Legendre quadrature over a quadrilateral region:

$$I = \sum_{i=0}^{n} \sum_{j=0}^{n} A_i A_j \, f\left[x(\xi_i, \eta_j), \, y(\xi_i, \eta_j)\right] \left|J(\xi_i, \eta_j)\right| \tag{6.46}$$

The ξ and η-coordinates of the integration points and the weights can again be obtained from Table 6.3.

■ gaussQuad2

The function gaussQuad2 in this module computes $\int \int_A f(x, y) \, dx \, dy$ over a quadrilateral element with Gauss-Legendre quadrature of integration order m. The quadrilateral is defined by the arrays **x** and **y**, which contain the coordinates of the four corners ordered in a *counterclockwise direction* around the element. The determinant of the Jacobian matrix is obtained by calling the function jac; mapping is performed by map. The weights and the values of ξ and η at the integration points are computed by gaussNodes listed in the previous section (note that ξ and η appear as s and t in the listing).

```
## module gaussQuad2
''' I = gaussQuad2(f,xc,yc,m).
    Gauss-Legendre integration of f(x,y) over a
    quadrilateral using integration order m.
    {xc},{yc} are the corner coordinates of the quadrilateral.
'''
from gaussNodes import *
import numpy as np
def gaussQuad2(f,x,y,m):

    def jac(x,y,s,t):
        J = np.zeros((2,2))
        J[0,0] = -(1.0 - t)*x[0] + (1.0 - t)*x[1] \
                 + (1.0 + t)*x[2] - (1.0 + t)*x[3]
        J[0,1] = -(1.0 - t)*y[0] + (1.0 - t)*y[1] \
                 + (1.0 + t)*y[2] - (1.0 + t)*y[3]
        J[1,0] = -(1.0 - s)*x[0] - (1.0 + s)*x[1] \
                 + (1.0 + s)*x[2] + (1.0 - s)*x[3]
        J[1,1] = -(1.0 - s)*y[0] - (1.0 + s)*y[1] \
                 + (1.0 + s)*y[2] + (1.0 - s)*y[3]
        return (J[0,0]*J[1,1] - J[0,1]*J[1,0])/16.0
```

```
def map(x,y,s,t):
    N = np.zeros(4)
    N[0] = (1.0 - s)*(1.0 - t)/4.0
    N[1] = (1.0 + s)*(1.0 - t)/4.0
    N[2] = (1.0 + s)*(1.0 + t)/4.0
    N[3] = (1.0 - s)*(1.0 + t)/4.0
    xCoord = np.dot(N,x)
    yCoord = np.dot(N,y)
    return xCoord,yCoord

s,A = gaussNodes(m)
sum = 0.0
for i in range(m):
    for j in range(m):
        xCoord,yCoord = map(x,y,s[i],s[j])
        sum = sum + A[i]*A[j]*jac(x,y,s[i],s[j])  \
                   *f(xCoord,yCoord)
return sum
```

EXAMPLE 6.13

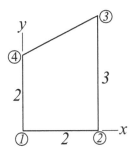

Evaluate the integral

$$I = \int \int_A (x^2 + y)\, dx\, dy$$

analytically by first transforming it from the quadrilateral region A shown to the "standard" rectangle.

Solution. The corner coordinates of the quadrilateral are

$$\mathbf{x}^T = \begin{bmatrix} 0 & 2 & 2 & 0 \end{bmatrix} \qquad \mathbf{y}^T = \begin{bmatrix} 0 & 0 & 3 & 2 \end{bmatrix}$$

The mapping is

$$x(\xi, \eta) = \sum_{k=1}^{4} N_k(\xi, \eta) x_k$$

$$= 0 + \frac{(1+\xi)(1-\eta)}{4}(2) + \frac{(1+\xi)(1+\eta)}{4}(2) + 0$$

$$= 1 + \xi$$

$$y(\xi, \eta) = \sum_{k=1}^{4} N_k(\xi, \eta) y_k$$

$$= 0 + 0 + \frac{(1+\xi)(1+\eta)}{4}(3) + \frac{(1-\xi)(1+\eta)}{4}(2)$$

$$= \frac{(5+\xi)(1+\eta)}{4}$$

which yields for the Jacobian matrix

$$J(\xi, \eta) = \begin{bmatrix} \dfrac{\partial x}{\partial \xi} & \dfrac{\partial y}{\partial \xi} \\[2mm] \dfrac{\partial x}{\partial \eta} & \dfrac{\partial y}{\partial \eta} \end{bmatrix} = \begin{bmatrix} 1 & \dfrac{1+\eta}{4} \\[2mm] 0 & \dfrac{5+\xi}{4} \end{bmatrix}$$

Thus the area scale factor is

$$|J(\xi, \eta)| = \frac{5+\xi}{4}$$

Now we can map the integral from the quadrilateral to the standard rectangle. Referring to Eq. (6.45), we obtain

$$I = \int_{-1}^{1} \int_{-1}^{1} \left[\left(\frac{1+\xi}{2} \right)^2 + \frac{(5+\xi)(1+\eta)}{4} \right] \frac{5+\xi}{4} d\xi \, d\eta$$

$$= \int_{-1}^{1} \int_{-1}^{1} \left(\frac{15}{8} + \frac{21}{16}\xi + \frac{1}{2}\xi^2 + \frac{1}{16}\xi^3 + \frac{25}{16}\eta + \frac{5}{8}\xi\eta + \frac{1}{16}\xi^2\eta \right) d\xi \, d\eta$$

Noting that only even powers of ξ and η contribute to the integral, the integral simplifies to

$$I = \int_{-1}^{1} \int_{-1}^{1} \left(\frac{15}{8} + \frac{1}{2}\xi^2 \right) d\xi \, d\eta = \frac{49}{6}$$

EXAMPLE 6.14
Evaluate the integral

$$\int_{-1}^{1} \int_{-1}^{1} \cos \frac{\pi x}{2} \cos \frac{\pi y}{2} \, dx \, dy$$

by Gauss-Legendre quadrature of order three.

Solution. From the quadrature formula in Eq. (6.40), we have

$$I = \sum_{i=0}^{2} \sum_{j=0}^{2} A_i A_j \cos \frac{\pi x_i}{2} \cos \frac{\pi y_j}{2}$$

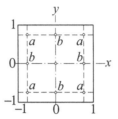

The integration points are shown in the figure; their coordinates and the corresponding weights are listed in Table 6.3. Note that the integrand, the integration points, and the weights are all symmetric about the coordinate axes. It follows that the points labeled a contribute equal amounts to I; the same is true for the points labeled b. Therefore,

$$I = 4(0.555\,556)^2 \cos^2 \frac{\pi\,(0.774\,597)}{2}$$

$$+4(0.555\,556)(0.888\,889) \cos \frac{\pi\,(0.774\,597)}{2} \cos \frac{\pi\,(0)}{2}$$

$$+(0.888\,889)^2 \cos^2 \frac{\pi\,(0)}{2}$$

$$= 1.623\,391$$

The exact value of the integral is $16/\pi^2 \approx 1.621\,139$.

EXAMPLE 6.15

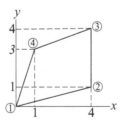

Use gaussQuad2 to evaluate $I = \int \int_A f(x, y)\,dx\,dy$ over the quadrilateral shown, where

$$f(x, y) = (x - 2)^2(y - 2)^2$$

Use enough integration points for an "exact" answer.

Solution. The required integration order is determined by the integrand in Eq. (6.45):

$$I = \int_{-1}^{1} \int_{-1}^{1} f\,[x(\xi, \eta), y(\xi, \eta)]\, |J(\xi, \eta)|\, d\xi\, d\eta \tag{a}$$

We note that $|J(\xi, \eta)|$, defined in Eqs. (6.44), is biquadratic. Because the specified $f(x, y)$ is also biquadratic, the integrand in Eq. (a) is a polynomial of degree 4 in both ξ and η. Thus third-order integration is sufficient for an "exact" result.

```
#!/usr/bin/python
## example 6_15
from gaussQuad2 import *
import numpy as np
def f(x,y): return ((x - 2.0)**2)*((y - 2.0)**2)

x = np.array([0.0, 4.0, 4.0, 1.0])
y = np.array([0.0, 1.0, 4.0, 3.0])
m = eval(input("Integration order ==> "))
print("Integral =", gaussQuad2(f,x,y,m))
input("\nPress return to exit")
```

Running this program produced the following result:

```
Integration order ==> 3
Integral = 11.3777777778
```

Quadrature over a Triangular Element

A triangle may be viewed as a degenerate quadrilateral with two of its corners occupying the same location, as illustrated in Figure 6.8. Therefore, the integration formulas over a quadrilateral region can also be used for a triangular element. However, it is computationally advantageous to use integration formulas specially developed for triangles, which we present without derivation.[6]

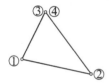

Figure 6.8. Degenerate quadrilateral.

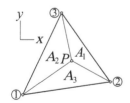

Figure 6.9. Triangular element.

Consider the triangular element in Figure 6.9. Drawing straight lines from the point P in the triangle to each of the corners, we divide the triangle into three parts with areas A_1, A_2, and A_3. The so-called *area coordinates* of P are defined as

$$\alpha_i = \frac{A_i}{A}, \quad i = 1, 2, 3 \tag{6.47}$$

---

[6] The triangle formulas are extensively used in the finite method analysis. See, for example, Zienkiewicz, O.C. and Taylor, R.L., *The Finite Element Method*, Vol. 1, 4th ed., McGraw-Hill, 1989.

where A is the area of the element. Because $A_1 + A_2 + A_3 = A$, the area coordinated are related by

$$\alpha_1 + \alpha_2 + \alpha_3 = 1 \tag{6.48}$$

Note that α_i ranges from 0 (when P lies on the side opposite to corner i) to 1 (when P is at corner i).

A convenient formula of computing A from the corner coordinates (x_i, y_i) is

$$A = \frac{1}{2} \begin{vmatrix} 1 & 1 & 1 \\ x_1 & x_2 & x_3 \\ y_1 & y_2 & y_3 \end{vmatrix} \tag{6.49}$$

The area coordinates are mapped onto the Cartesian coordinates by

$$x(\alpha_1, \alpha_2, \alpha_3) = \sum_{i=1}^{3} \alpha_i x_i \qquad y(\alpha_1, \alpha_2, \alpha_3) = \sum_{i=1}^{3} \alpha_i y_i \tag{6.50}$$

The integration formula over the element is

$$\int\int_A f[x(\boldsymbol{\alpha}), y(\boldsymbol{\alpha})] \, dA = A \sum_k W_k f[x(\boldsymbol{\alpha}_k), y(\boldsymbol{\alpha}_k)] \tag{6.51}$$

where $\boldsymbol{\alpha}_k$ represents the area coordinates of the integration point k, and W_k are the weights. The locations of the integration points are shown in Figure 6.10, and the corresponding values of $\boldsymbol{\alpha}_k$ and W_k are listed in Table 6.7. The quadrature in Eq. (6.51) is exact if $f(x, y)$ is a polynomial of the degree indicated.

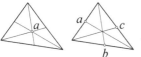

Figure 6.10. Integration points for triangular elements.

(a) Linear   (b) Quadratic   (c) Cubic

Degree of $f(x, y)$	Point	α_k	W_k
(a) Linear	a	1/3, 1/3, 1/3	1
(b) Quadratic	a	1/2, 0, 1/2	1/3
	b	1/2, 1/2, 0	1/3
	c	0, 1/2, 1/2	1/3
(c) Cubic	a	1/3, 1/3, 1/3	−27/48
	b	1/5, 1/5, 3/5	25/48
	c	3/5. 1/5, 1/5	25/48
	d	1/5, 3/5, 1/5	25/48

Table 6.7. Nodes and weights for quadrature over a triangle.

■ triangleQuad

The function `triangleQuad` computes $\int \int_A f(x, y)\, dx\, dy$ over a triangular region using the cubic formula—case (c) in Figure 6.10. The triangle is defined by its corner coordinate arrays `xc` and `yc`, where the coordinates are listed in a *counterclockwise* order around the triangle.

```
## module triangleQuad
''' integral = triangleQuad(f,xc,yc).
    Integration of f(x,y) over a triangle using
    the cubic formula.
    {xc},{yc} are the corner coordinates of the triangle.
'''
import numpy as np

def triangleQuad(f,xc,yc):
    alpha = np.array([[1/3, 1/3.0, 1/3], \
                      [0.2, 0.2, 0.6],                 \
                      [0.6, 0.2, 0.2],                 \
                      [0.2, 0.6, 0.2]])
    W = np.array([-27/48,25/48,25/48,25/48])
    x = np.dot(alpha,xc)
    y = np.dot(alpha,yc)
    A = (xc[1]*yc[2] - xc[2]*yc[1]         \
        - xc[0]*yc[2] + xc[2]*yc[0]         \
        + xc[0]*yc[1] - xc[1]*yc[0])/2.0
    sum = 0.0
    for i in range(4):
        sum = sum + W[i] * f(x[i],y[i])
    return A*sum
```

EXAMPLE 6.16

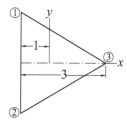

Evaluate $I = \int \int_A f(x, y)\, dx\, dy$ over the equilateral triangle shown, where[7]

$$f(x, y) = \frac{1}{2}(x^2 + y^2) - \frac{1}{6}(x^3 - 3xy^2) - \frac{2}{3}$$

Use the quadrature formulas for (1) a quadrilateral and (2) a triangle.

---

[7] This function is identical to the Prandtl stress function for torsion of a bar with the cross section shown; the integral is related to the torsional stiffness of the bar. See, for example Timoshenko, S.P and Goodier, J.N., *Theory of Elasticity*, 3rd ed., McGraw-Hill, 1970.

Solution of Part (1). Let the triangle be formed by collapsing corners 3 and 4 of a quadrilateral. The corner coordinates of this quadrilateral are $\mathbf{x} = \begin{bmatrix} -1, & -1, & 2, & 2 \end{bmatrix}^T$ and $\mathbf{y} = \begin{bmatrix} \sqrt{3}, & -\sqrt{3}, & 0, & 0 \end{bmatrix}^T$. To determine the minimum required integration order for an exact result, we must examine $f[x(\xi, \eta), y(\xi, \eta)] |J(\xi, \eta)|$, the integrand in Eqs. (6.44). Since $|J(\xi, \eta)|$ is biquadratic, and $f(x, y)$ is cubic in x, the integrand is a polynomial of degree five in x. Therefore, third-order integration is sufficient. The program used for the computations is similar to the one in Example 6.15:

```python
#!/usr/bin/python
## example6_16a
from gaussQuad2 import *
import numpy as np
import math

def f(x,y):
    return (x**2 + y**2)/2.0            \
           - (x**3 - 3.0*x*y**2)/6.0  \
           - 2.0/3.0

x = np.array([-1.0,-1.0,2.0,2.0])
y = np.array([math.sqrt(3.0),-math.sqrt(3.0),0.0,0.0])
m = eval(input("Integration order ==> "))
print("Integral =", gaussQuad2(f,x,y,m))
input("\nPress return to exit")
```

Here is the output:

```
Integration order ==> 3
Integral = -1.55884572681
```

Solution of Part (2). The following program uses `triangleQuad`:

```python
#!/usr/bin/python
## example6_16b
import numpy as np
import math
from triangleQuad import *

def f(x,y):
    return (x**2 + y**2)/2.0            \
           -(x**3 - 3.0*x*y**2)/6.0 \
           -2.0/3.0

xCorner = np.array([-1.0, -1.0, 2.0])
yCorner = np.array([math.sqrt(3.0), -math.sqrt(3.0), 0.0])
print("Integral =",triangleQuad(f,xCorner,yCorner))
input("Press return to  exit")
```

Because the integrand is a cubic, this quadrature is also exact, the result being

```
Integral = -1.55884572681
```

Note that only four function evaluations were required when using the triangle formulas. In contrast, the function had to be evaluated at nine points in part (1).

EXAMPLE 6.17

The corner coordinates of a triangle are $(0, 0)$, $(16, 10)$, and $(12, 20)$. Compute $\int \int_A (x^2 - y^2) \, dx \, dy$ over this triangle.

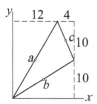

Solution. Because $f(x, y)$ is quadratic, quadrature over the three integration points shown in Figure 6.10(b) is sufficient for an "exact" result. Noting that the integration points lie in the middle of each side, their coordinates are $(6, 10)$, $(8, 5)$, and $(14, 15)$. The area of the triangle is obtained from Eq. (6.49):

$$A = \frac{1}{2} \begin{vmatrix} 1 & 1 & 1 \\ x_1 & x_2 & x_3 \\ y_1 & y_2 & y_3 \end{vmatrix} = \frac{1}{2} \begin{vmatrix} 1 & 1 & 1 \\ 0 & 16 & 12 \\ 0 & 10 & 20 \end{vmatrix} = 100$$

From Eq. (6.51) we get

$$I = A \sum_{k=a}^{c} W_k f(x_k, y_k)$$

$$= 100 \left[\frac{1}{3} f(6, 10) + \frac{1}{3} f(8, 5) + \frac{1}{3} f(14, 15) \right]$$

$$= \frac{100}{3} \left[(6^2 - 10^2) + (8^2 - 5^2) + (14^2 - 15^2) \right] = 1800$$

PROBLEM SET 6.3

1. Use Gauss-Legendre quadrature to compute

$$\int_{-1}^{1} \int_{-1}^{1} (1 - x^2)(1 - y^2) \, dx \, dy$$

2. Evaluate the following integral with Gauss-Legendre quadrature:

$$\int_{y=0}^{2} \int_{x=0}^{3} x^2 y^2 \, dx \, dy$$

3. Compute the approximate value of

$$\int_{-1}^{1} \int_{-1}^{1} e^{-(x^2+y^2)} \, dx \, dy$$

with Gauss-Legendre quadrature. Use integration order (a) two and (b) three. (The "exact" value of the integral is 2.230 985.)

4. Use third-order Gauss-Legendre quadrature to obtain an approximate value of

$$\int_{-1}^{1}\int_{-1}^{1} \cos \frac{\pi (x - y)}{2} \, dx \, dy$$

(The "exact" value of the integral is 1.621 139.)

5.

Map the integral $\int \int_A xy \, dx \, dy$ from the quadrilateral region shown to the "standard" rectangle and then evaluate it analytically.

6.

Compute $\int \int_A x \, dx \, dy$ over the quadrilateral region shown by first mapping it into the "standard" rectangle and then integrating analytically.

7.

Use quadrature to compute $\int \int_A x^2 \, dx \, dy$ over the triangle shown.

8. Evaluate $\int \int_A x^3 \, dx \, dy$ over the triangle shown in Prob. 7.

9.

Use quadrature to evaluate $\int \int_A (3 - x) y \, dx \, dy$ over the region shown. Treat the region as (a) a triangular element and (b) a degenerate quadrilateral.

10. Evaluate $\int \int_A x^2 y \, dx \, dy$ over the triangle shown in Prob. 9.

11. ■

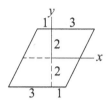

Evaluate $\int \int_A xy(2 - x^2)(2 - xy)\, dx\, dy$ over the region shown.

12. ■ Compute $\int \int_A xy \exp(-x^2)\, dx\, dy$ over the region shown in Prob. 11 to four decimal places.

13. ■

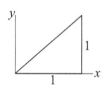

Evaluate $\int \int_A (1 - x)(y - x)y\, dx\, dy$ over the triangle shown.

14. ■ Estimate $\int \int_A \sin \pi x\, dx\, dy$ over the region shown in Prob. 13. Use the cubic integration formula for a triangle. (The exact integral is $1/\pi$.)

15. ■ Compute $\int \int_A \sin \pi x \sin \pi (y - x)\, dx\, dy$ to six decimal places, where A is the triangular region shown in Prob. 13. Consider the triangle as a degenerate quadrilateral.

16. ■

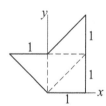

Write a program to evaluate $\int \int_A f(x, y)\, dx\, dy$ over an irregular region that has been divided into several triangular elements. Use the program to compute $\int \int_A xy(y - x)\, dx\, dy$ over the region shown.

7 Initial Value Problems

$$\boxed{\text{Solve } \mathbf{y'} = \mathbf{F}(x, \mathbf{y}) \text{ with the auxiliary conditions } \mathbf{y}(a) = \alpha}$$

7.1 Introduction

The general form of a *first-order differential equation* is

$$y' = f(x, y) \tag{7.1a}$$

where $y' = dy/dx$ and $f(x, y)$ is a given function. The solution of this equation contains an arbitrary constant (the constant of integration). To find this constant, we must know a point on the solution curve; that is, y must be specified at some value of x, say at $x = a$. We write this auxiliary condition as

$$y(a) = \alpha \tag{7.1b}$$

An ordinary differential equation of order n

$$y^{(n)} = f\left(x, y, y', \ldots, y^{(n-1)}\right) \tag{7.2}$$

can always be transformed into n first-order equations. Using the notation

$$y_0 = y \quad y_1 = y' \quad y_2 = y'' \quad \ldots \quad y_{n-1} = y^{(n-1)} \tag{7.3}$$

the equivalent first-order equations are

$$y_0' = y_1 \quad y_1' = y_2 \quad y_2' = y_3 \quad \ldots \quad y_n' = f(x, y_0, y_1, \ldots, y_{n-1}) \tag{7.4a}$$

The solution now requires the knowledge of n auxiliary conditions. If these conditions are specified at the same value of x, the problem is said to be an *initial value problem*. Then the auxiliary conditions, called *initial conditions*, have the form

$$y_0(a) = \alpha_0 \quad y_1(a) = \alpha_1 \quad \ldots \quad y_{n-1}(a) = \alpha_{n-1} \tag{7.4b}$$

If y_i are specified at different values of x, the problem is called a *boundary value problem*.

For example,

$$y'' = -y \qquad y(0) = 1 \qquad y'(0) = 0$$

is an initial value problem because both auxiliary conditions imposed on the solution are given at $x = 0$. In contrast,

$$y'' = -y \qquad y(0) = 1 \qquad y(\pi) = 0$$

is a boundary value problem because the two conditions are specified at different values of x.

In this chapter we consider only initial value problems. Boundary value problems, which are more difficult to solve, are discussed in the next chapter. We also make extensive use of vector notation, which allows us to manipulate sets of first-order equations in a concise form. For example, Eqs. (7.4) are written as

$$\mathbf{y}' = \mathbf{F}(x, \mathbf{y}) \qquad \mathbf{y}(a) = \boldsymbol{\alpha} \tag{7.5a}$$

where

$$\mathbf{F}(x, \mathbf{y}) = \begin{bmatrix} y_1 \\ y_2 \\ \vdots \\ f(x, \mathbf{y}) \end{bmatrix} \tag{7.5b}$$

A numerical solution of differential equations is essentially a table of x- and $\mathbf{y}$-values listed at discrete intervals of x.

7.2   Euler's Method

Euler's method of solution is conceptually simple. Its basis is the truncated Taylor series of $\mathbf{y}$ about x:

$$\mathbf{y}(x + h) \approx \mathbf{y}(x) + \mathbf{y}'(x)h \tag{7.6}$$

Because Eq. (7.6) predicts $\mathbf{y}$ at $x + h$ from the information available at x, it can be used to move the solution forward in steps of h, starting with the given initial values of x and $\mathbf{y}$.

The error in Eq. (7.6) caused by truncation of the Taylor series is given by Eq. (A4):

$$\mathbf{E} = \frac{1}{2}\mathbf{y}''(\xi)h^2 = \mathcal{O}(h^2), \quad x < \xi < x + h \tag{7.7}$$

A rough idea of the accumulated error $\mathbf{E}_{acc}$ can be obtained by assuming that per-step error is constant over the period of integration. Then after n integration steps covering the interval x_0 to x_n we have

$$\mathbf{E}_{acc} = n\mathbf{E} = \frac{x_n - x_0}{h}\mathbf{E} = \mathcal{O}(h) \tag{7.8}$$

Hence the accumulated error is one order less than the per-step error.

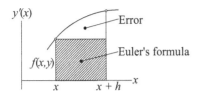

Figure 7.1. Graphical representation of Euler's formula.

Let us now take a look at the graphical interpretation of Euler's equation. For the sake of simplicity, we assume that there is a single dependent variable y, so that the differential equation is $y' = f(x, y)$. The change in the solution y between x and $x + h$ is

$$y(x + h) - y(h) = \int_x^{x+h} y' \, dx = \int_x^{x+h} f(x, y)dx$$

which is the area of the panel under the $y'(x)$ plot, shown in Figure 7.1. Euler's formula approximates this area by the area of the cross-hatched rectangle. The area between the rectangle and the plot represents the truncation error. Clearly, the truncation error is proportional to the slope of the plot; that is, proportional to $(y')' = y''(x)$.

Euler's method is seldom used in practice because of its computational inefficiency. Suppressing the truncation error to an acceptable level requires a very small h, resulting in many integration steps accompanied by an increase in the roundoff error. The value of the method lies mainly in its simplicity, which facilitates the discussion of certain important topics, such as stability.

■ euler

This function implements Euler's method of integration. It can handle any number of first-order differential equations. The user is required to supply the function F(x,y) that specifies the differential equations in the form of the array

$$\mathbf{F}(x, \mathbf{y}) = \begin{bmatrix} y_0' \\ y_1' \\ y_2' \\ \vdots \end{bmatrix}$$

The function returns the arrays X and Y that contain the values of x and $\mathbf{y}$ at intervals h.

```
## module euler
''' X,Y = integrate(F,x,y,xStop,h).
    Euler's method for solving the
    initial value problem {y}' = {F(x,{y})}, where
    {y} = {y[0],y[1],...y[n-1]}.
    x,y   = initial conditions
    xStop = terminal value of x
    h     = increment of x used in integration
```

```
      F     = user-supplied function that returns the
              array F(x,y) = {y'[0],y'[1],...,y'[n-1]}.
'''

import numpy as np
def integrate(F,x,y,xStop,h):
    X = []
    Y = []
    X.append(x)
    Y.append(y)
    while x < xStop:
        h = min(h,xStop - x)
        y = y + h*F(x,y)
        x = x + h
        X.append(x)
        Y.append(y)
    return np.array(X),np.array(Y)
```

■ printSoln

We use this function to print X and Y obtained from numerical integration. The amount of data is controlled by the parameter freq. For example, if freq = 5, every fifth integration step would be displayed. If freq = 0, only the initial and final values will be shown.

```
## module printSoln
''' printSoln(X,Y,freq).
    Prints X and Y returned from the differential
    equation solvers using printout frequency 'freq'.
        freq = n prints every nth step.
        freq = 0 prints initial and final values only.
'''
def printSoln(X,Y,freq):

    def printHead(n):
        print("\n        x   ",end=" ")
        for i in range (n):
            print("      y[",i,"] ",end=" ")
        print()

    def printLine(x,y,n):
        print("{:13.4e}".format(x),end=" ")
        for i in range (n):
            print("{:13.4e}".format(y[i]),end=" ")
        print()
```

```
m = len(Y)
try: n = len(Y[0])
except TypeError: n = 1
if freq == 0: freq = m
printHead(n)
for i in range(0,m,freq):
    printLine(X[i],Y[i],n)
if i != m - 1: printLine(X[m - 1],Y[m - 1],n)
```

EXAMPLE 7.1

Integrate the initial value problem

$$y' + 4y = x^2 \qquad y(0) = 1$$

in steps of $h = 0.01$ from $x = 0$ to 0.03. Also compute the analytical solution

$$y = \frac{31}{32}e^{-4x} + \frac{1}{4}x^2 - \frac{1}{8}x + \frac{1}{32}$$

and the accumulated truncation error at each step.

Solution. It is convenient to use the notation

$$x_i = ih \qquad y_i = y(x_i)$$

so that Euler's formula takes the form

$$y_{i+1} = y_i + y_i'h$$

where

$$y_i' = x_i^2 - 4y_i$$

Step 1. ($x_0 = 0$ to $x_1 = 0.01$):

$$y_0 = 0$$
$$y_0' = x_0^2 - 4y_0 = 0^2 - 4(1) = -4$$
$$y_1 = y_0 + y_0'h = 1 + (-4)(0.01) = 0.96$$
$$(y_1)_{\text{exact}} = \frac{31}{32}e^{-4(0.01)} + \frac{1}{4}0.01^2 - \frac{1}{8}0.01 + \frac{1}{32} = 0.9608$$
$$E_{\text{acc}} = 0.96 - 0.9608 = -0.0008$$

Step 2. ($x_1 = 0.01$ to $x_2 = 0.02$):

$$y_1' = x_1^2 - 4y_1 = 0.01^2 - 4(0.96) = -3.840$$
$$y_2 = y_1 + y_1'h = 0.96 + (-3.840)(0.01) = 0.9216$$
$$(y_2)_{\text{exact}} = \frac{31}{32}e^{-4(0.02)} + \frac{1}{4}0.02^2 - \frac{1}{8}0.02 + \frac{1}{32} = 0.9231$$
$$E_{\text{acc}} = 0.9216 - 0.9231 = -0.0015$$

Step 3. $(x_2 = 0.02 \text{ to } x_3 = 0.03)$:

$$y_2' = x_2^2 - 4y_2 = 0.02^2 - 4(0.9216) = -3.686$$

$$y_3 = y_2 + y_2'h = 0.9216 + (-3.686)(0.01) = 0.8847$$

$$(y_3)_{\text{exact}} = \frac{31}{32}e^{-4(0.03)} + \frac{1}{4}0.03^2 - \frac{1}{8}0.03 + \frac{1}{32} = 0.8869$$

$$E_{\text{acc}} = 0.8847 - 0.8869 = -0.0022$$

We note that the magnitude of the per-step error is roughly constant at 0.008. Thus after 10 integration steps the accumulated error would be approximately 0.08, thereby reducing the solution to one significant figure accuracy. After 100 steps all significant figures would be lost.

EXAMPLE 7.2

Integrate the initial value problem

$$y'' = -0.1y' - x \qquad y(0) = 0 \qquad y'(0) = 1$$

from $x = 0$ to 2 with Euler's method using $h = 0.05$. Plot the computed y together with the analytical solution,

$$y = 100x - 5x^2 + 990(e^{-0.1x} - 1)$$

Solution. With the notation $y_0 = y$ and $y_1 = y'$ the equivalent first-order equations and the initial conditions are.

$$\mathbf{F}(x, \mathbf{y}) = \begin{bmatrix} y_0' \\ y_1' \end{bmatrix} = \begin{bmatrix} y_1 \\ -0.1y_1 - x \end{bmatrix} \qquad \mathbf{y}(0) = \begin{bmatrix} 0 \\ 1 \end{bmatrix}$$

Here is a program that uses the function `euler`:

```
#!/usr/bin/python
## example7_2
import numpy as np
from euler import *
import matplotlib.pyplot as plt

def F(x,y):
    F = np.zeros(2)
    F[0] = y[1]
    F[1] = -0.1*y[1] - x
    return F

x = 0.0                      # Start of integration
xStop = 2.0                  # End of integration
y = np.array([0.0, 1.0])     # Initial values of {y}
h = 0.05                     # Step size
```

```
X,Y = integrate(F,x,y,xStop,h)
yExact = 100.0*X - 5.0*X**2 + 990.0*(np.exp(-0.1*X) - 1.0)
plt.plot(X,Y[:,0],'o',X,yExact,'-')
plt.grid(True)
plt.xlabel('x'); plt.ylabel('y')
plt.legend(('Numerical','Exact'),loc=0)
plt.show()
input("Press return to exit")
```

The resulting plot is

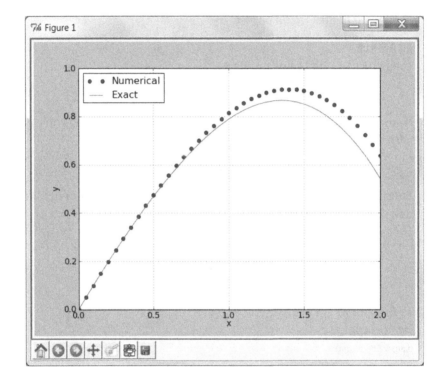

The initial portion of the plot is almost a straight line. Because the truncation error in the numerical solution is proportional to y'', the discrepancy between the two solutions is small. As the curvature of the plot increases, so does the truncation error.

You may want to see the effect of changing h by running the program with $h = 0.025$. Doing so should halve the truncation error.

7.3   Runge-Kutta Methods

Euler's method is classified as a first-order method because its cumulative truncation error behaves as $\mathcal{O}(h)$. Its basis was the truncated Taylor series

$$\mathbf{y}(x + h) = \mathbf{y}(x) + \mathbf{y}'(x)h$$

The accuracy of numerical integration can be greatly improved by keeping more terms of the series. Thus a nth order method would use the truncated Taylor series

$$\mathbf{y}(x+h) = \mathbf{y}(x) + \mathbf{y}'(x)h + \frac{1}{2!}\mathbf{y}''(x)h^2 + \cdots + \frac{1}{n!}\mathbf{y}^{(n)}(x)h^n$$

But now we must derive the expressions for $\mathbf{y}'', \mathbf{y}''', \ldots \mathbf{y}^{(n)}$ by repeatedly differentiating $\mathbf{y}' = \mathbf{F}(x, \mathbf{y})$ and write subroutines to evaluate them. This extra work can be avoided by using *Runge-Kutta methods* that are also based on truncated Taylor series, but do not require computation of higher derivatives of $\mathbf{y}(x)$.

Second-Order Runge-Kutta Method

To arrive at the second-order Runge-Kutta method, we assume an integration formula of the form

$$\mathbf{y}(x+h) = \mathbf{y}(x) + c_0\mathbf{F}(x, \mathbf{y})h + c_1\mathbf{F}\left[x + ph, \mathbf{y} + qh\mathbf{F}(x, \mathbf{y})\right]h \tag{a}$$

and attempt to find the parameters c_0, c_1, p, and q by matching Eq. (a) to the Taylor series:

$$\mathbf{y}(x+h) = \mathbf{y}(x) + \mathbf{y}'(x)h + \frac{1}{2!}\mathbf{y}''(x)h^2$$

$$= \mathbf{y}(x) + \mathbf{F}(x, \mathbf{y})h + \frac{1}{2}\mathbf{F}'(x, \mathbf{y})h^2 \tag{b}$$

Noting that

$$\mathbf{F}'(x, \mathbf{y}) = \frac{\partial \mathbf{F}}{\partial x} + \sum_{i=0}^{n-1} \frac{\partial \mathbf{F}}{\partial y_i}y_i' = \frac{\partial \mathbf{F}}{\partial x} + \sum_{i=0}^{n-1} \frac{\partial \mathbf{F}}{\partial y_i}F_i(x, \mathbf{y})$$

where n is the number of first-order equations, Eq.(b) can be written as

$$\mathbf{y}(x+h) = \mathbf{y}(x) + \mathbf{F}(x, \mathbf{y})h + \frac{1}{2}\left(\frac{\partial \mathbf{F}}{\partial x} + \sum_{i=0}^{n-1} \frac{\partial \mathbf{F}}{\partial y_i}F_i(x, \mathbf{y})\right)h^2) \tag{c}$$

Returning to Eq. (a), we can rewrite the last term by applying Taylor series in several variables,

$$\mathbf{F}\left[x + ph, \mathbf{y} + qh\mathbf{F}(x, \mathbf{y})\right] = \mathbf{F}(x, \mathbf{y}) + \frac{\partial \mathbf{F}}{\partial x}ph + qh\sum_{i=1}^{n-1} \frac{\partial \mathbf{F}}{\partial y_i}F_i(x, \mathbf{y})$$

so that Eq. (a) becomes

$$\mathbf{y}(x+h) = \mathbf{y}(x) + (c_0 + c_1)\mathbf{F}(x, \mathbf{y})h + c_1\left[\frac{\partial \mathbf{F}}{\partial x}ph + qh\sum_{i=1}^{n-1} \frac{\partial \mathbf{F}}{\partial y_i}F_i(x, \mathbf{y})\right]h \tag{d}$$

Comparing Eqs. (c) and (d), we find that they are identical if

$$c_0 + c_1 = 1 \qquad c_1 p = \frac{1}{2} \qquad c_1 q = \frac{1}{2} \tag{e}$$

Because Eqs. (e) represent three equation in four unknown parameters, we can assign any value to one of the parameters. Some of the popular choices and the names associated with the resulting formulas are as follows:

$$
\begin{array}{llll}
c_0 = 0 & c_1 = 1 & p = 1/2 & q = 1/2 \quad \text{Modified Euler's method} \\
c_0 = 1/2 & c_1 = 1/2 & p = 1 & q = 1 \quad \text{Heun's method} \\
c_0 = 1/3 & c_1 = 2/3 & p = 3/4 & q = 3/4 \quad \text{Ralston's method}
\end{array}
$$

All these formulas are classified as second-order Runge-Kutta methods, with no formula having a numerical superiority over the others. Choosing the *modified Euler's method*, substitution of the corresponding parameters into Eq. (a) yields

$$
\mathbf{y}(x + h) = \mathbf{y}(x) + \mathbf{F}\left[x + \frac{h}{2}, \mathbf{y} + \frac{h}{2}\mathbf{F}(x, \mathbf{y})\right] h \tag{f}
$$

This integration formula can conveniently evaluated by the following sequence of operations:

$$
\mathbf{K}_0 = h\mathbf{F}(x, \mathbf{y})
$$

$$
\mathbf{K}_1 = h\mathbf{F}\left(x + \frac{h}{2}, \mathbf{y} + \frac{1}{2}\mathbf{K}_0\right) \tag{7.9}
$$

$$
\mathbf{y}(x + h) = \mathbf{y}(x) + \mathbf{K}_1
$$

Second-order methods are not popular in computer application. Most programmers prefer integration formulas of order four, which achieve a given accuracy with less computational effort.

$y'(x)$

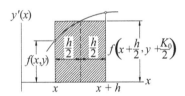

Figure 7.2. Graphical representation of modified Euler's formula.

Figure 7.2 displays the graphical interpretation of the modified Euler's formula for a single differential equation $y' = f(x, y)$. The first of Eqs. (7.9) yields an estimate of y at the midpoint of the panel by Euler's formula: $y(x + h/2) = y(x) + f(x, y)h/2 = y(x) + K_0/2$. The second equation then approximates the area of the panel by the area K_1 of the cross-hatched rectangle. The error here is proportional to the curvature $(y')'' = y'''$ of the plot.

Fourth-Order Runge-Kutta Method

The fourth-order Runge-Kutta method is obtained from the Taylor series along the same lines as the second-order method. Because the derivation is rather long and not very instructive, we shall skip it. The final form of the integration formula again depends on the choice of the parameters; that is, there is no unique Runge-Kutta fourth-order formula. The most popular version, which is known simply as the *Runge-Kutta method*, entails the following sequence of operations:

$$\mathbf{K}_0 = h\mathbf{F}(x, \mathbf{y})$$

$$\mathbf{K}_1 = h\mathbf{F}\left(x + \frac{h}{2}, \mathbf{y} + \frac{\mathbf{K}_0}{2}\right)$$

$$\mathbf{K}_2 = h\mathbf{F}\left(x + \frac{h}{2}, \mathbf{y} + \frac{\mathbf{K}_1}{2}\right) \tag{7.10}$$

$$\mathbf{K}_3 = h\mathbf{F}(x + h, \mathbf{y} + \mathbf{K}_2)$$

$$\mathbf{y}(x + h) = \mathbf{y}(x) + \frac{1}{6}(\mathbf{K}_0 + 2\mathbf{K}_1 + 2\mathbf{K}_2 + \mathbf{K}_3)$$

The main drawback of this method is that is does not lend itself to an estimate of the truncation error. Therefore, we must guess the integration step size h or determine it by trial and error. In contrast, the so-called *adaptive methods* can evaluate the truncation error in each integration step and adjust the value of h accordingly (but at a higher cost of computation). One such adaptive method is introduced in the next section.

■ run_kut4

The function `integrate` in this module implements the Runge-Kutta method of order four. The user must provide `integrate` with the function `F(x,y)` that defines the first-order differential equations $\mathbf{y}' = \mathbf{F}(x, \mathbf{y})$.

```
## module run_kut4
''' X,Y = integrate(F,x,y,xStop,h).
    4th-order Runge-Kutta method for solving the
    initial value problem {y}' = {F(x,{y})}, where
    {y} = {y[0],y[1],...y[n-1]}.
    x,y   = initial conditions
    xStop = terminal value of x
    h     = increment of x used in integration
    F     = user-supplied function that returns the
            array F(x,y) = {y'[0],y'[1],...,y'[n-1]}.
'''
import numpy as np
def integrate(F,x,y,xStop,h):

    def run_kut4(F,x,y,h):
        K0 = h*F(x,y)
        K1 = h*F(x + h/2.0, y + K0/2.0)
        K2 = h*F(x + h/2.0, y + K1/2.0)
        K3 = h*F(x + h, y + K2)
        return (K0 + 2.0*K1 + 2.0*K2 + K3)/6.0
```

```
X = []
Y = []
X.append(x)
Y.append(y)
while x < xStop:
    h = min(h,xStop - x)
    y = y + run_kut4(F,x,y,h)
    x = x + h
    X.append(x)
    Y.append(y)
return np.array(X),np.array(Y)
```

EXAMPLE 7.3
Use the second-order Runge-Kutta method to integrate

$$y' = \sin y \qquad y(0) = 1$$

from $x = 0$ to 0.5 in steps of $h = 0.1$. Keep four decimal places in the computations.

Solution. In this problem we have

$$F(x, y) = \sin y$$

so that the integration formulas in Eqs. (7.9) are

$$K_0 = hF(x, y) = 0.1 \sin y$$

$$K_1 = hF\left(x + \frac{h}{2}, y + \frac{1}{2}K_0\right) = 0.1 \sin\left(y + \frac{1}{2}K_0\right)$$

$$y(x + h) = y(x) + K_1$$

Noting that $y(0) = 1$, the integration then proceeds as follows:

$$K_0 = 0.1 \sin 1.0000 = 0.0841$$

$$K_1 = 0.1 \sin\left(1.0000 + \frac{0.0841}{2}\right) = 0.0863$$

$$y(0.1) = 1.0 + 0.0863 = 1.0863$$

$$K_0 = 0.1 \sin 1.0863 = 0.0885$$

$$K_1 = 0.1 \sin\left(1.0863 + \frac{0.0885}{2}\right) = 0.0905$$

$$y(0.2) = 1.0863 + 0.0905 = 1.1768$$

and so on. A summary of the computations is shown in the following table.

x	y	K_0	K_1
0.0	1.0000	0.0841	0.0863
0.1	1.0863	0.0885	0.0905
0.2	1.1768	0.0923	0.0940
0.3	1.2708	0.0955	0.0968
0.4	1.3676	0.0979	0.0988
0.5	1.4664		

The exact solution can be shown to be

$$x(y) = \ln(\csc y - \cot y) + 0.604582$$

which yields $x(1.4664) = 0.5000$. Therefore, up to this point the numerical solution is accurate to four decimal places. However, it is unlikely that this precision would be maintained if we were to continue the integration. Because the errors (due to truncation and roundoff) tend to accumulate, longer integration ranges require better integration formulas and more significant figures in the computations.

EXAMPLE 7.4
Integrate the initial value problem

$$y'' = -0.1y' - x \qquad y(0) = 0 \qquad y'(0) = 1$$

with the fourth-order Runge-Kutta method from $x = 0$ to 2 in increments of $h = 0.2$. Plot the computed y together with the analytical solution

$$y = 100x - 5x^2 + 990(e^{-0.1x} - 1)$$

(This problem was solved by Euler's method in Example 7.2.)

Solution. Letting $y_0 = y$ and $y_1 = y'$, the equivalent first-order equations are

$$\mathbf{F}(x, \mathbf{y}) = \begin{bmatrix} y_0' \\ y_1' \end{bmatrix} = \begin{bmatrix} y_1 \\ -0.1y_1 - x \end{bmatrix}$$

Except for two statements, the program shown next is identical to that used in Example 7.2.

```
#!/usr/bin/python
## example7_4
import numpy as np
from printSoln import *
from run_kut4 import *
import matplotlib.pyplot as plt

def F(x,y):
    F = np.zeros(2)
    F[0] = y[1]
```

```
    F[1] = -0.1*y[1] - x
    return F

x = 0.0                  # Start of integration
xStop = 2.0              # End of integration
y = np.array([0.0, 1.0])  # Initial values of {y}
h = 0.2                  # Step size

X,Y = integrate(F,x,y,xStop,h)
yExact = 100.0*X - 5.0*X**2 + 990.0*(np.exp(-0.1*X) - 1.0)
plt.plot(X,Y[:,0],'o',X,yExact,'-')
plt.grid(True)
plt.xlabel('x'); plt.ylabel('y')
plt.legend(('Numerical','Exact'),loc=0)
plt.show()
input("Press return to exit"))
```

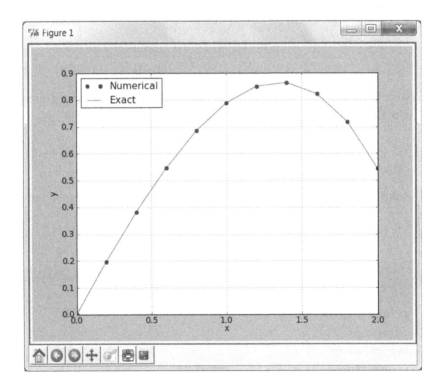

Comparing this plot with that in Example 7.2, we see the clear superiority of the fourth-order Runge-Kutta method over Euler's method. In this problem, the two methods involve about the same amount of computation. Euler's method requires one evaluation of $\mathbf{F}(x, \mathbf{y})$ per step, whereas the Runge-Kutta method involves four evaluations. Yet Euler's method has four times as many steps. Considering the large difference in the accuracy of the results, we conclude that the fourth-order Runge-Kutta method offers a much "bigger bang for a buck."

EXAMPLE 7.5
Use the fourth-order Runge-Kutta method to integrate

$$y' = 3y - 4e^{-x} \qquad y(0) = 1$$

from $x = 0$ to 10 in steps of $h = 0.1$. Compare the result with the analytical solution $y = e^{-x}$.

Solution. We used the program shown next. Recalling that run_kut4 expects y to be an array, we specified the initial value as y = np.array([1.0]) rather than y = 1.0.

```
#!/usr/bin/python
## example7_5
import numpy as np
from run_kut4 import *
from printSoln import *
from math import exp

def F(x,y):
    F = np.zeros(1)
    F[0] = 3.0*y[0] - 4.0*exp(-x)
    return F

x = 0.0                 # Start of integration
xStop = 10.0            # End of integration
y = np.array([1.0])     # Initial values of {y}
h = 0.1                 # Step size
freq = 20               # Printout frequency

X,Y = integrate(F,x,y,xStop,h)
printSoln(X,Y,freq)
input("\nPress return to exit")
```

Running the program produced the following output (every 20th integration step is shown):

x	y[0]
0.0000e+000	1.0000e+000
2.0000e+000	1.3250e-001
4.0000e+000	-1.1237e+000
6.0000e+000	-4.6056e+002
8.0000e+000	-1.8575e+005
1.0000e+001	-7.4912e+007

It is clear that something went wrong. According to the analytical solution, y should approach zero with increasing x, but the output shows the opposite trend: After an initial decrease, the magnitude of y increases dramatically. The explanation is found by taking a closer look at the analytical solution. The general solution of the given differential equation is

$$y = Ce^{3x} + e^{-x}$$

which can be verified by substitution. The initial condition $y(0) = 1$ yields $C = 0$, so that the solution to the problem is indeed $y = e^{-x}$.

The cause of the trouble in the numerical solution is the dormant term Ce^{3x}. Suppose that the initial condition contains a small error ε, so that we have $y(0) = 1 + \varepsilon$. This changes the analytical solution to

$$y = \varepsilon e^{3x} + e^{-x}$$

We now see that the term containing the error ε becomes dominant as x is increased. Because errors inherent in the numerical solution have the same effect as small changes in initial conditions, we conclude that our numerical solution is the victim of *numerical instability* due to sensitivity of the solution to initial conditions. The lesson is not to blindly trust the results of numerical integration.

EXAMPLE 7.6

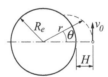

A spacecraft is launched at the altitude $H = 772$ km above sea level with the speed $v_0 = 6700$ m/s in the direction shown. The differential equations describing the motion of the spacecraft are

$$\ddot{r} = r\dot{\theta}^2 - \frac{GM_e}{r^2} \qquad \ddot{\theta} = -\frac{2\dot{r}\dot{\theta}}{r}$$

where r and θ are the polar coordinates of the spacecraft. The constants involved in the motion are

$$G = 6.672 \times 10^{-11} \text{ m}^3 \text{ kg}^{-1}\text{s}^{-2} = \text{universal gravitational constant}$$

$$M_e = 5.9742 \times 10^{24} \text{ kg} = \text{mass of the earth}$$

$$R_e = 6378.14 \text{ km} = \text{radius of the earth at sea level}$$

(1) Derive the first-order differential equations and the initial conditions of the form $\dot{\mathbf{y}} = \mathbf{F}(t, \mathbf{y})$, $\mathbf{y}(0) = \mathbf{b}$. (2) Use the fourth-order Runge-Kutta method to integrate the equations from the time of launch until the spacecraft hits the earth. Determine θ at the impact site.

Solution of Part (1). We have

$$GM_e = \left(6.672 \times 10^{-11}\right)\left(5.9742 \times 10^{24}\right) = 3.9860 \times 10^{14} \text{ m}^3 \text{ s}^{-2}$$

Letting

$$\mathbf{y} = \begin{bmatrix} y_0 \\ y_1 \\ y_2 \\ y_3 \end{bmatrix} = \begin{bmatrix} r \\ \dot{r} \\ \theta \\ \dot{\theta} \end{bmatrix}$$

the equivalent first-order equations become

$$\mathbf{F}(t, \mathbf{y}) = \begin{bmatrix} \dot{y}_0 \\ \dot{y}_1 \\ \dot{y}_2 \\ \dot{y}_3 \end{bmatrix} = \begin{bmatrix} y_1 \\ y_0 y_3^2 - 3.9860 \times 10^{14}/y_0^2 \\ y_3 \\ -2y_1 y_3/y_0 \end{bmatrix}$$

and the initial conditions are

$$r(0) = R_e + H = (6378.14 + 772) \times 10^3 = 7.15014 \times 10^6 \text{ m}$$

$$\dot{r}(0) = 0$$

$$\theta(0) = 0$$

$$\dot{\theta}(0) = v_0/r(0) = (6700)/(7.15014 \times 10^6) = 0.937045 \times 10^{-3} \text{ rad/s}$$

Therefore,

$$\mathbf{y}(0) = \begin{bmatrix} 7.\,15014 \times 10^6 \\ 0 \\ 0 \\ 0.937045 \times 10^{-3} \end{bmatrix}$$

Solution of Part (2). The program used for numerical integration is listed next. Note that the independent variable t is denoted by $\mathbf{x}$. The period of integration xStop (the time when the spacecraft hits) was estimated from a previous run of the program.

```
#!/usr/bin/python
## example7_6
import numpy as np
from run_kut4 import *
from printSoln import *

def F(x,y):
    F = np.zeros(4)
    F[0] = y[1]
    F[1] = y[0]*(y[3]**2) - 3.9860e14/(y[0]**2)
    F[2] = y[3]
```

```
    F[3] = -2.0*y[1]*y[3]/y[0]
    return F

x = 0.0
xStop = 1200.0
y = np.array([7.15014e6, 0.0, 0.0, 0.937045e-3])
h = 50.0
freq = 2

X,Y = integrate(F,x,y,xStop,h)
printSoln(X,Y,freq)
input("\nPress return to exit")
```

Here is the output:

x	y[0]	y[1]	y[2]	y[3]
0.0000e+000	7.1501e+006	0.0000e+000	0.0000e+000	9.3704e-004
1.0000e+002	7.1426e+006	-1.5173e+002	9.3771e-002	9.3904e-004
2.0000e+002	7.1198e+006	-3.0276e+002	1.8794e-001	9.4504e-004
3.0000e+002	7.0820e+006	-4.5236e+002	2.8292e-001	9.5515e-004
4.0000e+002	7.0294e+006	-5.9973e+002	3.7911e-001	9.6951e-004
5.0000e+002	6.9622e+006	-7.4393e+002	4.7697e-001	9.8832e-004
6.0000e+002	6.8808e+006	-8.8389e+002	5.7693e-001	1.0118e-003
7.0000e+002	6.7856e+006	-1.0183e+003	6.7950e-001	1.0404e-003
8.0000e+002	6.6773e+006	-1.1456e+003	7.8520e-001	1.0744e-003
9.0000e+002	6.5568e+006	-1.2639e+003	8.9459e-001	1.1143e-003
1.0000e+003	6.4250e+006	-1.3708e+003	1.0083e+000	1.1605e-003
1.1000e+003	6.2831e+006	-1.4634e+003	1.1269e+000	1.2135e-003
1.2000e+003	6.1329e+006	-1.5384e+003	1.2512e+000	1.2737e-003

The spacecraft hits the earth when r equals $R_e = 6.378\,14 \times 10^6$ m. This occurs between $t = 1000$ and 1100 s. A more accurate value of t can be obtained by polynomial interpolation. If no great precision is needed, linear interpolation will do. Letting $1000 + \Delta t$ be the time of impact, we can write

$$r(1000 + \Delta t) = R_e$$

Expanding r in a two-term Taylor series, we get

$$r(1000) + r'(1000)\Delta t = R_e$$
$$6.4250 \times 10^6 + \left(-1.3708 \times 10^3\right) x = 6378.14 \times 10^3$$

from which

$$\Delta t = 34.184 \text{ s}$$

The coordinate θ of the impact site can be estimated in a similar manner. Using again two terms of the Taylor series, we have

$$\theta(1000 + \Delta t) = \theta(1000) + \theta'(1000)\Delta t$$

$$= 1.0083 + \left(1.1605 \times 10^{-3}\right)(34.184)$$

$$= 1.0480 \text{ rad} = 60.00°$$

PROBLEM SET 7.1

1. Given

$$y' + 4y = x^2 \qquad y(0) = 1$$

   compute $y(0.03)$ using two steps of the second-order Runge-Kutta method. Compare the result with the analytical solution given in Example 7.1.
2. Solve Prob. 1 with one step of the fourth-order Runge-Kutta method.
3. Integrate

$$y' = \sin y \qquad y(0) = 1$$

   from $x = 0$ to 0.5 with Euler's method using $h = 0.1$. Compare the result with Example 7.3.
4. ■ Verify that the problem

$$y' = y^{1/3} \qquad y(0) = 0$$

   has two solutions: $y = 0$ and $y = (2x/3)^{3/2}$. Which of the solutions would be reproduced by numerical integration if the initial condition is set at (a) $y = 0$ and (b) $y = 10^{-16}$? Verify your conclusions by integrating with any numerical method.
5. Convert the following differential equations into first-order equations of the form $\mathbf{y}' = \mathbf{F}(x, \mathbf{y})$:

   (a) $\quad \ln y' + y = \sin x$
   (b) $\quad y''y - xy' - 2y^2 = 0$
   (c) $\quad y^{(4)} - 4y''\sqrt{1 - y^2} = 0$
   (d) $\quad (y'')^2 = |32y'x - y^2|$

6. In the following sets of coupled differential equations t is the independent variable. Convert these equations into first-order equations of the form $\dot{\mathbf{y}} = \mathbf{F}(t, \mathbf{y})$:

   (a) $\quad \ddot{y} = x - 2y \qquad\qquad\qquad \ddot{x} = y - x$
   (b) $\quad \ddot{y} = -y\left(\dot{y}^2 + \dot{x}^2\right)^{1/4} \qquad \ddot{x} = -x\left(\dot{y}^2 + \dot{x}\right)^{1/4} - 32$
   (c) $\quad \ddot{y}^2 + t\sin y = 4\dot{x} \qquad\qquad x\ddot{x} + t\cos y = 4\dot{y}$

7. ■ The differential equation for the motion of a simple pendulum is

$$\frac{d^2\theta}{dt^2} = -\frac{g}{L}\sin\theta$$

where

$$\theta = \text{angular displacement from the vertical}$$

$$g = \text{gravitational acceleration}$$

$$L = \text{length of the pendulum}$$

With the transformation $\tau = t\sqrt{g/L}$ the equation becomes

$$\frac{d^2\theta}{d\tau^2} = -\sin\theta$$

Use numerical integration to determine the period of the pendulum if the amplitude is $\theta_0 = 1$ rad. Note that for small amplitudes ($\sin\theta \approx \theta$) the period is $2\pi\sqrt{L/g}$.

8. ■ A skydiver of mass m in a vertical free fall experiences an aerodynamic drag force $F_D = c_D\dot{y}^2$, where y is measured downward from the start of the fall. The differential equation describing the fall is

$$\ddot{y} = g - \frac{c_D}{m}\dot{y}^2$$

Determine the time of a 5000 m fall. Use $g = 9.80665$ m/s^2, $C_D = 0.2028$ kg/m, and $m = 80$ kg.

9. ■

$$P(t) = \begin{cases} 10t\text{ N} & \text{when } t < 2\text{ s} \\ 20\text{ N} & \text{when } t \geq 2\text{ s} \end{cases}$$

The spring-mass system is at rest when the force $P(t)$ is applied, where

The differential equation of the ensuing motion is

$$\ddot{y} = \frac{P(t)}{m} - \frac{k}{m}y$$

Determine the maximum displacement of the mass. Use $m = 2.5$ kg and $k = 75$ N/m.

10. ■

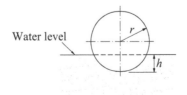

Water level

The equilibrium position of the floating cylinder is $h = r$. If the cylinder is displaced to the position $h = 1.5r$ and released, the differential equation describing its motion is

$$\ddot{y} = \frac{2}{\pi}\left[\tan^{-1}\frac{1-y}{\sqrt{2y-y^2}} + (1-y)\sqrt{2y-y^2}\right]$$

where $y = h/r$. Plot h/r from $t = 0$ to 6 s. Use the plot to estimate the period of the motion.

11. ■ Solve the differential equation

$$y' = \sin xy$$

with the initial conditions $y(0) = 2.0, 2.5, 3.0$, and 3.5. Plot the solutions in the range $0 \le x \le 10$.

12. ■

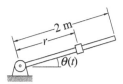

The system consisting of a sliding mass and a guide rod is at rest with the mass at $r = 0.75$ m. At time $t = 0$ a motor is turned on that imposes the motion $\theta(t) = (\pi/12)\cos\pi t$ on the rod. The differential equation describing the resulting motion of the slider is

$$\ddot{r} = \left(\frac{\pi^2}{12}\right)^2 r\sin^2\pi t - g\sin\left(\frac{\pi}{12}\cos\pi t\right)$$

Determine the time when the slider reaches the tip of the rod. Use $g = 9.806\,65$ m/s^2.

13. ■

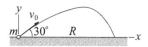

A ball of mass $m = 0.25$ kg is launched with the velocity $v_0 = 50$ m/s in the direction shown. Assuming that the aerodynamic drag force acting on the ball is $F_D = C_D v^{3/2}$, the differential equations describing the motion are

$$\ddot{x} = -\frac{C_D}{m}\dot{x}v^{1/2} \qquad \ddot{y} = -\frac{C_D}{m}\dot{y}v^{1/2} - g$$

where $v = \sqrt{\dot{x}^2 + \dot{y}^2}$. Determine the time of flight and the range R. Use $C_D = 0.03$ kg/(m·s)$^{1/2}$ and $g = 9.806\,65$ m/s^2.

14. ■ The differential equation describing the angular position θ of a mechanical arm is

$$\ddot{\theta} = \frac{a(b - \theta) - \theta\dot{\theta}^2}{1 + \theta^2}$$

where $a = 100 \text{ s}^{-2}$ and $b = 15$. If $\theta(0) = 2\pi$ and $\dot{\theta}(0) = 0$, compute θ and $\dot{\theta}$ when $t = 0.5$ s.

15. ■

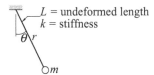

L = undeformed length
k = stiffness

The mass m is suspended from an elastic cord with an extensional stiffness k and undeformed length L. If the mass is released from rest at $\theta = 60°$ with the cord unstretched, find the length r of the cord when the position $\theta = 0$ is reached for the first time. The differential equations describing the motion are

$$\ddot{r} = r\dot{\theta}^2 + g\cos\theta - \frac{k}{m}(r - L)$$

$$\ddot{\theta} = \frac{-2\dot{r}\dot{\theta} - g\sin\theta}{r}$$

Use $g = 9.80665 \text{ m/s}^2$, $k = 40 \text{ N/m}$, $L = 0.5$ m, and $m = 0.25$ kg.

16. ■ Solve Prob. 15 if the mass is released from the position $\theta = 60°$ with the cord stretched by 0.075 m.

17. ■

Consider the mass-spring system where dry friction is present between the block and the horizontal surface. The frictional force has a constant magnitude μmg (μ is the coefficient of friction) and always opposes the motion. The differential equation for the motion of the block can be expressed as

$$\ddot{y} = -\frac{k}{m}y - \mu g\frac{\dot{y}}{|\dot{y}|}$$

where y is measured from the position where the spring is unstretched. If the block is released from rest at $y = y_0$, verify by numerical integration that the next positive peak value of y is $y_0 - 4\mu mg/k$ (this relationship can be derived analytically). Use $k = 3000 \text{ N/m}$, $m = 6$ kg, $\mu = 0.5$, $g = 9.80665 \text{ m/s}^2$, and $y_0 = 0.1$ m.

18. ■ Integrate the following problems from $x = 0$ to 20 and plot y vs. x:

$$\text{(a)} \quad y'' + 0.5(y^2 - 1) + y = 0 \qquad y(0) = 1 \qquad y'(0) = 0$$
$$\text{(b)} \quad y'' = y \cos 2x \qquad\qquad\qquad\quad y(0) = 0 \qquad y'(0) = 1$$

These differential equations arise in nonlinear vibration analysis.

19. ■ The solution of the problem

$$y'' + \frac{1}{x}y' + \left(1 - \frac{1}{x^2}\right)y \qquad y(0) = 0 \qquad y'(0) = 1$$

is the Bessel function $J_1(x)$. Use numerical integration to compute $J_1(5)$ and compare the result with $-0.327\,579$, the value listed in mathematical tables. *Hint:* To avoid singularity at $x = 0$, start the integration at $x = 10^{-12}$.

20. ■ Consider the following initial value problem:

$$y'' = 16.81y \qquad y(0) = 1.0 \qquad y'(0) = -4.1$$

(a) Derive the analytical solution. (b) Do you anticipate difficulties in numerical solution of this problem? (c) Try numerical integration from $x = 0$ to 8 to see if your concerns were justified.

21. ■

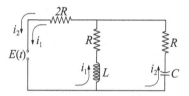

Kirchoff's equations for the circuit shown are

$$L\frac{di_1}{dt} + Ri_1 + 2R(i_1 + i_2) = E(t) \tag{a}$$

$$\frac{q_2}{C} + Ri_2 + 2R(i_2 + i_1) = E(t) \tag{b}$$

where i_1 and i_2 are the loop currents, and q_2 is the charge of the condenser. Differentiating Eq. (b) and substituting the charge-current relationship $dq_2/dt = i_2$, we get

$$\frac{di_1}{dt} = \frac{-3\,Ri_1 - 2\,Ri_2 + E(t)}{L} \tag{c}$$

$$\frac{di_2}{dt} = -\frac{2}{3}\frac{di_1}{dt} - \frac{i_2}{3\,RC} + \frac{1}{3\,R}\frac{dE}{dt} \tag{d}$$

We could substitute di_1/dt from Eq. (c) into Eq. (d), so that the latter would assume the usual form $di_2/dt = f(t, i_1, i_2)$, but it is more convenient to leave the equations as they are. Assuming that the voltage source is turned on at time $t = 0$, plot the loop currents i_1 and i_2 from $t = 0$ to 0.05 s. Use $E(t) = 240\sin(120\pi t)$ V, $R = 1.0\,\Omega$, $L = 0.2 \times 10^{-3}$ H, and $C = 3.5 \times 10^{-3}$ F.

22. ■

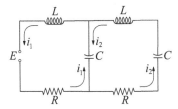

The constant voltage source of the circuit shown is turned on at $t = 0$, causing transient currents i_1 and i_2 in the two loops that last about 0.05 s. Plot these currents from $t = 0$ to 0.05 s, using the following data: $E = 9$ V, $R = 0.25\,\Omega$, $L = 1.2 \times 10^{-3}$ H, and $C = 5 \times 10^{-3}$ F. Kirchoff's equations for the two loops are

$$L\frac{di_1}{dt} + Ri_1 + \frac{q_1 - q_2}{C} = E$$

$$L\frac{di_2}{dt} + Ri_2 + \frac{q_2 - q_1}{C} + \frac{q_2}{C} = 0$$

Two additional equations are the current-charge relationships:

$$\frac{dq_1}{dt} = i_1 \qquad \frac{dq_2}{dt} = i_2$$

23. Print a table of the sine integral

$$\text{Si}(x) = \int_0^x \frac{\sin t}{t}\,dt$$

from $x = 0$ to 3.6 in increments of 0.2.

7.4 Stability and Stiffness

Loosely speaking, a method of numerical integration is said to be stable if the effects of local errors do not accumulate catastrophically; that is, if the global error remains bounded. If the method is unstable, the global error will increase exponentially, eventually causing numerical overflow. Stability has nothing to do with accuracy; in fact, an inaccurate method can be very stable.

Stability is determined by three factors: the differential equations, the method of solution, and the value of the increment h. Unfortunately, it is not easy to determine stability beforehand, unless the differential equation is linear.

Stability of Euler's Method

As a simple illustration of stability, consider the linear problem

$$y' = -\lambda y \qquad y(0) = \beta \tag{7.11}$$

where λ is a positive constant. The analytical solution of this problem is

$$y(x) = \beta e^{-\lambda x}$$

Let us now investigate what happens when we attempt to solve Eq. (7.11) numerically with Euler's formula,

$$y(x + h) = y(x) + hy'(x) \tag{7.12}$$

Substituting $y'(x) = -\lambda y(x)$, we get

$$y(x + h) = (1 - \lambda h)y(x)$$

If $|1 - \lambda h| > 1$, the method is clearly unstable because $|y|$ increases in every integration step. Thus Euler's method is stable only if $|1 - \lambda h| \le 1$, or

$$h \le 2/\lambda \tag{7.13}$$

The results can be extended to a system of n differential equations of the form

$$\mathbf{y}' = -\mathbf{\Lambda}\mathbf{y} \tag{7.14}$$

where $\mathbf{\Lambda}$ is a constant matrix with the positive eigenvalues λ_i, $\;\; i = 1, 2, \ldots, n$. It can be shown that Euler's method of integration is stable if

$$h < 2/\lambda_{\max} \tag{7.15}$$

where $\lambda_{\max}$ is the largest eigenvalue of $\mathbf{\Lambda}$.

Stiffness

An initial value problem is called *stiff* if some terms in the solution vector $\mathbf{y}(x)$ vary much more rapidly with x than others. Stiffness can be easily predicted for the differential equations $\mathbf{y}' = -\mathbf{\Lambda}\mathbf{y}$ with constant coefficient matrix $\mathbf{\Lambda}$. The solution of these equations is $\mathbf{y}(x) = \sum_i C_i \mathbf{v}_i \exp(-\lambda_i x)$, where λ_i are the eigenvalues of $\mathbf{\Lambda}$ and $\mathbf{v}_i$ are the corresponding eigenvectors. It is evident that the problem is stiff if there is a large disparity in the magnitudes of the positive eigenvalues.

Numerical integration of stiff equations requires special care. The step size h needed for stability is determined by the largest eigenvalue $\lambda_{\max}$, even if the terms $\exp(-\lambda_{\max}x)$ in the solution decay very rapidly and become insignificant as we move away from the origin.

For example, consider the differential equation[1]

$$y'' + 1001y' + 1000y = 0 \tag{7.16}$$

Using $y_0 = y$ and $y_1 = y_1$, the equivalent first-order equations are

$$\mathbf{y}' = \begin{bmatrix} y_1 \\ -1000y_0 - 1001y_1 \end{bmatrix}$$

---

[1] This example is taken from C.E. Pearson, *Numerical Methods in Engineering and Science*, van Nostrand and Reinhold, 1986.

In this case

$$\mathbf{\Lambda} = \begin{bmatrix} 0 & -1 \\ 1000 & 1001 \end{bmatrix}$$

The eigenvalues of $\mathbf{\Lambda}$ are the roots of

$$|\mathbf{\Lambda} - \lambda\mathbf{I}| = \begin{vmatrix} -\lambda & -1 \\ 1000 & 1001 - \lambda \end{vmatrix} = 0$$

Expanding the determinant we get

$$-\lambda(1001 - \lambda) + 1000 = 0$$

which has the solutions $\lambda_1 = 1$ and $\lambda_2 = 1000$. These equations are clearly stiff. According to Eq. (7.15) we would need $h \leq 2/\lambda_2 = 0.002$ for Euler's method to be stable. The Runge-Kutta method would have approximately the same limitation on the step size.

When the problem is very stiff, the usual methods of solution, such as the Runge-Kutta formulas, become impractical because of the very small h required for stability. These problems are best solved with methods that are specially designed for stiff equations. Stiff problem solvers, which are outside the scope of this text, have much better stability characteristics; some are even unconditionally stable. However, the higher degree of stability comes at a cost—the general rule is that stability can be improved only by reducing the order of the method (and thus increasing the truncation error).

EXAMPLE 7.7
(1) Show that the problem

$$y'' = -\frac{19}{4}y - 10y' \qquad y(0) = -9 \qquad y'(0) = 0$$

is moderately stiff and estimate h_{max}, the largest value of h for which the Runge-Kutta method would be stable. (2) Confirm the estimate by computing $y(10)$ with $h \approx h_{max}/2$ and $h \approx 2h_{max}$.

Solution of Part (1). With the notation $y = y_0$ and $y' = y_1$ the equivalent first-order differential equations are

$$\mathbf{y}' = \begin{bmatrix} y_1 \\ -\dfrac{19}{4}y_0 - 10y_1 \end{bmatrix} = -\mathbf{\Lambda}\begin{bmatrix} y_0 \\ y_1 \end{bmatrix}$$

where

$$\mathbf{\Lambda} = \begin{bmatrix} 0 & -1 \\ \dfrac{19}{4} & 10 \end{bmatrix}$$

The eigenvalues of $\mathbf{\Lambda}$ are given by

$$|\mathbf{\Lambda} - \lambda\mathbf{I}| = \begin{vmatrix} -\lambda & -1 \\ \dfrac{19}{4} & 10 - \lambda \end{vmatrix} = 0$$

which yields $\lambda_1 = 1/2$ and $\lambda_2 = 19/2$. Because λ_2 is quite a bit larger than λ_1, the equations are moderately stiff.

Solution of Part (2). An estimate for the upper limit of the stable range of h can be obtained from Eq. (7.15):

$$h_{\text{max}} = \frac{2}{\lambda_{\text{max}}} = \frac{2}{19/2} = 0.2153$$

Although this formula is strictly valid for Euler's method, it is usually not too far off for higher order integration formulas.

Here are the results from the Runge-Kutta method with $h = 0.1$ (by specifying `freq = 0` in `printSoln`, only the initial and final values were printed):

```
      x            y[ 0 ]           y[ 1 ]
 0.0000e+000   -9.0000e+000     0.0000e+000
 1.0000e+001   -6.4011e-002     3.2005e-002
```

The analytical solution is

$$y(x) = -\frac{19}{2}e^{-x/2} + \frac{1}{2}e^{-19x/2}$$

yielding $y(10) = -0.064011$, which agrees with the value obtained numerically.

With $h = 0.5$ we encountered instability, as expected:

```
      x            y[ 0 ]           y[ 1 ]
 0.0000e+000   -9.0000e+000     0.0000e+000
 1.0000e+001    2.7030e+020    -2.5678e+021
```

7.5 Adaptive Runge-Kutta Method

Determination of a suitable step size h can be a major headache in numerical integration. If h is too large, the truncation error may be unacceptable; if h is too small, we are squandering computational resources. Moreover, a constant step size may not be appropriate for the entire range of integration. For example, if the solution curve starts off with rapid changes before becoming smooth (as in a stiff problem), we should use a small h at the beginning and increase it as we reach the smooth region. This is where *adaptive methods* come in. They estimate the truncation error at each integration step and automatically adjust the step size to keep the error within prescribed limits.

The adaptive Runge-Kutta methods use so-called *embedded integration formulas*. These formulas come in pairs: One formula has the integration order m, and the other one is of order $m + 1$. The idea is to use both formulas to advance the solution from x to $x + h$. Denoting the results by $\mathbf{y}_m(x + h)$ and $\mathbf{y}_{m+1}(x + h)$, an estimate of the truncation error in the formula of order m is obtained from

$$\mathbf{E}(h) = \mathbf{y}_{m+1}(x + h) - \mathbf{y}_m(x + h) \tag{7.17}$$

What makes the embedded formulas attractive is that they share the points where $\mathbf{F}(x, \mathbf{y})$ is evaluated. This means that once $\mathbf{y}_{m+1}(x + h)$ has been computed, relatively small additional effort is required to calculate $\mathbf{y}_m(x + h)$.

Here are Runge-Kutta formulas of order five:

$$\mathbf{K}_0 = h\mathbf{F}(x, y)$$

$$\mathbf{K}_i = h\mathbf{F}\left(x + A_i h, \mathbf{y} + \sum_{j=0}^{i-1} B_{ij}\mathbf{K}_j\right), \quad i = 1, 2, \ldots, 6 \tag{7.18}$$

$$\mathbf{y}_5(x + h) = \mathbf{y}(x) + \sum_{i=0}^{6} C_i\mathbf{K}_i \tag{7.19a}$$

The embedded fourth-order formula is

$$\mathbf{y}_4(x + h) = \mathbf{y}(x) + \sum_{i=0}^{6} D_i\mathbf{K}_i \tag{7.19b}$$

The coefficients appearing in these formulas are not unique. Table 7.1 gives the coefficients proposed by Dormand and Prince.[2] They are claimed to provide superior error prediction than alternative values.

The solution is advanced with the fifth-order formula in Eq. (7.19a). The fourth-order formula is used only implicitly in estimating the truncation error

$$\mathbf{E}(h) = \mathbf{y}_5(x + h) - \mathbf{y}_4(x + h) = \sum_{i=0}^{6}(C_i - D_i)\mathbf{K}_i \tag{7.20}$$

Because Eq. (7.20) actually applies to the fourth-order formula, it tends to overestimate the error in the fifth-order formula.

Note that $\mathbf{E}(h)$ is a vector, its components $E_i(h)$ representing the errors in the dependent variables y_i. This brings up the question: What is the error measure $e(h)$ that we wish to control? There is no single choice that works well in all problems. If we want to control the largest component of $\mathbf{E}(h)$, the error measure would be

$$e(h) = \max_i |E_i(h)| \tag{7.21}$$

We could also control some gross measure of the error, such as the root-mean-square error defined by

$$\bar{E}(h) = \sqrt{\frac{1}{n}\sum_{i=0}^{n-1} E_i^2(h)} \tag{7.22}$$

[2] Dormand, R.R. and Prince, P.J., *Journal of Computational and Applied Mathematics*, Vol. 6, p. 1980.

i	A_i	B_{ij}						C_i	D_i
0	—	—	—	—	—	—		$\dfrac{35}{384}$	$\dfrac{5179}{57600}$
1	$\dfrac{1}{5}$	$\dfrac{1}{5}$	—	—	—	—		0	0
2	$\dfrac{3}{10}$	$\dfrac{3}{40}$	$\dfrac{9}{40}$	—	—	—		$\dfrac{500}{1113}$	$\dfrac{7571}{16695}$
3	$\dfrac{4}{5}$	$\dfrac{44}{45}$	$-\dfrac{56}{15}$	$\dfrac{32}{9}$	—	—		$\dfrac{125}{192}$	$\dfrac{393}{640}$
4	$8/9$	$\dfrac{19372}{6561}$	$-\dfrac{25360}{2187}$	$\dfrac{64448}{6561}$	$-\dfrac{212}{729}$	—		$-\dfrac{2187}{6784}$	$-\dfrac{92097}{339200}$
5	1	$\dfrac{9017}{3168}$	$-\dfrac{355}{33}$	$\dfrac{46732}{5247}$	$\dfrac{49}{176}$	$-\dfrac{5103}{18656}$		$\dfrac{11}{84}$	$\dfrac{187}{2100}$
6	1	$\dfrac{35}{384}$	0	$\dfrac{500}{1113}$	$\dfrac{125}{192}$	$-\dfrac{2187}{6784}$	$\dfrac{11}{84}$	0	$\dfrac{1}{40}$

Table 7.1. Dormand-Prince coefficients.

where n is the number of first-order equations. Then we would use

$$e(h) = \bar{E}(h) \tag{7.23}$$

for the error measure. Because the root-mean-square error is easier to handle, we adopt it for our program.

Error control is achieved by adjusting the increment h so that the per-step error $e(h)$ is approximately equal to a prescribed tolerance ε. Noting that the truncation error in the fourth-order formula is $\mathcal{O}(h^5)$, we conclude that

$$\frac{e(h_1)}{e(h_2)} \approx \left(\frac{h_1}{h_2}\right)^5 \tag{a}$$

Let us suppose that we performed an integration step with h_1 that resulted in the error $e(h_1)$. The step size h_2 that we should have used can now be obtained from Eq. (a) by setting $e(h_2) = \varepsilon$:

$$h_2 = h_1 \left[\frac{\varepsilon}{e(h_1)}\right]^{1/5} \tag{b}$$

If $h_2 \geq h_1$, we could repeat the integration step with h_2, but because the error was below the tolerance, that would be a waste of a perfectly good result. So we accept the current step and try h_2 in the next step. However, if $h_2 < h_1$, we must scrap the current step and repeat it with h_2.

Because Eq. (b) is only a crude approximation, it is prudent to incorporate a small margin of safety. In our program we use the formula

$$h_2 = 0.9 h_1 \left[\frac{\varepsilon}{e(h_1)} \right]^{1/5} \tag{7.24}$$

We also prevent excessively large changes in h by applying the constraints

$$0.1 \leq h_2 / h_1 \leq 10$$

Recall that $e(h)$ applies to a single integration step; that is, it is a measure of the local truncation error. The all-important global truncation error is caused by the accumulation of the local errors. What should ε be set at to achieve a global error tolerance $\varepsilon_{\text{global}}$? Because $e(h)$ is a conservative estimate of the actual error, setting $\varepsilon = \varepsilon_{\text{global}}$ is usually adequate. If the number of integration steps is very large, it is advisable to decrease ε accordingly.

Is there any reason to use the nonadaptive methods at all? Usually no—however, there are special cases where adaptive methods break down. For example, adaptive methods generally do not work if $\mathbf{F}(x, \mathbf{y})$ contains discontinuities. Because the error behaves erratically at the point of discontinuity, the program can get stuck in an infinite loop trying to find the appropriate value of h. Nonadaptive methods are also handy if the output is to have evenly spaced values of x.

■ run_kut5

This module is compatible with run_kut4 listed in the previous section. Any program that calls integrate can choose between the adaptive and the nonadaptive methods by importing either run_kut5 or run_kut4. The input argument h is the trial value of the increment for the first integration step.

Note that $\mathbf{K}_0$ is computed from scratch only in the first integration step. Subsequently, we use

$$(\mathbf{K}_0)_{m+1} = \frac{h_{m+1}}{h_m} (\mathbf{K}_6)_m \tag{c}$$

if the mth step was accepted, and

$$(\mathbf{K}_0)_{m+1} = \frac{h_{m+1}}{h_m} (\mathbf{K}_0)_m \tag{d}$$

if step m was rejected due to excessive truncation error.

To prove Eq. (c), we let $i = 6$ in Eq. (7.18), obtaining

$$(\mathbf{K}_6)_m = h_m \mathbf{F} \left[x_m + A_6 h_m, \, \mathbf{y}_m + \sum_{i=0}^{5} B_{6i} (\mathbf{K}_i)_m \right]$$

Table 6.1 shows that the last row of B-coefficients is identical to the C-coefficients (i.e., $B_{6i} = C_i$). Also note that $A_6 = 1$. Therefore,

$$(\mathbf{K}_6)_m = h_m \mathbf{F} \left[x_m + h_m, \, \mathbf{y}_m + \sum_{i=0}^{5} C_i (\mathbf{K}_i)_m \right] \tag{e}$$

But according to Eq. (7.19a) the fifth-order formula is

$$\mathbf{y}_{m+1} = \mathbf{y}_m + \sum_{i=0}^{6} C_i (\mathbf{K}_i)_m$$

Since $C_6 = 0$ (see Table 7.1), we can reduce the upper limit of the summation from 6 to 5. Therefore, Eq. (e) becomes

$$(\mathbf{K}_6)_m = h_m \mathbf{F}(x_{m+1}, \mathbf{y}_{m+1}) = \frac{h_m}{h_{m+1}} (\mathbf{K}_0)_{m+1}$$

which completes the proof.

The validity of Eq. (d) is rather obvious by inspection of the first equation of Eqs. (7.18). Because step $m + 1$ repeats step m with a different value of h, we have

$$(\mathbf{K}_0)_m = h_m \mathbf{F}(x_m, \mathbf{y}_m) \qquad (\mathbf{K}_0)_{m+1} = h_{m+1} \mathbf{F}(x_m, \mathbf{y}_m)$$

which leads directly to Eq. (d).

```
## module run_kut5
''' X,Y = integrate(F,x,y,xStop,h,tol=1.0e-6).
    Adaptive Runge-Kutta method with Dormand-Prince
    coefficients for solving the
    initial value problem {y}' = {F(x,{y})}, where
    {y} = {y[0],y[1],...y[n-1]}.

    x,y   = initial conditions
    xStop = terminal value of x
    h     = initial increment of x used in integration
    tol   = per-step error tolerance
    F     = user-supplied function that returns the
            array F(x,y) = {y'[0],y'[1],...,y'[n-1]}.
'''
import math
import numpy as np

def integrate(F,x,y,xStop,h,tol=1.0e-6):

    a1 = 0.2; a2 = 0.3; a3 = 0.8; a4 = 8/9; a5 = 1.0
    a6 = 1.0

    c0 = 35/384; c2 = 500/1113; c3 = 125/192
    c4 = -2187/6784; c5 = 11/84

    d0 = 5179/57600; d2 = 7571/16695; d3 = 393/640
    d4 = -92097/339200; d5 = 187/2100; d6 = 1/40

    b10 = 0.2
```

```
b20 = 0.075; b21 = 0.225
b30 = 44/45; b31 = -56/15; b32 = 32/9
b40 = 19372/6561; b41 = -25360/2187; b42 = 64448/6561
b43 = -212/729
b50 = 9017/3168; b51 =-355/33; b52 = 46732/5247
b53 = 49/176; b54 = -5103/18656
b60 = 35/384; b62 = 500/1113; b63 = 125/192;
b64 = -2187/6784; b65 = 11/84


X = []
Y = []
X.append(x)
Y.append(y)
stopper = 0   # Integration stopper(0 = off, 1 = on)
k0 = h*F(x,y)

for i in range(500):
    k1 = h*F(x + a1*h, y + b10*k0)
    k2 = h*F(x + a2*h, y + b20*k0 + b21*k1)
    k3 = h*F(x + a3*h, y + b30*k0 + b31*k1 + b32*k2)
    k4 = h*F(x + a4*h, y + b40*k0 + b41*k1 + b42*k2 + b43*k3)
    k5 = h*F(x + a5*h, y + b50*k0 + b51*k1 + b52*k2 + b53*k3 \
            + b54*k4)
    k6 = h*F(x + a6*h, y + b60*k0 + b62*k2 + b63*k3 + b64*k4 \
            + b65*k5)

    dy = c0*k0 + c2*k2 + c3*k3 + c4*k4 + c5*k5
    E = (c0 - d0)*k0 + (c2 - d2)*k2 + (c3 - d3)*k3  \
            + (c4 - d4)*k4 + (c5 - d5)*k5 - d6*k6
    e = math.sqrt(np.sum(E**2)/len(y))
    hNext = 0.9*h*(tol/e)**0.2

  # Accept integration step if error e is within tolerance
    if  e <= tol:
        y = y + dy
        x = x + h
        X.append(x)
        Y.append(y)
        if stopper == 1: break  # Reached end of x-range
        if abs(hNext) > 10.0*abs(h): hNext = 10.0*h

      # Check if next step is the last one; if so, adjust h
        if (h > 0.0) == ((x + hNext) >= xStop):
            hNext = xStop - x
```

```
            stopper = 1
        k0 = k6*hNext/h
    else:
        if abs(hNext) < 0.1*abs(h): hNext = 0.1*h
        k0 = k0*hNext/h

    h = hNext
return np.array(X),np.array(Y)
```

EXAMPLE 7.8

The aerodynamic drag force acting on a certain object in free fall can be approximated by

$$F_D = av^2 e^{-by}$$

where

$$v = \text{velocity of the object in m/s}$$

$$y = \text{elevation of the object in meters}$$

$$a = 7.45 \text{ kg/m}$$

$$b = 10.53 \times 10^{-5} \text{ m}^{-1}$$

The exponential term accounts for the change of air density with elevation. The differential equation describing the fall is

$$m\ddot{y} = -mg + F_D$$

where $g = 9.80665$ m/s^2 and $m = 114$ kg is the mass of the object. If the object is released at an elevation of 9 km, determine its elevation and speed after a 10-s fall with the adaptive Runge-Kutta method.

Solution. The differential equation and the initial conditions are

$$\ddot{y} = -g + \frac{a}{m}\dot{y}^2 \exp(-by)$$

$$= -9.80665 + \frac{7.45}{114}\dot{y}^2 \exp(-10.53 \times 10^{-5}y)$$

$$y(0) = 9000 \text{ m} \qquad \dot{y}(0) = 0$$

Letting $y_0 = y$ and $y_1 = \dot{y}$, the equivalent first-order equations become

$$\dot{\mathbf{y}} = \begin{bmatrix} \dot{y}_0 \\ \dot{y}_1 \end{bmatrix} = \begin{bmatrix} y_1 \\ -9.80665 + (65.351 \times 10^{-3}) \, y_1^2 \exp(-10.53 \times 10^{-5}y_0) \end{bmatrix}$$

$$\mathbf{y}(0) = \begin{bmatrix} 9000 \text{ m} \\ 0 \end{bmatrix}$$

The driver program for `run_kut5` is listed next. We specified a per-step error toler-ance of 10^{-2} in `integrate`. Considering the magnitude of **y**, this should be enough for five decimal point accuracy in the solution.

```
#!/usr/bin/python
## example7_8
import numpy as np
import math
from run_kut5 import *
from printSoln import *

def F(x,y):
    F = np.zeros(2)
    F[0] = y[1]
    F[1] = -9.80665 + 65.351e-3 * y[1]**2 * math.exp(-10.53e-5*y[0])
    return F

x = 0.0
xStop = 10.0
y = np.array([9000, 0.0])
h = 0.5
freq = 1
X,Y = integrate(F,x,y,xStop,h,1.0e-2)
printSoln(X,Y,freq)
input("\nPress return to exit")
```

Running the program resulted in the following output:

```
      x            y[ 0 ]         y[ 1 ]
0.0000e+00     9.0000e+03     0.0000e+00
5.0000e-01     8.9988e+03    -4.8043e+00
2.4229e+00     8.9763e+03    -1.6440e+01
3.4146e+00     8.9589e+03    -1.8388e+01
4.6318e+00     8.9359e+03    -1.9245e+01
5.9739e+00     8.9098e+03    -1.9501e+01
7.6199e+00     8.8777e+03    -1.9549e+01
9.7063e+00     8.8369e+03    -1.9524e+01
1.0000e+01     8.8312e+03    -1.9519e+01
```

The first step was carried out with the prescribed trial value $h = 0.5$ s. Apparently the error was well within the tolerance, so that the step was accepted. Subsequent step sizes, determined from Eq. (7.24), were considerably larger.

Inspecting the output, we see that at $t = 10$ s the object is moving with the speed $v = -\dot{y} = 19.52$ m/s at an elevation of $y = 8831$ m.

EXAMPLE 7.9

Integrate the moderately stiff problem

$$y'' = -\frac{19}{4}y - 10y' \qquad y(0) = -9 \qquad y'(0) = 0$$

from $x = 0$ to 10 with the adaptive Runge-Kutta method and plot the results (this problem also appeared in Example 7.7).

Solution. Because we use an adaptive method, there is no need to worry about the stable range of h, as we did in Example 7.7. As long as we specify a reasonable tolerance for the per-step error (in this case the default value 10^{-6} is fine), the algorithm will find the appropriate step size. Here is the program and its numerical output:

```
## example7_9
import numpy as np
import matplotlib.pyplot as plt
from run_kut5 import *
from printSoln import *

def F(x,y):
    F = np.zeros(2)
    F[0] = y[1]
    F[1] = -4.75*y[0] - 10.0*y[1]
    return F

x = 0.0
xStop = 10.0
y = np.array([-9.0, 0.0])
h = 0.1
freq = 4
X,Y = integrate(F,x,y,xStop,h)
printSoln(X,Y,freq)
plt.plot(X,Y[:,0],'o-',X,Y[:,1],'^-')
plt.xlabel('x')
plt.legend(('y','dy/dx'),loc=0)
plt.grid(True)
plt.show()
input("\nPress return to exit")
```

The following printout displays every fourth integration step:

x	y[0]	y[1]
0.0000e+00	-9.0000e+00	0.0000e+00
7.7774e-02	-8.8988e+00	2.2999e+00
1.6855e-01	-8.6314e+00	3.4083e+00
2.7656e-01	-8.2370e+00	3.7933e+00

```
4.0945e-01    -7.7311e+00    3.7735e+00
5.8108e-01    -7.1027e+00    3.5333e+00
8.2045e-01    -6.3030e+00    3.1497e+00
1.2036e+00    -5.2043e+00    2.6021e+00
2.0486e+00    -3.4110e+00    1.7055e+00
3.5357e+00    -1.6216e+00    8.1081e-01
4.9062e+00    -8.1724e-01    4.0862e-01
6.3008e+00    -4.0694e-01    2.0347e-01
7.7202e+00    -2.0012e-01    1.0006e-01
9.1023e+00    -1.0028e-01    5.0137e-02
1.0000e+01    -6.4010e-02    3.2005e-02
```

The results are in agreement with the analytical solution.

The plots of y and y' are shown in the following figure. Note the high density of points near $x = 0$ where y' changes rapidly. As the y'-curve becomes smoother, the distance between the points increases.

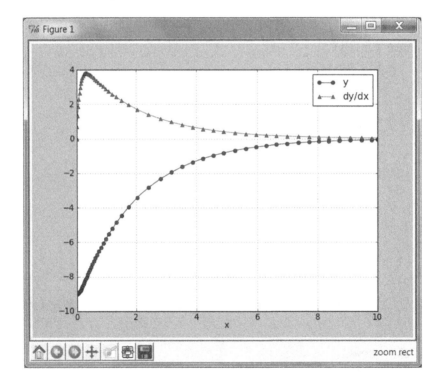

7.6   Bulirsch-Stoer Method

Midpoint Method

The midpoint formula of numerical integration of $\mathbf{y}' = \mathbf{F}(x, \mathbf{y})$ is

$$\mathbf{y}(x + h) = \mathbf{y}(x - h) + 2h\mathbf{F}\left[x, \mathbf{y}(x)\right] \tag{7.25}$$

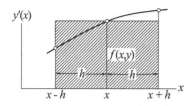

Figure 7.3. Graphical representation of the midpoint formula.

It is a second-order formula, like the modified Euler's formula. We discuss it here because it is the basis of the powerful *Bulirsch-Stoer method*, which is the technique of choice in problems where high accuracy is required.

Figure 7.3 illustrates the midpoint formula for a single differential equation $y' = f(x, y)$. The change in y over the two panels shown is

$$y(x + h) - y(x - h) = \int_{x-h}^{x+h} y'(x)\,dx$$

which equals the area under the $y'(x)$ curve. The midpoint method approximates this area by the area $2hf(x, y)$ of the cross-hatched rectangle.

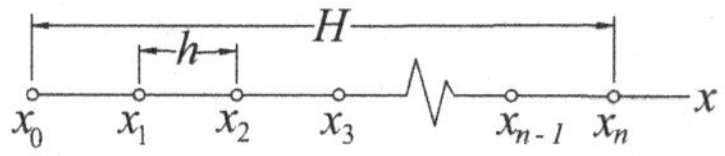

Figure 7.4. Mesh used in the midpoint method.

Consider now advancing the solution of $\mathbf{y}'(x) = \mathbf{F}(x, \mathbf{y})$ from $x = x_0$ to $x_0 + H$ with the midpoint formula. We divide the interval of integration into n steps of length $h = H/n$ each, as shown in Figure 7.4, and then carry out the following computations:

$$\mathbf{y}_1 = \mathbf{y}_0 + h\mathbf{F}_0$$

$$\mathbf{y}_2 = \mathbf{y}_0 + 2h\mathbf{F}_1$$

$$\mathbf{y}_3 = \mathbf{y}_1 + 2h\mathbf{F}_2 \qquad (7.26)$$

$$\vdots$$

$$\mathbf{y}_n = \mathbf{y}_{n-2} + 2h\mathbf{F}_{n-1}$$

Here we used the notation $\mathbf{y}_i = \mathbf{y}(x_i)$ and $\mathbf{F}_i = \mathbf{F}(x_i, \mathbf{y}_i)$. The first of Eqs. (7.26) uses Euler's formula to "seed" the midpoint method; the other equations are midpoint formulas. The final result is obtained by averaging $\mathbf{y}_n$ in Eq. (7.26) and the estimate $\mathbf{y}_n \approx \mathbf{y}_{n-1} + h\mathbf{F}_n$ available from the Euler formula. The result is

$$\mathbf{y}(x_0 + H) = \frac{1}{2}\left[\mathbf{y}_n + \left(\mathbf{y}_{n-1} + h\mathbf{F}_n\right)\right] \qquad (7.27)$$

Richardson Extrapolation

It can be shown that the error in Eq. (7.27) is

$$\mathbf{E} = \mathbf{c}_1 h^2 + \mathbf{c}_2 h^4 + \mathbf{c}_3 h^6 + \cdots$$

Herein lies the great utility of the midpoint method: We can eliminate as many of the leading error terms as we wish by Richarson's extrapolation. For example, we could compute $\mathbf{y}(x_0 + H)$ with a certain value of h and then repeat the process with $h/2$. Denoting the corresponding results by $\mathbf{g}(h)$ and $\mathbf{g}(h/2)$, Richardson's extrapolation—see Eq. (5.9)—then yields the improved result

$$\mathbf{y}_{\text{better}}(x_0 + H) = \frac{4\mathbf{g}(h/2) - \mathbf{g}(h)}{3}$$

which is fourth-order accurate. Another round of integration with $h/4$ followed by Richardson's extrapolation get us sixth-order accuracy, and so on Rather than halving the interval in successive integrations, we use the sequence $h/2$, $h/4$, $h/6$, $h/8$, $h/10$, ... which has been found to be more economical.

The $\mathbf{y}$'s in Eqs. (7.26) should be viewed as a temporary variables, because unlike $\mathbf{y}(x_0 + H)$, they cannot be refined by Richardson's extrapolation.

■ midpoint

The function `integrate` in this module combines the midpoint method with Richardson extrapolation. The first application of the midpoint method uses two integration steps. The number of steps is increased by two in successive integrations, each integration being followed by Richardson extrapolation. The procedure is stopped when two successive solutions differ (in the root-mean-square sense) by less than a prescribed tolerance.

```
## module midpoint
''' yStop = integrate (F,x,y,xStop,tol=1.0e-6)
    Modified midpoint method for solving the
    initial value problem y' = F(x,y}.
    x,y   = initial conditions
    xStop = terminal value of x
    yStop = y(xStop)
    F     = user-supplied function that returns the
            array F(x,y) = {y'[0],y'[1],...,y'[n-1]}.
'''

import numpy as np
import math
def integrate(F,x,y,xStop,tol):

    def midpoint(F,x,y,xStop,nSteps):

  # Midpoint formulas
        h = (xStop - x)/nSteps
        y0 = y
        y1 = y0 + h*F(x,y0)
```

```
        for i in range(nSteps-1):
            x = x + h
            y2 = y0 + 2.0*h*F(x,y1)
            y0 = y1
            y1 = y2
        return 0.5*(y1 + y0 + h*F(x,y2))

    def richardson(r,k):
  # Richardson's extrapolation
        for j in range(k-1,0,-1):
            const = (k/(k - 1.0))**(2.0*(k-j))
            r[j] = (const*r[j+1] - r[j])/(const - 1.0)
        return

    kMax = 51
    n = len(y)
    r = np.zeros((kMax,n))
  # Start with two integration steps
    nSteps = 2
    r[1] = midpoint(F,x,y,xStop,nSteps)
    r_old = r[1].copy()
  # Increase the number of integration points by 2
  # and refine result by Richardson extrapolation
    for k in range(2,kMax):
        nSteps = 2*k
        r[k] = midpoint(F,x,y,xStop,nSteps)
        richardson(r,k)
      # Compute RMS change in solution
        e = math.sqrt(np.sum((r[1] - r_old)**2)/n)
      # Check for convergence
        if e < tol: return r[1]
        r_old = r[1].copy()
    print("Midpoint method did not converge")
```

Bulirsch-Stoer Algorithm

The fundamental idea behind the Bulirsch-Stroer method is simple: Apply the midpoint method in a piecewise fashion. That is, advance the solution in stages of length H, using the midpoint method with Richardson extrapolation to perform the integration in each stage. The value of H can be quite large, because the precision of the result is determined by the step length h in the midpoint method, not by H. However, if H is too large, the the midpoint method may not converge. If this happens, try a smaller value of H or a larger error tolerance.

The original Bulirsch and Stoer technique[3] is a complex procedure that incorporates many refinements missing in our algorithm, such as determining the optimal value of H. However, the function bulStoer shown next retains the essential ideas of Bulirsch and Stoer.

What are the relative merits of adaptive Runge-Kutta and Bulirsch-Stoer methods? The Runge-Kutta method is more robust, having higher tolerance for nonsmooth functions and stiff problems. The Bulirsch-Stoer algorithm (in its original form) is used mainly in problems where high accuracy is of paramount importance. Our simplified version is no more accurate than the adaptive Runge-Kutta method, but it is useful if the output is to appear at equally spaced values of x.

■ bulStoer

This function contains a simplified algorithm for the Bulirsch-Stoer method.

```
## module bulStoer
''' X,Y = bulStoer(F,x,y,xStop,H,tol=1.0e-6).
    Simplified Bulirsch-Stoer method for solving the
    initial value problem {y}' = {F(x,{y})}, where
    {y} = {y[0],y[1],...y[n-1]}.
    x,y   = initial conditions
    xStop = terminal value of x
    H     = increment of x at which results are stored
    F     = user-supplied function that returns the
            array F(x,y) = {y'[0],y'[1],...,y'[n-1]}.
'''
import numpy as np
from midpoint import *

def bulStoer(F,x,y,xStop,H,tol=1.0e-6):
    X = []
    Y = []
    X.append(x)
    Y.append(y)
    while x < xStop:
        H = min(H,xStop - x)
        y = integrate(F,x,y,x + H,tol) # Midpoint method
        x = x + H
        X.append(x)
        Y.append(y)
    return np.array(X),np.array(Y)
```

---

[3] Stoer, J. and Bulirsch, R., *Introduction to Numerical Analysis*, Springer, 1980.

EXAMPLE 7.10

Compute the solution of the initial value problem

$$y' = \sin y \qquad y(0) = 1$$

at $x = 0.5$ with the midpoint formulas using $n = 2$ and $n = 4$, followed by Richardson extrapolation (this problem was solved with the second-order Runge-Kutta method in Example 7.3).

Solution. With $n = 2$ the step length is $h = 0.25$. The midpoint formulas, Eqs. (7.26) and (7.27) yield

$$y_1 = y_0 + hf_0 = 1 + 0.25 \sin 1.0 = 1.210\,368$$

$$y_2 = y_0 + 2hf_1 = 1 + 2(0.25)\sin 1.210\,368 = 1.467\,87\,3$$

$$y_h(0.5) = \frac{1}{2}(y_1 + y_0 + hf_2)$$

$$= \frac{1}{2}(1.210\,368 + 1.467\,87\,3 + 0.25 \sin 1.467\,87\,3)$$

$$= 1.463\,459$$

Using $n = 4$ we have $h = 0.125$, and the midpoint formulas become

$$y_1 = y_0 + hf_0 = 1 + 0.125 \sin 1.0 = 1.105\,184$$

$$y_2 = y_0 + 2hf_1 = 1 + 2(0.125)\sin 1.105\,184 = 1.223\,387$$

$$y_3 = y_1 + 2hf_2 = 1.105\,184 + 2(0.125)\sin 1.223\,387 = 1.340\,248$$

$$y_4 = y_2 + 2hf_3 = 1.223\,387 + 2(0.125)\sin 1.340\,248 = 1.466\,772$$

$$y_{h/2}(0.5) = \frac{1}{2}(y_4 + y_3 + hf_4)$$

$$= \frac{1}{2}(1.466\,772 + 1.340\,248 + 0.125 \sin 1.466\,772)$$

$$= 1.465\,672$$

Richardson extrapolation results in

$$y(0.5) = \frac{4y_{h/2}(0.5) - y_h(0.5)}{3} = \frac{4(1.465\,672) - 1.463\,459}{3} = 1.466\,410$$

which compares favorably with the "true" solution, $y(0.5) = 1.466\,404$.

EXAMPLE 7.11

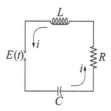

The differential equations governing the loop current i and the charge q on the capacitor of the electric circuit shown are

$$L\frac{di}{dt} + Ri + \frac{q}{C} = E(t) \qquad \frac{dq}{dt} = i$$

If the applied voltage E is suddenly increased from zero to 9 V, plot the resulting loop current during the first 10 s. Use $R = 1.0\,\Omega$, $L = 2\,H$, and $C = 0.45\,F$.

Solution. Letting

$$\mathbf{y} = \begin{bmatrix} y_0 \\ y_1 \end{bmatrix} = \begin{bmatrix} q \\ i \end{bmatrix}$$

and substituting the given data, the differential equations become

$$\dot{\mathbf{y}} = \begin{bmatrix} \dot{y}_0 \\ \dot{y}_1 \end{bmatrix} = \begin{bmatrix} y_1 \\ (-Ry_1 - y_0/C + E)/L \end{bmatrix}$$

The initial conditions are

$$\mathbf{y}(0) = \begin{bmatrix} 0 \\ 0 \end{bmatrix}$$

We solved the problem with the function $\mathtt{bul\_stoer}$ with the increment $H = 0.25$ s:

```
#!/usr/bin/python
## example7_11
from bulStoer import *
import numpy as np
import matplotlib.pyplot as plt

def F(x,y):
    F = np.zeros(2)
    F[0] = y[1]
    F[1] =(-y[1] - y[0]/0.45 + 9.0)/2.0
    return F

H = 0.25
xStop = 10.0
x = 0.0
y = np.array([0.0, 0.0])
X,Y = bulStoer(F,x,y,xStop,H)
plt.plot(X,Y[:,1],'o-')
plt.xlabel('Time (s)')
plt.ylabel('Current (A)')
plt.grid(True)
plt.show()
input("\nPress return to exit")
```

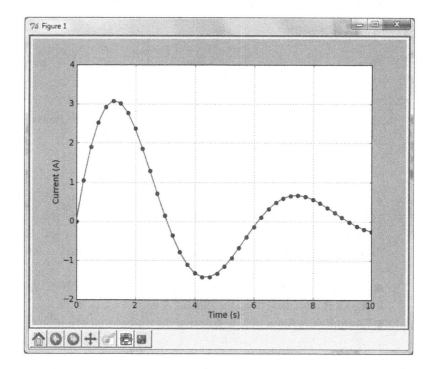

Recall that in each interval H (the spacing of points in the plot), the integration was performed by the modified midpoint method and refined by Richardson's extrapolation.

PROBLEM SET 7.2

1. Derive the analytical solution of the problem

$$y'' + y' - 380y = 0 \qquad y(0) = 1 \qquad y'(0) = -20$$

Would you expect difficulties in solving this problem numerically?

2. Consider the problem

$$y' = x - 10y \qquad y(0) = 10$$

(a) Verify that the analytical solution is $y(x) = 0.1x - 0.001 + 10.01e^{-10x}$. (b) Determine the step size h that you would use in numerical solution with the (non-adaptive) Runge-Kutta method.

3. ■ Integrate the initial value problem in Prob. 2 from $x = 0$ to 5 with the Runge-Kutta method using (a) $h = 0.1$, (b) $h = 0.25$, and (c) $h = 0.5$. Comment on the results.

4. ■ Integrate the initial value problem in Prob. 2 from $x = 0$ to 10 with the adaptive Runge-Kutta method and plot the result.

5. ■

The differential equation describing the motion of the mass-spring-dashpot system is

$$\ddot{y} + \frac{c}{m}\dot{y} + \frac{k}{m}y = 0$$

where $m = 2$ kg, $c = 460$ N·s/m, and $k = 450$ N/m. The initial conditions are $y(0) = 0.01$ m and $\dot{y}(0) = 0$. (a) Show that this is a stiff problem and determine a value of h that you would use in numerical integration with the nonadaptive Runge-Kutta method. (b) Carry out the integration from $t = 0$ to 0.2 s with the chosen h and plot $\dot{y}$ vs. t.

6. ■ Integrate the initial value problem specified in Prob. 5 with the adaptive Runge-Kutta method from $t = 0$ to 0.2 s, and plot $\dot{y}$ vs. t.

7. ■ Compute the numerical solution of the differential equation

$$y'' = 16.81y$$

from $x = 0$ to 2 with the adaptive Runge-Kutta method and plot the results. Use the initial conditions (a) $y(0) = 1.0$, $y'(0) = -4.1$; and (b) $y(0) = 1.0$, $y'(0) = -4.11$. Explain the large difference in the two solutions. *Hint*: Derive and plot the analytical solutions.

8. ■ Integrate

$$y'' + y' - y^2 = 0 \qquad y(0) = 1 \qquad y'(0) = 0$$

from $x = 0$ to 3.5. Is the sudden increase in y near the upper limit real or an artifact caused by instability?

9. ■ Solve the stiff problem—see Eq. (7.16)—

$$y'' + 1001y' + 1000y = 0 \qquad y(0) = 1 \qquad y'(0) = 0$$

from $x = 0$ to 0.2 with the adaptive Runge-Kutta method and plot y' vs. x.

10. ■ Solve

$$y'' + 2y' + 3y = 0 \qquad y(0) = 0 \qquad y'(0) = \sqrt{2}$$

with the adaptive Runge-Kutta method from $x = 0$ to 5 (the analytical solution is $y = e^{-x}\sin\sqrt{2}x$).

11. ■ Solve the differential equation

$$y'' = 2yy'$$

from $x = 0$ to 10 with the initial conditions $y(0) = 1$, $y'(0) = -1$. Plot y vs. x.

12. ■ Repeat Prob. 11 with the initial conditions $y(0) = 0$, $y'(0) = 1$ and the integration range $x = 0$ to 1.5.

13. ■ Use the adaptive Runge-Kutta method to integrate

$$y' = \left(\frac{9}{y} - y\right)x \qquad y(0) = 5$$

from $x = 0$ to 4 and plot y vs. x.

14. ■ Solve Prob. 13 with the Bulirsch-Stoer method using $H = 0.25$.

15. ■ Integrate

$$x^2 y'' + xy' + y = 0 \qquad y(1) = 0 \qquad y'(1) = -2$$

from $x = 1$ to 20, and plot y and y' vs. x. Use the Bulirsch-Stoer method.

16. ■

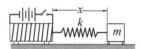

The magnetized iron block of mass m is attached to a spring of stiffness k and free length L. The block is at rest at $x = L$ when the electromagnet is turned on, exerting the repulsive force $F = c/x^2$ on the block. The differential equation of the resulting motion is

$$m\ddot{x} = \frac{c}{x^2} - k(x - L)$$

Determine the period of the ensuing motion by numerical integration with the adaptive Runge-Kutta method. Use $c = 5$ N·m^2, $k = 120$ N/m, $L = 0.2$ m, and $m = 1.0$ kg.

17. ■

The bar ABC is attached to the vertical rod with a horizontal pin. The assembly is free to rotate about the axis of the rod. Neglecting friction, the equations of motion of the system are

$$\ddot{\theta} = \dot{\phi}^2 \sin\theta \cos\theta \qquad \ddot{\phi} = -2\dot{\theta}\dot{\phi} \cot\theta$$

The system is set into motion with the initial conditions $\theta(0) = \pi/12$ rad, $\dot{\theta}(0) = 0$, $\phi(0) = 0$, and $\dot{\phi}(0) = 20$ rad/s. Obtain a numerical solution with the adaptive Runge-Kutta method from $t = 0$ to 1.5 s and plot $\dot{\phi}$ vs. t.

18. ■ Solve the circuit problem in Example 7.11 if $R = 0$ and

$$E(t) = \begin{cases} 0 \text{ when } t < 0 \\ 9 \sin \pi t \text{ when } t \geq 0 \end{cases}$$

19. ■ Solve Prob. 21 in Problem Set 1 if $E = 240\,\text{V}$ (constant).
20. ■

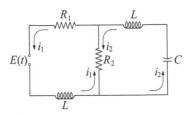

Kirchoff's equations for the circuit in the figure are

$$L\frac{di_1}{dt} + R_1 i_1 + R_2(i_1 - i_2) = E(t)$$

$$L\frac{di_2}{dt} + R_2(i_2 - i_1) + \frac{q_2}{C} = 0$$

where

$$\frac{dq_2}{dt} = i_2$$

Using the data $R_1 = 4\,\Omega$, $R_2 = 10\,\Omega$, $L = 0.032\,\text{H}$, $C = 0.53\,\text{F}$, and

$$E(t) = \begin{cases} 20\,\text{V if } 0 < t < 0.005\,\text{s} \\ 0 \text{ otherwise} \end{cases}$$

plot the transient loop currents i_1 and i_2 from $t = 0$ to $0.05\,\text{s}$.

21. ■ Consider a closed biological system populated by M number of prey and N number of predators. Volterra postulated that the two populations are related by the differential equations

$$\dot{M} = aM - bMN$$

$$\dot{N} = -cN + dMN$$

where a, b, c, and d are constants. The steady-state solution is $M_0 = c/d$, $N_0 = a/b$; if numbers other than these are introduced into the system, the populations undergo periodic fluctuations. Introducing the notation

$$y_0 = M/M_0 \qquad y_1 = N/N_0$$

allows us to write the differential equations as

$$\dot{y}_0 = a(y_0 - y_0 y_1)$$

$$\dot{y}_1 = b(-y_1 + y_0 y_1)$$

Using $a = 1.0/\text{year}$, $b = 0.2/\text{year}$, $y_0(0) = 0.1$, and $y_1(0) = 1.0$, plot the two populations from $t = 0$ to 5 years.

22. ■ The equations

$$\dot{u} = -au + av$$

$$\dot{v} = cu - v - uw$$

$$\dot{w} = -bw + uv$$

known as the Lorenz equations, are encountered in the theory of fluid dynamics. Letting $a = 5.0$, $b = 0.9$, and $c = 8.2$, solve these equations from $t = 0$ to 10 with the initial conditions $u(0) = 0$, $v(0) = 1.0$, $w(0) = 2.0$, and plot $u(t)$. Repeat the solution with $c = 8.3$. What conclusions can you draw from the results?

23. ■

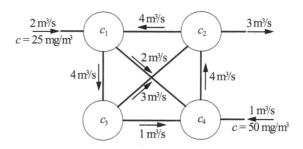

Four mixing tanks are connected by pipes. The fluid in the system is pumped through the pipes at the rates shown in the figure. The fluid entering the system contains a chemical of concentration c as indicated. The rate at which the mass of the chemical changes in tank i is

$$V_i \frac{dc_i}{dt} = \Sigma\,(Qc)_{\text{in}} - \Sigma(Qc)_{\text{out}}$$

where V_i is the volume of the tank and Q represents the flow rate in the pipes connected to it. Applying this equation to each tank, we obtain

$$V_1 \frac{dc_i}{dt} = -6c_1 + 4c_2 + 2(25)$$

$$V_2 \frac{dc_2}{dt} = -7c_2 + 3c_3 + 4c_4$$

$$V_3 \frac{dc_3}{dt} = 4c_1 - 4c_3$$

$$V_4 \frac{dc_4}{dt} = 2c_1 + c_3 - 4c_4 + 50$$

Plot the concentration of the chemical in tanks 1 and 2 vs. time t from $t = 0$ to 100 s. Let $V_1 = V_2 = V_3 = V_4 = 10$ m^3, and assume that the concentration in each tank is zero at $t = 0$. The steady-state version of this problem was solved in Prob. 21, Problem Set 2.2.

7.7   Other Methods

The methods described so far belong to a group known as *single-step methods*. The name stems from the fact that the information at a single point on the solution curve is sufficient to compute the next point. There are also *multistep methods* that use several points on the curve to extrapolate the solution at the next step. Well-known members of this group are the methods of Adams, Milne, Hamming, and Gere. These methods were popular once, but have lost some of their luster in the last few years. Multistep methods have two shortcomings that complicate their implementation:

1. The methods are not self-starting, but must be provided with the solution at the first few points by a single-step method.
2. The integration formulas assume equally spaced steps, which makes it difficult to change the step size.

Both of these hurdles can be overcome, but the price is complexity of the algorithm, which increases with the sophistication of the method. The benefits of multistep methods are minimal—the best of them can outperform their single-step counterparts in certain problems, but these occasions are rare.

8 Two-Point Boundary Value Problems

$$\boxed{\text{Solve } y'' = f(x, y, y'), \quad y(a) = \alpha, \quad y(b) = \beta.}$$

8.1 Introduction

In two-point boundary value problems the auxiliary conditions associated with the differential equation, called the *boundary conditions*, are specified at two different values of x. This seemingly small departure from initial value problems has a major repercussion—it makes boundary value problems considerably more difficult to solve. In an initial value problem we were able to start at the point where the initial values were given and march the solution forward as far as needed. This technique does not work for boundary value problems, because there are not enough starting conditions available at either endpoint to produce a unique solution.

One way to overcome the lack of starting conditions is to guess the missing values. The resulting solution is very unlikely to satisfy boundary conditions at the other end, but by inspecting the discrepancy we can estimate what changes to make to the initial conditions before integrating again. This iterative procedure is known as the *shooting method*. The name is derived from analogy with target shooting—take a shot and observe where it hits the target; then correct the aim and shoot again.

Another means of solving two-point boundary value problems is the *finite difference method*, where the differential equations are approximated by finite differences at evenly spaced mesh points. As a consequence, a differential equation is transformed into a set of simultaneous algebraic equations.

The two methods have a common problem: They give rise to nonlinear sets of equations if the differential equations are not linear. As we noted in Chapter 2, all methods of solving nonlinear equations are iterative procedures that can consume a lot of computational resources. Thus solving nonlinear boundary value problems is not cheap. Another complication is that iterative methods need reasonably good starting values to converge. Since there is no set formula for determining these starting values, an algorithm for solving nonlinear boundary value problems requires informed input; it cannot be treated as a "black box."

8.2 Shooting Method

Second-Order Differential Equation

The simplest two-point boundary value problem is a second-order differential equation with one condition specified at $x = a$ and another one at $x = b$. Here is an example of such a problem:

$$y'' = f(x, y, y'), \quad y(a) = \alpha, \quad y(b) = \beta \qquad (8.1)$$

Let us now attempt to turn Eqs. (8.1) into the initial value problem

$$y'' = f(x, y, y'), \quad y(a) = \alpha, \quad y'(a) = u \qquad (8.2)$$

The key to success is finding the correct value of u. This could be done by trial and error: Guess u and solve the initial value problem by marching from $x = a$ to b. If the solution agrees with the prescribed boundary condition $y(b) = \beta$, we are done; otherwise we have to adjust u and try again. Clearly, this procedure is very tedious.

More systematic methods become available to us if we realize that the determination of u is a root-finding problem. Because the solution of the initial value problem depends on u, the computed value of $y(b)$ is a function of u; that is

$$y(b) = \theta(u)$$

Hence u is a root of

$$r(u) = \theta(u) - \beta = 0 \qquad (8.3)$$

where $r(u)$ is the *boundary residual* (the difference between the computed and specified boundary value at $x = b$). Equation (8.3) can be solved by one of the root-finding methods discussed in Chapter 4. We reject the method of bisection because it involves too many evaluations of $\theta(u)$. In the Newton-Raphson method we run into the problem of having to compute $d\theta/du$, which can be done but not easily. That leaves Ridder's algorithm as our method of choice.

Here is the procedure we use in solving nonlinear boundary value problems:

> Specify the starting values u_1 and u_2 that *must bracket the root u* of Eq. (8.3).
> Apply Ridder's method to solve Eq. (8.3) for u. Note that each iteration requires evaluation of $\theta(u)$ by solving the differential equation as an initial value problem.
> Having determined the value of u, solve the differential equations once more and record the results.

If the differential equation is linear, any root-finding method will need only one interpolation to determine u. Because Ridder's method uses three points (u_1, u_2, and u_3), it is wasteful compared with linear interpolation, which uses only two points

(u_1 and u_2). Therefore, we replace Ridder's method with linear interpolation whenever the differential equation is linear.

■ linInterp

Here is the algorithm we use for linear interpolation:

```
## module linInterp
''' root = linInterp(f,x1,x2).
    Finds the zero of the linear function f(x) by straight
    line interpolation based on x = x1 and x2.
'''
def linInterp(f,x1,x2):
    f1 = f(x1)
    f2 = f(x2)
    return = x2 - f2*(x2 - x1)/(f2 - f1)
```

EXAMPLE 8.1

Solve the boundary value problem

$$y'' + 3yy' = 0 \qquad y(0) = 0 \qquad y(2) = 1$$

Solution. The equivalent first-order equations are

$$\mathbf{y}' = \begin{bmatrix} y'_0 \\ y'_1 \end{bmatrix} = \begin{bmatrix} y_1 \\ -3y_0 y_1 \end{bmatrix}$$

with the boundary conditions

$$y_0(0) = 0 \qquad y_0(2) = 1$$

Now comes the daunting task of determining the trial values of $y'(0)$. We could always pick two numbers at random and hope for the best. However, it is possible to reduce the element of chance with a little detective work. We start by making the reasonable assumption that y is smooth (does not wiggle) in the interval $0 \leq x \leq 2$. Next we note that y has to increase from 0 to 1, which requires $y' > 0$. Because both y and y' are positive, we conclude that y'' must be negative to satisfy the differential equation. Now we are in a position to make a rough sketch of y:

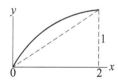

Looking at the sketch it is clear that $y'(0) > 0.5$, so that $y'(0) = 1$ and 2 appear to be reasonable values for the brackets of $y'(0)$; if they are not, Ridder's method will display an error message.

In the program listed next we chose the fourth-order Runge-Kutta method for integration. It can be replaced by the adaptive version by substituting `run_kut5` for `run_kut4` in the `import` statement. Note that three user-supplied functions are needed to describe the problem at hand. Apart from the function `F(x,y)` that defines the differential equations, we also need the functions `initCond(u)` to specify the initial conditions for integration, and `r(u)` to provide Ridder's method with the boundary condition residual. By changing a few statements in these functions, the program can be applied to any second-order boundary value problem. It also works for third-order equations if integration is started at the end where two of the three boundary conditions are specified.

```python
#!/usr/bin/python
## example8_1
import numpy as np
from run_kut4 import *
from ridder import *
from printSoln import *

def initCond(u):  # Init. values of [y,y']; use 'u' if unknown
    return np.array([0.0, u])

def r(u):          # Boundary condition residual--see Eq. (8.3)
    X,Y = integrate(F,xStart,initCond(u),xStop,h)
    y = Y[len(Y) - 1]
    r = y[0] - 1.0
    return r

def F(x,y):        # First-order differential equations
    F = np.zeros(2)
    F[0] = y[1]
    F[1] = -3.0*y[0]*y[1]
    return F

xStart = 0.0        # Start of integration
xStop = 2.0         # End of integration
u1 = 1.0            # 1st trial value of unknown init. cond.
u2 = 2.0            # 2nd trial value of unknown init. cond.
h = 0.1             # Step size
freq = 2            # Printout frequency
u = ridder(r,u1,u2) # Compute the correct initial condition
X,Y = integrate(F,xStart,initCond(u),xStop,h)
printSoln(X,Y,freq)
input("\nPress return to exit")
```

Here is the solution:

x	y[0]	y[1]
0.0000e+00	0.0000e+00	1.5145e+00
2.0000e-01	2.9404e-01	1.3848e+00
4.0000e-01	5.4170e-01	1.0743e+00
6.0000e-01	7.2187e-01	7.3287e-01
8.0000e-01	8.3944e-01	4.5752e-01
1.0000e+00	9.1082e-01	2.7013e-01
1.2000e+00	9.5227e-01	1.5429e-01
1.4000e+00	9.7572e-01	8.6471e-02
1.6000e+00	9.8880e-01	4.7948e-02
1.8000e+00	9.9602e-01	2.6430e-02
2.0000e+00	1.0000e+00	1.4522e-02

Note that $y'(0) = 1.5145$, so that our starting values of 1.0 and 2.0 were on the mark.

EXAMPLE 8.2

Numerical integration of the initial value problem

$$y'' + 4y = 4x \qquad y(0) = 0 \qquad y'(0) = 0$$

yielded $y'(2) = 1.653\,64$. Use this information to determine the value of $y'(0)$ that would result in $y'(2) = 0$.

Solution. We use linear interpolation

$$u = u_2 - \theta(u_2)\frac{u_2 - u_1}{\theta(u_2) - \theta(u_1)}$$

where in our case $u = y'(0)$ and $\theta(u) = y'(2)$. So far we are given $u_1 = 0$ and $\theta(u_1) = 1.653\,64$. To obtain the second point, we need another solution of the initial value problem. An obvious solution is $y = x$, which gives us $y(0) = 0$ and $y'(0) = y'(2) = 1$. Thus the second point is $u_2 = 1$ and $\theta(u_2) = 1$. Linear interpolation now yields

$$y'(0) = u = 1 - (1)\frac{1 - 0}{1 - 1.653\,64} = 2.529\,89$$

EXAMPLE 8.3

Solve the third-order boundary value problem

$$y''' = 2y'' + 6xy \qquad y(0) = 2 \qquad y(5) = y'(5) = 0$$

and plot y and y' vs. x.

Solution. The first-order equations and the boundary conditions are

$$\mathbf{y}' = \begin{bmatrix} y_0' \\ y_1' \\ y_2' \end{bmatrix} = \begin{bmatrix} y_1 \\ y_2 \\ 2y_2 + 6xy_0 \end{bmatrix}$$

$$y_0(0) = 2 \qquad y_0(5) = y_1(5) = 0$$

The program listed next is based on `example8_1`. Because two of the three boundary conditions are specified at the right end, we start the integration at $x = 5$ and proceed with negative h toward $x = 0$. Two of the three initial conditions are prescribed— $y_0(5) = y_1(5) = 0$—whereas the third condition $y_2(5)$ is unknown. Because the differential equation is linear, we replaced `ridder` with `linInterp`. In linear interpolation the two guesses for $y_2(5)$ (u_1 and u_2) are not important, so we left them as they were in Example 8.1. The adaptive Runge-Kutta method (`run_kut5`) was chosen for the integration.

```python
#!/usr/bin/python
## example8_3
import matplotlib.pyplot as plt
import numpy as np
from run_kut5 import *
from linInterp import *

def initCond(u):   # Initial values of [y,y',y"];
                   # use 'u' if unknown
    return np.array([0.0, 0.0, u])

def r(u):   # Boundary condition residual--see Eq. (8.3)
    X,Y = integrate(F,xStart,initCond(u),xStop,h)
    y = Y[len(Y) - 1]
    r = y[0] - 2.0
    return r

def F(x,y):   # First-order differential equations
    F = np.zeros(3)
    F[0] = y[1]
    F[1] = y[2]
    F[2] = 2.0*y[2] + 6.0*x*y[0]
    return F

xStart = 5.0        # Start of integration
xStop = 0.0         # End of integration
u1 = 1.0            # 1st trial value of unknown init. cond.
u2 = 2.0            # 2nd trial value of unknown init. cond.
h = -0.1            # initial step size
freq = 2            # printout frequency
u = linInterp(r,u1,u2)
X,Y = integrate(F,xStart,initCond(u),xStop,h)

plt.plot(X,Y[:,0],'o-',X,Y[:,1],'^-')
plt.xlabel('x')
```

```
plt.legend(('y','dy/dx'),loc = 3)
plt.grid(True)
plt.show()
input("\nPress return to exit")
```

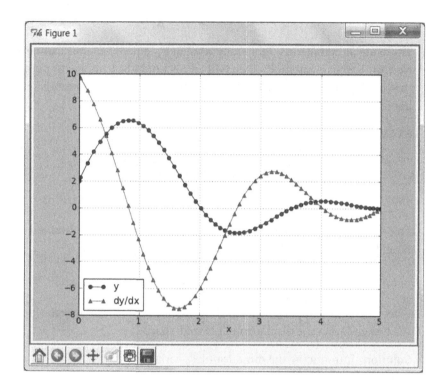

Higher Order Equations

Let us consider the fourth-order differential equation

$$y^{(4)} = f(x, y, y', y'', y''') \tag{8.4a}$$

with the boundary conditions

$$y(a) = \alpha_1 \qquad y''(a) = \alpha_2 \qquad y(b) = \beta_1 \qquad y''(b) = \beta_2 \tag{8.4b}$$

To solve Eqs. (8.4) with the shooting method, we need four initial conditions at $x = a$, only two of which are specified. Denoting the unknown initial values by u_1 and u_2, the set of initial conditions is

$$y(a) = \alpha_1 \qquad y'(a) = u_1 \qquad y''(a) = \alpha_2 \qquad y'''(a) = u_2 \tag{8.5}$$

If Eq. (8.4a) is solved with the shooting method using the initial conditions in Eq. (8.5), the computed boundary values at $x = b$ depend on the choice of u_1 and u_2. We denote this dependence as

$$y(b) = \theta_1(u_1, u_2) \qquad y''(b) = \theta_2(u_1, u_2) \tag{8.6}$$

The correct values u_1 and u_2 satisfy the given boundary conditions at $x = b$,

$$\theta_1(u_1, u_2) = \beta_1 \qquad \theta_2(u_1, u_2) = \beta_2$$

or, using vector notation

$$\boldsymbol{\theta}(\mathbf{u}) = \boldsymbol{\beta} \tag{8.7}$$

These are simultaneous, (generally nonlinear) equations that can be solved by the Newton-Raphson method discussed in Section 4.6. It must be pointed out again that intelligent estimates of u_1 and u_2 are needed if the differential equation is not linear.

EXAMPLE 8.4

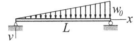

The displacement v of the simply supported beam can be obtained by solving the boundary value problem

$$\frac{d^4 v}{dx^4} = \frac{w_0}{EI}\frac{x}{L} \qquad v = \frac{d^2 v}{dx^2} = 0 \text{ at } x = 0 \text{ and } x = L$$

where EI is the bending rigidity. Determine by numerical integration the slopes at the two ends and the displacement at mid-span.

Solution. Introducing the dimensionless variables

$$\xi = \frac{x}{L} \qquad y = \frac{EI}{w_0 L^4}v$$

the problem is transformed to

$$\frac{d^4 y}{d\xi^4} = \xi \qquad y = \frac{d^2 y}{d\xi^2} = 0 \text{ at } \xi = 0 \text{ and } 1$$

The equivalent first-order equations and the boundary conditions are (the prime denotes $d/d\xi$)

$$\mathbf{y}' = \begin{bmatrix} y_0' \\ y_1' \\ y_2' \\ y_3' \end{bmatrix} = \begin{bmatrix} y_1 \\ y_2 \\ y_3 \\ \xi \end{bmatrix}$$

$$y_0(0) = y_2(0) = y_0(1) = y_2(1) = 0$$

The program listed next is similar to the one in Example 8.1. With appropriate changes in functions F(x,y), initCond(u), and r(u) the program can solve boundary value problems of any order greater than two. For the problem at hand we chose the Bulirsch-Stoer algorithm to do the integration because it gives us control over the printout (we need y precisely at mid-span). The nonadaptive Runge-Kutta method could also be used here, but we would have to guess a suitable step size h.

Because the differential equation is linear, the solution requires only one iteration with the Newton-Raphson method. In this case the initial values $u_1 = dy/d\xi|_{x=0}$ and $u_2 = d^3y/d\xi^3|_{x=0}$ are irrelevant; convergence always occurs in one iteration.

```python
#!/usr/bin/python
## example8_4
import numpy as np
from bulStoer import *
from newtonRaphson2 import *
from printSoln import *

def initCond(u):   # Initial values of [y,y',y",y"'];
                   # use 'u' if unknown
    return np.array([0.0, u[0], 0.0, u[1]])

def r(u):   # Boundary condition residuals--see Eq. (8.7)
    r = np.zeros(len(u))
    X,Y = bulStoer(F,xStart,initCond(u),xStop,H)
    y = Y[len(Y) - 1]
    r[0] = y[0]
    r[1] = y[2]
    return r

def F(x,y):   # First-order differential equations
    F = np.zeros(4)
    F[0] = y[1]
    F[1] = y[2]
    F[2] = y[3]
    F[3] = x
    return F

xStart = 0.0                # Start of integration
xStop = 1.0                 # End of integration
u = np.array([0.0, 1.0])    # Initial guess for {u}
H = 0.5                     # Printout increment
freq = 1                    # Printout frequency
u = newtonRaphson2(r,u,1.0e-4)
X,Y = bulStoer(F,xStart,initCond(u),xStop,H)
printSoln(X,Y,freq)
input("\nPress return to exit")
```

Here is the output:

x	y[0]	y[1]	y[2]	y[3]
0.0000e+00	0.0000e+00	1.9444e-02	0.0000e+00	-1.6667e-01
5.0000e-01	6.5104e-03	1.2150e-03	-6.2500e-02	-4.1667e-02
1.0000e+00	7.6881e-12	-2.2222e-02	7.5002e-11	3.3333e-01

Noting that

$$\frac{dv}{dx} = \frac{dv}{d\xi}\frac{d\xi}{dx} = \left(\frac{w_0 L^4}{EI}\frac{dy}{d\xi}\right)\frac{1}{L} = \frac{w_0 L^3}{EI}\frac{dy}{d\xi}$$

we obtain

$$\left.\frac{dv}{dx}\right|_{x=0} = 19.444 \times 10^{-3}\frac{w_0 L^3}{EI}$$

$$\left.\frac{dv}{dx}\right|_{x=L} = -22.222 \times 10^{-3}\frac{w_0 L^3}{EI}$$

$$v|_{x=0.5L} = 6.5104 \times 10^{-3}\frac{w_0 L^4}{EI}$$

which agree with the analytical solution (easily obtained by direct integration of the differential equation).

EXAMPLE 8.5
Consider the differential equation

$$y^{(4)} + \frac{4}{x}y^3 = 0$$

with the boundary conditions

$$y(0) = y'(0) = 0 \qquad y''(1) = 0 \qquad y'''(1) = 1$$

Use numerical integration to determine $y(1)$.

Solution. Our first task is to handle the indeterminacy of the differential equation at the origin, where $x = y = 0$. The problem is resolved by applying L'Hospital's rule: $4y^3/x \to 12y^2 y'$ as $x \to 0$. Thus the equivalent first-order equations and the boundary conditions that we use in the solution are

$$\mathbf{y}' = \begin{bmatrix} y_0' \\ y_1' \\ y_2' \\ y_3' \end{bmatrix} = \begin{bmatrix} y_1 \\ y_2 \\ y_3 \\ \begin{cases} -12y_0^2 y_1 & \text{if } x = 0 \\ -4y_0^3/x & \text{otherwise} \end{cases} \end{bmatrix}$$

$$y_0(0) = y_1(0) = 0 \qquad y_2(1) = 0 \qquad y_3(1) = 1$$

Because the problem is nonlinear, we need reasonable estimates for $y''(0)$ and $y'''(0)$. Based on the boundary conditions $y''(1) = 0$ and $y'''(1) = 1$, the plot of y'' is likely to look something like this:

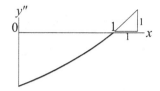

If we are correct, then $y''(0) < 0$ and $y'''(0) > 0$. Based on this rather scanty information, we try $y''(0) = -1$ and $y'''(0) = 1$.

The following program uses the adaptive Runge-Kutta method for integration:

```python
#!/usr/bin/python
## example8_5
import numpy as np
from run_kut5 import *
from newtonRaphson2 import *
from printSoln import *

def initCond(u):  # Initial values of [y,y',y",y"'];
                  # use 'u' if unknown
    return np.array([0.0, 0.0, u[0], u[1]])

def r(u):  # Boundary condition residuals--see Eq. (8.7)
    r = np.zeros(len(u))
    X,Y = integrate(F,x,initCond(u),xStop,h)
    y = Y[len(Y) - 1]
    r[0] = y[2]
    r[1] = y[3] - 1.0
    return r

def F(x,y):  # First-order differential equations
    F = np.zeros(4)
    F[0] = y[1]
    F[1] = y[2]
    F[2] = y[3]
    if x == 0.0: F[3] = -12.0*y[1]*y[0]**2
    else:        F[3] = -4.0*(y[0]**3)/x
    return F

x = 0.0                     # Start of integration
xStop = 1.0                 # End of integration
u = np.array([-1.0, 1.0])   # Initial guess for u
h = 0.1                     # Initial step size
freq = 0                    # Printout frequency
u = newtonRaphson2(r,u,1.0e-5)
```

```
X,Y = integrate(F,x,initCond(u),xStop,h)
printSoln(X,Y,freq)
input("\nPress return to exit")
```

The results are as follows:

```
    x          y[ 0 ]         y[ 1 ]         y[ 2 ]        y[ 3 ]
0.0000e+00   0.0000e+00    0.0000e+00    -9.7607e-01   9.7132e-01
1.0000e+00  -3.2607e-01   -4.8975e-01    -6.7408e-11   1.0000e+00
```

According to the printout, we have $y(1) = -0.326\,07$. Note that by good fortune, our initial estimates $y''(0) = -1$ and $y'''(0) = 1$ were very close to the final values.

PROBLEM SET 8.1

1. Numerical integration of the initial value problem

$$y'' + y' - y = 0 \qquad y(0) = 0 \qquad y'(0) = 1$$

yielded $y(1) = 0.741028$. What is the value of $y'(0)$ that would result in $y(1) = 1$, assuming that $y(0)$ is unchanged?

2. The solution of the differential equation

$$y''' + y'' + 2y' = 6$$

with the initial conditions $y(0) = 2$, $y'(0) = 0$, and $y''(0) = 1$ yielded $y(1) = 3.03765$. When the solution was repeated with $y''(0) = 0$ (the other conditions being unchanged), the result was $y(1) = 2.72318$. Determine the value of $y''(0)$ so that $y(1) = 0$.

3. Roughly sketch the solution of the following boundary value problems. Use the sketch to estimate $y'(0)$ for each problem.

   (a) $y'' = -e^{-y}$ $\qquad$ $y(0) = 1$ $\qquad$ $y(1) = 0.5$

   (b) $y'' = 4y^2$ $\qquad$ $y(0) = 10$ $\qquad$ $y'(1) = 0$

   (c) $y'' = \cos(xy)$ $\qquad$ $y(0) = 0$ $\qquad$ $y(1) = 2$

4. Using a rough sketch of the solution, estimate $y(0)$ for the following boundary value problems.

   (a) $y'' = y^2 + xy$ $\qquad$ $y'(0) = 0$ $\qquad$ $y(1) = 2$

   (b) $y'' = -\dfrac{2}{x}y' - y^2$ $\qquad$ $y'(0) = 0$ $\qquad$ $y(1) = 2$

   (c) $y'' = -x(y')^2$ $\qquad$ $y'(0) = 2$ $\qquad$ $y(1) = 1$

5. Obtain a rough estimate of $y''(0)$ for the following boundary value problem:

$$y''' + 5y''y^2 = 0$$

$$y(0) = 0 \qquad y'(0) = 1 \qquad y(1) = 0$$

6. Obtain rough estimates of $y''(0)$ and $y'''(0)$ for the boundary value problem

$$y^{(4)} + 2y'' + y' \sin y = 0$$

$$y(0) = y'(0) = 0 \qquad y(1) = 5 \qquad y'(1) = 0$$

7. Obtain rough estimates of $\dot{x}(0)$ and $\dot{y}(0)$ for the boundary value problem

$$\ddot{x} + 2x^2 - y = 0 \qquad x(0) = 1 \qquad x(1) = 0$$

$$\ddot{y} + y^2 - 2x = 1 \qquad y(0) = 0 \qquad y(1) = 1$$

8. ■ Solve the following boundary value problem:

$$y'' + (1 - 0.2x)\, y^2 = 0 \qquad y(0) = 0 \qquad y(\pi/2) = 1$$

9. ■ Solve the following boundary value problem:

$$y'' + 2y' + 3y^2 = 0 \qquad y(0) = 0 \qquad y(2) = -1$$

10. ■ Solve this boundary value problem:

$$y'' + \sin y + 1 = 0 \qquad y(0) = 0 \qquad y(\pi) = 0$$

11. ■ Solve the following boundary value problem:

$$y'' + \frac{1}{x}y' + y = 0 \qquad y(0) = 1 \qquad y'(2) = 0$$

and plot y vs. x. *Warning:* y changes very rapidly near $x = 0$.

12. ■ Solve the boundary value problem

$$y'' - \left(1 - e^{-x}\right) y = 0 \qquad y(0) = 1 \qquad y(\infty) = 0$$

and plot y vs. x. *Hint:* Replace the infinity by a finite value β. Check your choice of β by repeating the solution with 1.5β. If the results change, you must increase β.

13. ■ Solve the following boundary value problem:

$$y''' = -\frac{1}{x}y'' + \frac{1}{x^2}y' + 0.1(y')^3$$

$$y(1) = 0 \qquad y''(1) = 0 \qquad y(2) = 1$$

14. ■ Solve the following boundary value problem:

$$y''' + 4y'' + 6y' = 10$$

$$y(0) = y''(0) = 0 \qquad y(3) - y'(3) = 5$$

15. ■ Solve the following boundary value problem:

$$y''' + 2y'' + \sin y = 0$$

$$y(-1) = 0 \qquad y'(-1) = -1 \qquad y'(1) = 1$$

16. ■ Solve the differential equation in Prob. 15 with the boundary conditions

$$y(-1) = 0 \qquad y(0) = 0 \qquad y(1) = 1$$

(this is a three-point boundary value problem).

17. ■ Solve the following boundary value problem:

$$y^{(4)} = -xy^2$$

$$y(0) = 5 \qquad y''(0) = 0 \qquad y'(1) = 0 \qquad y'''(1) = 2$$

18. ■ Solve the following boundary value problem:

$$y^{(4)} = -2yy''$$

$$y(0) = y'(0) = 0 \qquad y(4) = 0 \qquad y'(4) = 1$$

19. ■

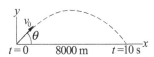

A projectile of mass m in free flight experiences the aerodynamic drag force $F_d = cv^2$, where v is the velocity. The resulting equations of motion are

$$\ddot{x} = -\frac{c}{m}v\dot{x} \qquad \ddot{y} = -\frac{c}{m}v\dot{y} - g$$

$$v = \sqrt{\dot{x}^2 + \dot{y}^2}$$

If the projectile hits a target 8 km away after a 10-s flight, determine the launch velocity v_0 and its angle of inclination θ. Use $m = 20$ kg, $c = 3.2 \times 10^{-4}$ kg/m, and $g = 9.80665$ m/s^2.

20. ■

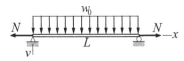

The simply supported beam carries a uniform load of intensity w_0 and the tensile force N. The differential equation for the vertical displacement v can be shown to be

$$\frac{d^4v}{dx^4} - \frac{N}{EI}\frac{d^2v}{dx^2} = \frac{w_0}{EI}$$

where EI is the bending rigidity. The boundary conditions are $v = d^2v/dx^2 = 0$ at $x = 0$ and L. Changing the variables to $\xi = \dfrac{x}{L}$ and $y = \dfrac{EI}{w_0 L^4}v$ transforms the problem to the dimensionless form

$$\frac{d^4 y}{d\xi^4} - \beta \frac{d^2 y}{d\xi^2} = 1 \qquad \beta = \frac{NL^2}{EI}$$

$$y|_{\xi=0} = \frac{d^2 y}{d\xi^2}\bigg|_{\xi=0} = y|_{\xi=0} = \frac{d^2 y}{d\xi^2}\bigg|_{x=1} = 0$$

Determine the maximum displacement if (a) $\beta = 1.65929$ and (b) $\beta = -1.65929$ (N is compressive).

21. ■ Solve the boundary value problem

$$y''' + yy'' = 0 \qquad y(0) = y'(0) = 0, \; y'(\infty) = 2$$

and plot $y(x)$ and $y'(x)$. This problem arises in determining the velocity profile of the boundary layer in incompressible flow (Blasius solution).

22. ■

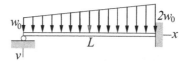

The differential equation that governs the displacement v of the beam shown is

$$\frac{d^4 v}{dx^4} = \frac{w_0}{EI}\left(1 + \frac{x}{L}\right)$$

The boundary conditions are

$$v = \frac{d^2 v}{dx^2} = 0 \text{ at } x = 0 \qquad v = \frac{dv}{dx} = 0 \text{ at } x = L$$

Integrate the differential equation numerically and plot the displacement. Follow the steps used in solving a similar problem in Example 8.4.

8.3   Finite Difference Method

In the finite difference method we divide the range of integration (a, b) into m equal subintervals of length h each, as shown in Figure 8.1. The values of the numerical solution at the mesh points are denoted by y_i, $i = 0, 1, \ldots, m$; the purpose of y_{-1} and y_{m+1} is explained shortly. We now make two approximations:

1. The derivatives of y in the differential equation are replaced by the finite difference expressions. It is common practice to use the first central difference approximations (see Chapter 5):

$$y_i' = \frac{y_{i+1} - y_{i-1}}{2h} \qquad y_i'' = \frac{y_{i-1} - 2y_i + y_{i+1}}{h^2} \qquad \text{etc.} \qquad (8.8)$$

2. The differential equation is enforced only at the mesh points.

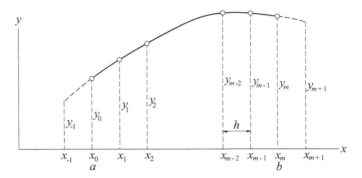

Figure 8.1. Finite difference mesh.

As a result, the differential equations are replaced by $m + 1$ simultaneous algebraic equations, the unknowns being y_i, $i = 0, 1, \ldots .m$. If the differential equation is nonlinear, the algebraic equations will also be nonlinear and must be solved by the Newton-Raphson method.

Since the truncation error in a first central difference approximation is $\mathcal{O}(h^2)$, the finite difference method is not nearly as accurate as the shooting method—recall that the Runge Kutta method has a truncation error of $\mathcal{O}(h^5)$. Therefore, the convergence criterion specified in the Newton-Raphson method should not be too severe.

Second-Order Differential Equation

Consider the second-order differential equation

$$y'' = f(x, y, y')$$

with the boundary conditions

$$y(a) = \alpha \quad \text{or} \quad y'(a) = \alpha$$

$$y(b) = \beta \quad \text{or} \quad y'(b) = \beta$$

Approximating the derivatives at the mesh points by finite differences, the problem becomes

$$\frac{y_{i-1} - 2y_i + y_{i+1}}{h^2} = f\left(x_i, y_i, \frac{y_{i+1} - y_{i-1}}{2h}\right), \quad i = 0, 1, \ldots, m \tag{8.9}$$

$$y_0 = \alpha \quad \text{or} \quad \frac{y_1 - y_{-1}}{2h} = \alpha \tag{8.10a}$$

$$y_m = \beta \quad \text{or} \quad \frac{y_{m+1} - y_{m-1}}{2h} = \beta \tag{8.10b}$$

Note the presence of y_{-1} and y_{m+1}, which are associated with points outside the solution domain (a, b). This "spillover" can be eliminated by using the boundary conditions. But before we do that, let us rewrite Eqs. (8.9) as

$$y_{-1} - 2y_0 + y_1 - h^2 f\left(x_0, y_0, \frac{y_1 - y_{-1}}{2h}\right) = 0 \tag{a}$$

$$y_{i-1} - 2y_i + y_{i+1} - h^2 f\left(x_i, y_i, \frac{y_{i+1} - y_{i-1}}{2h}\right) = 0, \quad i = 1, 2, \ldots, m-1 \tag{b}$$

$$y_{m-1} - 2y_m + y_{m+1} - h^2 f\left(x_m, y_i, \frac{y_{m+1} - y_{m-1}}{2h}\right) = 0 \tag{c}$$

The boundary conditions on y are easily dealt with: Eq. (a) is simply replaced by $y_0 - \alpha = 0$ and Eq. (c) is replaced by $y_m - \beta = 0$. If y' are prescribed, we obtain from Eqs. (8.10) $y_{-1} = y_1 - 2h\alpha$ and $y_{m+1} = y_{m-1} + 2h\beta$, which are then substituted into Eqs. (a) and (c), respectively. Hence we finish up with $m+1$ equations in the unknowns $y_0, y_1, \ldots, y_m$:

$$\left. \begin{aligned} y_0 - \alpha &= 0 && \text{if } y(a) = \alpha \\ -2y_0 + 2y_1 - h^2 f(x_0, y_0, \alpha) - 2h\alpha &= 0 && \text{if } y'(a) = \alpha \end{aligned} \right\} \tag{8.11a}$$

$$y_{i-1} - 2y_i + y_{i+1} - h^2 f\left(x_i, y_i, \frac{y_{i+1} - y_{i-1}}{2h}\right) = 0 \quad i = 1, 2, \ldots, m-1 \tag{8.11b}$$

$$\left. \begin{aligned} y_m - \beta &= 0 && \text{if } y(b) = \beta \\ 2y_{m-1} - 2y_m - h^2 f(x_m, y_m, \beta) + 2h\beta &= 0 && \text{if } y'(b) = \beta \end{aligned} \right\} \tag{8.11c}$$

EXAMPLE 8.6

Write out Eqs. (8.11) for the following linear boundary value problem for $m = 10$:

$$y'' = -4y + 4x \qquad y(0) = 0 \qquad y'(\pi/2) = 0$$

Solve these equations with a computer program.

Solution. In this case $\alpha = y(0) = 0$, $\beta = y'(\pi/2) = 0$, and $f(x, y, y') = -4y + 4x$. Hence Eqs. (8.11) are

$$y_0 = 0$$

$$y_{i-1} - 2y_i + y_{i+1} - h^2(-4y_i + 4x_i) = 0, \quad i = 1, 2, \ldots, m-1$$

$$2y_9 - 2y_{10} - h^2(-4y_{10} + 4x_{10}) = 0$$

or, using matrix notation,

$$\begin{bmatrix} 1 & 0 & & & & \\ 1 & -2+4h^2 & 1 & & & \\ & \ddots & \ddots & \ddots & & \\ & & 1 & -2+4h^2 & 1 & \\ & & & 2 & -2+4h^2 \end{bmatrix} \begin{bmatrix} y_0 \\ y_1 \\ \vdots \\ y_{m-1} \\ y_m \end{bmatrix} = \begin{bmatrix} 0 \\ 4h^2 x_1 \\ \vdots \\ 4h^2 x_{m-1} \\ 4h^2 x_m \end{bmatrix}$$

Note that the coefficient matrix is tridiagonal, so that the equations can be solved efficiently by the decomposition and back substitution routines in module LUde-comp3, described in Section 2.4. Recalling that in LUdecomp3 the diagonals of the coefficient matrix are stored in vectors **c**, **d** and **e**, we arrive at the following program:

```
#!/usr/bin/python
## example8_6
import numpy as np
from LUdecomp3 import *
import math

def equations(x,h,m): # Set up finite difference eqs.
    h2 = h*h
    d = np.ones(m + 1)*(-2.0 + 4.0*h2)
    c = np.ones(m)
    e = np.ones(m)
    b = np.ones(m+1)*4.0*h2*x
    d[0] = 1.0
    e[0] = 0.0
    b[0] = 0.0
    c[m-1] = 2.0
    return c,d,e,b

xStart = 0.0            # x at left end
xStop = math.pi/2.0    # x at right end
m = 10                 # Number of mesh spaces
h = (xStop - xStart)/m
x = np.arange(xStart,xStop + h,h)
c,d,e,b = equations(x,h,m)
c,d,e = LUdecomp3(c,d,e)
y = LUsolve3(c,d,e,b)
print("\n       x                    y")
for i in range(m + 1):
    print('{:14.5e} {:14.5e}'.format(x[i],y[i]))
input("\nPress return to exit")
```

The solution is

x	y
0.00000e+00	0.00000e+00
1.57080e-01	3.14173e-01
3.14159e-01	6.12841e-01
4.71239e-01	8.82030e-01
6.28319e-01	1.11068e+00

```
7.85398e-01      1.29172e+00
9.42478e-01      1.42278e+00
1.09956e+00      1.50645e+00
1.25664e+00      1.54995e+00
1.41372e+00      1.56451e+00
1.57080e+00      1.56418e+00
```

The exact solution of the problem is

$$y = x - \sin 2x$$

which yields $y(\pi/2) = \pi/2 = 1.57080$. Thus the error in the numerical solution is about 0.4%. More accurate results can be achieved by increasing m. For example, with $m = 100$, we would get $y(\pi/2) = 1.57073$, which is in error by only 0.0002%.

EXAMPLE 8.7

Solve the boundary value problem

$$y'' = -3yy' \qquad y(0) = 0 \qquad y(2) = 1$$

with the finite difference method. Use $m = 10$ and compare the output with the results of the shooting method in Example 8.1.

Solution. Because the problem is nonlinear, Eqs. (8.11) must be solved by the Newton-Raphson method. The program listed next can be used as a model for other second-order boundary value problems. The function residual(y) returns the residuals of the finite difference equations, which are the left-hand sides of Eqs. (8.11). The differential equation $y'' = f(x, y, y')$ is defined in the function F(x,y,yPrime). In this problem we chose for the initial solution $y_i = 0.5x_i$, which corresponds to the dashed straight line shown in the rough plot of y in Example 8.1. The starting values of $y_0, y_1, \ldots, y_m$ are specified by function startSoln(x). Note that we relaxed the convergence criterion in the Newton-Raphson method to 1.0×10^{-5}, which is more in line with the truncation error in the finite difference method.

```python
#!/usr/bin/python
## example8_7
import numpy as np
from newtonRaphson2 import *

def residual(y):   # Residuals of finite diff. Eqs. (8.11)
    r = np.zeros(m + 1)
    r[0] = y[0]
    r[m] = y[m] - 1.0
    for i in range(1,m):
        r[i] = y[i-1] - 2.0*y[i] + y[i+1]                      \
             - h*h*F(x[i],y[i],(y[i+1] - y[i-1])/(2.0*h))
    return r
```

```
def F(x,y,yPrime):        # Differential eqn. y" = F(x,y,y')
    F = -3.0*y*yPrime
    return F

def startSoln(x):         # Starting solution y(x)
    y = np.zeros(m + 1)
    for i in range(m + 1): y[i] = 0.5*x[i]
    return y

xStart = 0.0                  # x at left end
xStop = 2.0                   # x at right end
m = 10                        # Number of mesh intervals
h = (xStop - xStart)/m
x = np.arange(xStart,xStop + h,h)
y = newtonRaphson2(residual,startSoln(x),1.0e-5)
print "\n        x                    y"
for i in range(m + 1):
    print "%14.5e %14.5e" %(x[i],y[i])
raw_input("\nPress return to exit")
```

Here is the output from our program together with the solution obtained in
Example 8.1.

x	y	y from Ex. 8.1
0.00000e+000	0.00000e+000	0.00000e+000
2.00000e-001	3.02404e-001	2.94050e-001
4.00000e-001	5.54503e-001	5.41710e-001
6.00000e-001	7.34691e-001	7.21875e-001
8.00000e-001	8.49794e-001	8.39446e-001
1.00000e+000	9.18132e-001	9.10824e-001
1.20000e+000	9.56953e-001	9.52274e-001
1.40000e+000	9.78457e-001	9.75724e-001
1.60000e+000	9.90201e-001	9.88796e-001
1.80000e+000	9.96566e-001	9.96023e-001
2.00000e+000	1.00000e+000	1.00000e+000

The maximum discrepancy between the solutions is 1.8% occurring at $x = 0.6$.
Because the shooting method used in Example 8.1 is considerably more accurate
than the finite difference method, the discrepancy can be attributed to truncation
error in the finite difference solution. This error would be acceptable in many engineering problems. Again, accuracy can be increased by using a finer mesh. With
$m = 100$ we can reduce the error to 0.07%, but we must question whether the 10-fold
increase in computation time is really worth the extra precision.

Fourth-Order Differential Equation

For the sake of brevity we limit our discussion to the special case where y' and y''' do not appear explicitly in the differential equation; that is, we consider

$$y^{(4)} = f(x, y, y'')$$

We assume that two boundary conditions are prescribed at each end of the solution domain (a, b). Problems of this form are commonly encountered in beam theory.

Again we divide the solution domain into m intervals of length h each. Replacing the derivatives of y by finite differences at the mesh points, we get the finite difference equations

$$\frac{y_{i-2} - 4y_{i-1} + 6y_i - 4y_{i+1} + y_{i+2}}{h^4} = f\left(x_i, y_i, \frac{y_{i-1} - 2y_i + y_{i+1}}{h^2}\right) \tag{8.12}$$

where $i = 0, 1, \ldots, m$. It is more revealing to write these equations as

$$y_{-2} - 4y_{-1} + 6y_0 - 4y_1 + y_2 - h^4 f\left(x_0, y_0, \frac{y_{-1} - 2y_0 + y_1}{h^2}\right) = 0 \tag{8.13a}$$

$$y_{-1} - 4y_0 + 6y_1 - 4y_2 + y_3 - h^4 f\left(x_1, y_1, \frac{y_0 - 2y_1 + y_2}{h^2}\right) = 0 \tag{8.13b}$$

$$y_0 - 4y_1 + 6y_2 - 4y_3 + y_4 - h^4 f\left(x_2, y_2, \frac{y_1 - 2y_2 + y_3}{h^2}\right) = 0 \tag{8.13c}$$

$$\vdots$$

$$y_{m-3} - 4y_{m-2} + 6y_{m-1} - 4y_m + y_{m+1} - h^4 f\left(x_{m-1}, y_{m-1}, \frac{y_{m-2} - 2y_{m-1} + y_m}{h^2}\right) = 0 \tag{8.13d}$$

$$y_{m-2} - 4y_{m-1} + 6y_m - 4y_{m+1} + y_{m+2} - h^4 f\left(x_m, y_m, \frac{y_{m-1} - 2y_m + y_{m+1}}{h^2}\right) = 0 \tag{8.13e}$$

We now see that there are four unknowns— y_{-2}, y_{-1}, y_{m+1}, and y_{m+2}—that lie outside the solution domain and must be eliminated by applying the boundary conditions, a task that is facilitated by Table 8.1.

The astute observer may notice that some combinations of boundary conditions will not work in eliminating the "spillover." One such combination is clearly $y(a) = \alpha_1$ and $y'''(a) = \alpha_2$. The other one is $y'(a) = \alpha_1$ and $y''(a) = \alpha_2$. In the context of beam theory, this makes sense: We can impose either a displacement y or a shear force EIy''' at a point, but it is impossible to enforce both simultaneously. Similarly, it makes no physical sense to prescribe both the slope y' and the bending moment EIy'' at the same point.

Boundary cond.	Equivalent finite difference expression
$y(a) = \alpha$	$y_0 = \alpha$
$y'(a) = \alpha$	$y_{-1} = y_1 - 2h\alpha$
$y''(a) = \alpha$	$y_{-1} = 2y_0 - y_1 + h^2\alpha$
$y'''(a) = \alpha$	$y_{-2} = 2y_{-1} - 2y_1 + y_2 - 2h^3\alpha$
$y(b) = \beta$	$y_m = \beta$
$y'(b) = \beta$	$y_{m+1} = y_{m-1} + 2h\beta$
$y''(b) = \beta$	$y_{m+1} = 2y_m - y_{m-1} + h^2\beta$
$y'''(b) = \beta$	$y_{m+2} = 2y_{m+1} - 2y_{m-1} + y_{m-2} + 2h^3\beta$

Table 8.1. Finite difference approximations for boundary conditions.

EXAMPLE 8.8

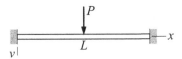

The uniform beam of length L and bending rigidity EI is attached to rigid supports at both ends. The beam carries a concentrated load P at its mid-span. If we utilize symmetry and model only the left half of the beam, the displacement v can be obtained by solving the boundary value problem

$$EI\frac{d^4v}{dx^4} = 0$$

$$v|_{x=0} = 0 \qquad \frac{dv}{dx}\bigg|_{x=0} = 0 \qquad \frac{dv}{dx}\bigg|_{x=L/2} = 0 \qquad EI\,\frac{d^3v}{dx^3}\bigg|_{x=L/2} = -P/2$$

Use the finite difference method to determine the displacement and the bending moment $M = -EI\,d^2v/dx^2$ at the mid-span (the exact values are $v = PL^3/(192EI)$ and $M = PL/8$).

Solution. By introducing the dimensionless variables

$$\xi = \frac{x}{L} \qquad y = \frac{EI}{PL^3}v$$

the problem becomes

$$\frac{d^4y}{d\xi^4} = 0$$

$$y|_{\xi=0} = 0 \qquad \frac{dy}{d\xi}\bigg|_{\xi=0} = 0 \qquad \frac{dy}{d\xi}\bigg|_{\xi=1/2} = 0 \qquad \frac{d^3y}{d\xi^3}\bigg|_{\xi=1/2} = -\frac{1}{2}$$

We now proceed to writing Eqs. (8.13), taking into account the boundary conditions. Referring to Table 8.1, the finite difference expressions of the boundary conditions at the left end are $y_0 = 0$ and $y_{-1} = y_1$. Hence Eqs. (8.13a) and (8.13b) become

$$y_0 = 0 \tag{a}$$

$$-4y_0 + 7y_1 - 4y_2 + y_3 = 0 \tag{b}$$

Equation (8.13c) is

$$y_0 - 4y_1 + 6y_2 - 4y_3 + y_4 = 0 \tag{c}$$

At the right end the boundary conditions are equivalent to $y_{m+1} = y_{m-1}$ and

$$y_{m+2} = 2y_{m+1} + y_{m-2} - 2y_{m-1} + 2h^3(-1/2) = y_{m-2} - h^3$$

Substitution into Eqs. (8.13d) and (8.13e) yields

$$y_{m-3} - 4y_{m-2} + 7y_{m-1} - 4y_m = 0 \tag{d}$$

$$2y_{m-2} - 8y_{m-1} + 6y_m = h^3 \tag{e}$$

The coefficient matrix of Eqs. (a)-(e) can be made symmetric by dividing Eq. (e) by 2. The result is

$$
\begin{bmatrix}
1 & 0 & 0 & & & & & \\
0 & 7 & -4 & 1 & & & & \\
0 & -4 & 6 & -4 & 1 & & & \\
 & \ddots & \ddots & \ddots & \ddots & \ddots & & \\
 & & 1 & -4 & 6 & -4 & 1 & \\
 & & & 1 & -4 & 7 & -4 & \\
 & & & & 1 & -4 & 3 &
\end{bmatrix}
\begin{bmatrix}
y_0 \\ y_1 \\ y_2 \\ \vdots \\ y_{m-2} \\ y_{m-1} \\ y_m
\end{bmatrix}
=
\begin{bmatrix}
0 \\ 0 \\ 0 \\ \vdots \\ 0 \\ 0 \\ 0.5h^3
\end{bmatrix}
$$

This system of equations can be solved with the decomposition and back substitution routines in module LUdecomp5—see Section 2.4. Recall that LUdecomp5 works with the vectors **d**, **e**, and **f** that form the diagonals of the upper half of the matrix. The constant vector is denoted by **b**. The program that sets up solves the equations is as follows:

```
#!/usr/bin/python
## example8_8
import numpy as np
from LUdecomp5 import *

def equations(x,h,m): # Set up finite difference eqs.
    h4 = h**4
    d = np.ones(m + 1)*6.0
    e = np.ones(m)*(-4.0)
    f = np.ones(m-1)
    b = np.zeros(m+1)
```

```
        d[0] = 1.0
        d[1] = 7.0
        e[0] = 0.0
        f[0] = 0.0
        d[m-1] = 7.0
        d[m] = 3.0
        b[m] = 0.5*h**3
        return d,e,f,b


xStart = 0.0         # x at left end
xStop = 0.5          # x at right end
m = 20               # Number of mesh spaces
h = (xStop - xStart)/m
x = np.arange(xStart,xStop + h,h)
d,e,f,b = equations(x,h,m)
d,e,f = LUdecomp5(d,e,f)
y = LUsolve5(d,e,f,b)
print('\n         x                     y')
print('{:14.5e} {:14.5e}'.format(x[m-1],y[m-1]))
print('{:14.5e} {:14.5e}'.format(x[m],y[m]))
input("\nPress return to exit")
```

We ran the program with $m = 20$, printing only the last two lines of the solution:

```
       x                y
  4.75000e-01    5.19531e-03
  5.00000e-01    5.23438e-03
```

Thus at the mid-span we have

$$v|_{x=0.5L} = \frac{PL^3}{EI} y|_{\xi=0.5} = 5.234\,38 \times 10^{-3} \frac{PL^3}{EI}$$

$$\frac{d^2 v}{dx^2}\bigg|_{x=0.5L} = \frac{PL^3}{EI} \left(\frac{1}{L^2} \frac{d^2 y}{d\xi^2}\bigg|_{\xi=0.5} \right) \approx \frac{PL}{EI} \frac{y_{m-1} - 2y_m + y_{m+1}}{h^2}$$

$$= \frac{PL}{EI} \frac{(5.19531 - 2(5.23438) + 5.19531) \times 10^{-3}}{0.025^2}$$

$$= -0.125\,024 \frac{PL}{EI}$$

$$M|_{x=0.5L} = -EI \frac{d^2 v}{dx^2}\bigg|_{\xi=0.5} = 0.125\,024\, PL$$

In comparison, the exact solution yields

$$v|_{x=0.5L} = 5.208\,33 \times 10^{-3}\frac{PL^3}{EI}$$

$$M|_{x=0.5L} = = 0.125\,000\,PL$$

PROBLEM SET 8.2

Problems 1–5 Use first central difference approximations to transform the boundary value problem shown into simultaneous equations $\mathbf{Ay} = \mathbf{b}$.

Problems 6–10 Solve the given boundary value problem with the finite difference method using $m = 20$.

1. $y'' = (2 + x)y$, $y(0) = 0$, $y'(1) = 5$.
2. $y'' = y + x^2$, $y(0) = 0$, $y(1) = 1$.
3. $y'' = e^{-x}y'$, $y(0) = 1$, $y(1) = 0$.
4. $y^{(4)} = y'' - y$, $y(0) = 0$, $y'(0) = 1$, $y(1) = 0$, $y'(1) = -1$.
5. $y^{(4)} = -9y + x$, $y(0) = y''(0) = 0$, $y'(1) = y'''(1) = 0$.
6. ■ $y'' = xy$, $y(1) = 1.5$, $y(2) = 3$.
7. ■ $y'' + 2y' + y = 0$, $y(0) = 0$, $y(1) = 1$. Exact solution is $y = xe^{1-x}$.
8. ■ $x^2 y'' + xy' + y = 0$, $y(1) = 0$, $y(2) = 0.638961$. Exact solution is $y = \sin(\ln x)$.
9. ■ $y'' = y^2 \sin y$, $y'(0) = 0$, $y(\pi) = 1$.
10. ■ $y'' + 2y(2xy' + y) = 0$, $y(0) = 1/2$, $y'(1) = -2/9$. Exact solution is $y = (2 + x^2)^{-1}$.
11. ■

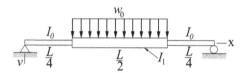

The simply supported beam consists of three segments with the moments of inertia I_0 and I_1 as shown. A uniformly distributed load of intensity w_0 acts over the middle segment. Modeling only the *left half* of the beam, the differential equation

$$\frac{d^2 v}{dx^2} = -\frac{M}{EI}$$

for the displacement v is

$$\frac{d^2 v}{dx^2} = -\frac{w_0 L^2}{4EI_0} \times \begin{cases} \dfrac{x}{L} & \text{in } 0 < x < \dfrac{L}{4} \\[3mm] \dfrac{I_0}{I_1}\left[\dfrac{x}{L} - 2\left(\dfrac{x}{L} - \dfrac{1}{4}\right)^2\right] & \text{in } \dfrac{L}{4} < x < \dfrac{L}{2} \end{cases}$$

Introducing the dimensionless variables

$$\xi = \frac{x}{L} \qquad y = \frac{EI_0}{w_0 L^4} v \qquad \gamma = \frac{I_1}{I_0}$$

the differential equation becomes

$$\frac{d^2 y}{d\xi^2} = \begin{cases} -\dfrac{1}{4}\xi & \text{in } 0 < \xi < \dfrac{1}{4} \\[3mm] -\dfrac{1}{4\gamma}\left[\xi - 2\left(\xi - \dfrac{1}{4}\right)^2\right] & \text{in } \dfrac{1}{4} < \xi < \dfrac{1}{2} \end{cases}$$

with the boundary conditions

$$y|_{\xi=0} = \frac{d^2 y}{d\xi^2}\bigg|_{\xi=0} = \frac{dy}{d\xi}\bigg|_{\xi=1/2} = \frac{d^3 y}{d\xi^3}\bigg|_{\xi=1/2} = 0$$

Use the finite difference method to determine the maximum displacement of the beam using $m = 20$ and $\gamma = 1.5$ and compare it with the exact solution

$$v_{max} = \frac{61}{9216}\frac{w_0 L^4}{EI_0}$$

12. ■

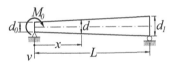

The simply supported, tapered beam has a circular cross section. A couple of magnitude M_0 is applied to the left end of the beam. The differential equation for the displacement v is

$$\frac{d^2 v}{dx^2} = -\frac{M}{EI} = -\frac{M_0(1 - x/L)}{EI_0(d/d_0)^4}$$

where

$$d = d_0\left[1 + \left(\frac{d_1}{d_0} - 1\right)\frac{x}{L}\right] \qquad I_0 = \frac{\pi d_0^4}{64}$$

Substituting

$$\xi = \frac{x}{L} \qquad y = \frac{EI_0}{M_0 L^2} v \qquad \delta = \frac{d_1}{d_0}$$

the differential equation becomes

$$\frac{d^2 y}{d\xi^2} = -\frac{1 - \xi}{[1 + (\delta - 1)\xi]^4}$$

with the boundary conditions

$$y|_{\xi=0} = \frac{d^2 y}{dx^2}\bigg|_{\xi=0} = y|_{\xi=1} = \frac{d^2 y}{dx^2}\bigg|_{\xi=1} = 0$$

Solve the problem with the finite difference method with $\delta = 1.5$ and $m = 20$; plot y vs. ξ. The exact solution is

$$y = -\frac{(3 + 2\delta\xi - 3\xi)\xi^2}{6(1 + \delta\xi - \xi^2)} + \frac{1}{3\delta}$$

13. ■ Solve Example 8.4 by the finite difference method with $m = 20$. *Hint*: Compute end slopes from second noncentral differences in Tables 5.3.

14. ■ Solve Prob. 20 in Problem Set 8.1 with the finite difference method. Use $m = 20$.

15. ■

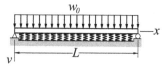

The simply supported beam of length L is resting on an elastic foundation of stiffness k N/m^2. The displacement v of the beam due to the uniformly distributed load of intensity w_0 N/m is given by the solution of the boundary value problem

$$EI\frac{d^4v}{dx^4} + kv = w_0, \quad v|_{x=0} = \frac{d^2y}{dx^2}\bigg|_{x=0} = v|_{x=L} = \frac{d^2v}{dx^2}\bigg|_{x=L} = 0$$

The nondimensional form of the problem is

$$\frac{d^2y}{d\xi^4} + \gamma y = 1, \quad y|_{\xi=0} = \frac{d^2y}{dx^2}\bigg|_{\xi-0} = y|_{\xi=1} = \frac{d^2y}{dx^2}\bigg|_{\xi=1} = 0$$

where

$$\xi = \frac{x}{L} \qquad y = \frac{EI}{w_0 L^4}v \qquad \gamma = \frac{kL^4}{EI}$$

Solve this problem by the finite difference method with $\gamma = 10^5$ and plot y vs. ξ.

16. ■ Solve Prob. 15 if the ends of the beam are free and the load is confined to the middle half of the beam. Consider only the left half of the beam, in which case the nondimensional form of the problem is

$$\frac{d^4y}{d\xi^4} + \gamma y = \begin{cases} 0 & \text{in } 0 < \xi < 1/4 \\ 1 & \text{in } 1/4 < \xi < 1/2 \end{cases}$$

$$\frac{d^2y}{d\xi^2}\bigg|_{\xi=0} = \frac{d^3y}{d\xi^3}\bigg|_{\xi=0} = \frac{dy}{d\xi}\bigg|_{\xi=1/2} = \frac{d^3y}{d\xi^3}\bigg|_{\xi=1/2} = 0$$

17. ■ The general form of a linear, second-order boundary value problem is

$$y'' = r(x) + s(x)y + t(x)y'$$

$$y(a) = \alpha \text{ or } y'(a) = \alpha$$

$$y(b) = \beta \text{ or } y'(b) = \beta$$

Write a program that solves this problem with the finite difference method for any user-specified $r(x)$, $s(x)$, and $t(x)$. Test the program by solving Prob. 8.

18. ■

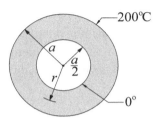

The thick cylinder conveys a fluid with a temperature of $0°$ C. At the same time the cylinder is immersed in a bath that is kept at $200°$ C. The differential equation and the boundary conditions that govern steady-state heat conduction in the cylinder are

$$\frac{d^2 T}{dr^2} = -\frac{1}{r}\frac{dT}{dr} \qquad T|_{r=a/2} = 0 \qquad T|_{r=a} = 200° \text{ C}$$

where T is the temperature. Determine the temperature profile through the thickness of the cylinder with the finite difference method, and compare it with the analytical solution

$$T = 200\left(1 - \frac{\ln r/a}{\ln 0.5}\right)$$

9 Symmetric Matrix Eigenvalue Problems

$$\boxed{\text{Find } \lambda \text{ for which nontrivial solutions of } \mathbf{Ax} = \lambda\mathbf{x} \text{ exist.}}$$

9.1 Introduction

The *standard form* of the matrix eigenvalue problem is

$$\mathbf{Ax} = \lambda\mathbf{x} \tag{9.1}$$

where $\mathbf{A}$ is a given $n \times n$ matrix. The problem is to find the scalar λ and the vector $\mathbf{x}$. Rewriting Eq. (9.1) in the form

$$(\mathbf{A} - \lambda\mathbf{I})\,\mathbf{x} = \mathbf{0} \tag{9.2}$$

it becomes apparent that we are dealing with a system of n homogeneous equations. An obvious solution is the trivial one $\mathbf{x} = \mathbf{0}$. A nontrivial solution can exist only if the determinant of the coefficient matrix vanishes; that is, if

$$|\mathbf{A} - \lambda\mathbf{I}| = \mathbf{0} \tag{9.3}$$

Expansion of the determinant leads to the polynomial equation, also known as the *characteristic equation*

$$a_0 + a_1\lambda + a_2\lambda^2 + \cdots + a_n\lambda^n = 0$$

which has the roots λ_i, $i = 1, 2, \ldots, n$, called the *eigenvalues* of the matrix $\mathbf{A}$. The solutions $\mathbf{x}_i$ of $(\mathbf{A} - \lambda_i\mathbf{I})\,\mathbf{x} = \mathbf{0}$ are known as the eigenvectors.

As an example, consider the matrix

$$\mathbf{A} = \begin{bmatrix} 1 & -1 & 0 \\ -1 & 2 & -1 \\ 0 & -1 & 1 \end{bmatrix} \tag{a}$$

The characteristic equation is

$$|\mathbf{A} - \lambda\mathbf{I}| = \begin{vmatrix} 1-\lambda & -1 & 0 \\ -1 & 2-\lambda & -1 \\ 0 & -1 & 1-\lambda \end{vmatrix} = -3\lambda + 4\lambda^2 - \lambda^3 = 0 \tag{b}$$

The roots of this equation are $\lambda_1 = 0$, $\lambda_2 = 1$, $\lambda_3 = 3$. To compute the eigenvector corresponding the λ_3, we substitute $\lambda = \lambda_3$ into Eq. (9.2), obtaining

$$\begin{bmatrix} -2 & -1 & 0 \\ -1 & -1 & -1 \\ 0 & -1 & -2 \end{bmatrix} \begin{bmatrix} x_1 \\ x_2 \\ x_3 \end{bmatrix} = \begin{bmatrix} 0 \\ 0 \\ 0 \end{bmatrix} \tag{c}$$

We know that the determinant of the coefficient matrix is zero, so that the equations are not linearly independent. Therefore, we can assign an arbitrary value to any one component of $\mathbf{x}$ and use two of the equations to compute the other two components. Choosing $x_1 = 1$, the first equation of Eq. (c) yields $x_2 = -2$, and from the third equation we get $x_3 = 1$. Thus the eigenvector associated with λ_3 is

$$\mathbf{x}_3 = \begin{bmatrix} 1 \\ -2 \\ 1 \end{bmatrix}$$

The other two eigenvectors

$$\mathbf{x}_2 = \begin{bmatrix} 1 \\ 0 \\ -1 \end{bmatrix} \qquad \mathbf{x}_1 = \begin{bmatrix} 1 \\ 1 \\ 1 \end{bmatrix}$$

can be obtained in the same manner.

It is sometimes convenient to display the eigenvectors as columns of a matrix $\mathbf{X}$. For the problem at hand, this matrix is

$$\mathbf{X} = \begin{bmatrix} \mathbf{x}_1 & \mathbf{x}_2 & \mathbf{x}_3 \end{bmatrix} = \begin{bmatrix} 1 & 1 & 1 \\ 1 & 0 & -2 \\ 1 & -1 & 1 \end{bmatrix}$$

It is clear from this example that the magnitude of an eigenvector is indeterminate; only its direction can be computed from Eq. (9.2). It is customary to *normalize* the eigenvectors by assigning a unit magnitude to each vector. Thus the normalized eigenvectors in our example are

$$\mathbf{X} = \begin{bmatrix} 1/\sqrt{3} & 1/\sqrt{2} & 1/\sqrt{6} \\ 1/\sqrt{3} & 0 & -2/\sqrt{6} \\ 1/\sqrt{3} & -1/\sqrt{2} & 1/\sqrt{6} \end{bmatrix}$$

Throughout this chapter we assume that the eigenvectors are normalized.

Here are some useful properties of eigenvalues and eigenvectors, given without proof:

- All the eigenvalues of a symmetric matrix are real.
- All eigenvalues of a symmetric, positive-definite matrix are real and positive.
- The eigenvectors of a symmetric matrix are orthonormal; that is, $\mathbf{X}^T\mathbf{X} = \mathbf{I}$.
- If the eigenvalues of $\mathbf{A}$ are λ_i, then the eigenvalues of $\mathbf{A}^{-1}$ are λ_i^{-1}.

Eigenvalue problems that originate from physical problems often end up with a symmetric **A**. This is fortunate, because symmetric eigenvalue problems are easier to solve than their nonsymmetric counterparts (which may have complex eigenvalues). In this chapter we largely restrict our discussion to eigenvalues and eigenvectors of symmetric matrices.

Common sources of eigenvalue problems are analyses of vibrations and stability. These problems often have the following characteristics:

- The matrices are large and sparse (e.g., have a banded structure).
- We need to know only the eigenvalues; if eigenvectors are required, only a few of them are of interest.

A useful eigenvalue solver must be able to use these characteristics to minimize the computations. In particular, it should be flexible enough to compute only what we need and no more.

EXAMPLE 9.1

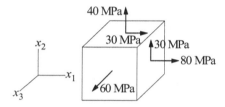

The stress matrix (tensor) corresponding to the state of stress shown is

$$\mathbf{S} = \begin{bmatrix} 80 & 30 & 0 \\ 30 & 40 & 0 \\ 0 & 0 & 60 \end{bmatrix} \text{MPa}$$

(each row of the matrix consists of the three stress components acting on a coordinate plane). It can be shown that the eigenvalues of **S** are the *principal stresses* and that the eigenvectors are normal to the *principal planes*. Determine the principal stresses and the eigenvectors.

Solution. The characteristic equation $|\mathbf{S} - \lambda\mathbf{I}| = \mathbf{0}$ is

$$\begin{vmatrix} 80 - \lambda & 30 & 0 \\ 30 & 40 - \lambda & 0 \\ 0 & 0 & 60 - \lambda \end{vmatrix} = 0$$

Expanding the determinant gives us

$$(60 - \lambda)\left[(80 - \lambda)(40 - \lambda) - 900\right] = 0$$

$$(60 - \lambda)(\lambda^2 - 120\lambda + 2300) = 0$$

which yields the principal stresses,

$$\lambda_1 = 23.944 \text{ MPa} \qquad \lambda_2 = 60 \text{ MPa} \qquad \lambda_3 = 96.056 \text{ MPa}$$

The first eigenvector is the solution of $(\mathbf{S} - \lambda_1\mathbf{I})\mathbf{x} = \mathbf{0}$, or

$$\begin{bmatrix} 56.056 & 30.0 & 0 \\ 30.0 & 16.056 & 0 \\ 0 & 0 & 36.056 \end{bmatrix} \begin{bmatrix} x_1 \\ x_2 \\ x_3 \end{bmatrix} = \begin{bmatrix} 0 \\ 0 \\ 0 \end{bmatrix}$$

Choosing $x_1 = 1$, we get $x_2 = -56.056/30 = -1.8685$ and $x_3 = 0$. Thus the normalized eigenvector is

$$\mathbf{x}_1 = \begin{bmatrix} 0.4719 & -0.8817 & 0 \end{bmatrix}^T$$

The second eigenvector is obtained from the equation $(\mathbf{S} - \lambda_2\mathbf{I})\mathbf{x} = \mathbf{0}$:

$$\begin{bmatrix} 20 & 30 & 0 \\ 30 & -20 & 0 \\ 0 & 0 & 0 \end{bmatrix} \begin{bmatrix} x_1 \\ x_2 \\ x_3 \end{bmatrix} = \begin{bmatrix} 0 \\ 0 \\ 0 \end{bmatrix}$$

This equation is satisfied with $x_1 = x_2 = 0$ and any value of x_3. Choosing $x_3 = 1$, the eigenvector becomes

$$\mathbf{x}_2 = \begin{bmatrix} 0 & 0 & 1 \end{bmatrix}^T$$

The third eigenvector is obtained by solving $(\mathbf{S} - \lambda_3\mathbf{I})\mathbf{x} = \mathbf{0}$:

$$\begin{bmatrix} -16.056 & 30 & 0 \\ 30 & -56.056 & 0 \\ 0 & 0 & -36.056 \end{bmatrix} \begin{bmatrix} x_1 \\ x_2 \\ x_3 \end{bmatrix} = \begin{bmatrix} 0 \\ 0 \\ 0 \end{bmatrix}$$

With the choice $x_1 = 1$, this yields $x_2 = 16.056/30 = 0.5352$ and $x_3 = 0$. After normalizing we get

$$\mathbf{x}_3 = \begin{bmatrix} 0.8817 & 0.4719 & 0 \end{bmatrix}^T$$

9.2 Jacobi Method

Jacobi method is a relatively simple iterative procedure that extracts *all* the eigenvalues and eigenvectors of a symmetric matrix. Its utility is limited to small matrices (say, less than 50 × 50), because the computational effort increases very rapidly with the size of the matrix. The main strength of the method is its robustness—it seldom fails to deliver.

Similarity Transformation and Diagonalization

Consider the standard matrix eigenvalue problem

$$\mathbf{Ax} = \lambda\mathbf{x} \qquad (9.4)$$

where $\mathbf{A}$ is symmetric. Let us now apply the transformation

$$\mathbf{x} = \mathbf{P}\mathbf{x}^* \tag{9.5}$$

where $\mathbf{P}$ is a nonsingular matrix. Substituting Eq. (9.5) into Eq. (9.4) and pre-multiplying each side by $\mathbf{P}^{-1}$, we get

$$\mathbf{P}^{-1}\mathbf{A}\mathbf{P}\mathbf{x}^* = \lambda\mathbf{P}^{-1}\mathbf{P}\mathbf{x}^*$$

or

$$\mathbf{A}^*\mathbf{x}^* = \lambda\mathbf{x}^* \tag{9.6}$$

where $\mathbf{A}^* = \mathbf{P}^{-1}\mathbf{A}\mathbf{P}$. Because λ was untouched by the transformation, the eigenvalues of $\mathbf{A}$ are also the eigenvalues of $\mathbf{A}^*$. Matrices that have the same eigenvalues are deemed to be *similar*, and the transformation between them is called a *similarity transformation*.

Similarity transformations are frequently used to change an eigenvalue problem to a form that is easier to solve. Suppose that we managed by some means to find a $\mathbf{P}$ that diagonalizes $\mathbf{A}^*$. Equations (9.6) then are

$$\begin{bmatrix} A_{11}^* - \lambda & 0 & \cdots & 0 \\ 0 & A_{22}^* - \lambda & \cdots & 0 \\ \vdots & \vdots & \ddots & \vdots \\ 0 & 0 & \cdots & A_{nn}^* - \lambda \end{bmatrix}\begin{bmatrix} x_1^* \\ x_2^* \\ \vdots \\ x_n^* \end{bmatrix} = \begin{bmatrix} 0 \\ 0 \\ \vdots \\ 0 \end{bmatrix}$$

which have the solutions

$$\lambda_1 = A_{11}^* \qquad \lambda_2 = A_{22}^* \qquad \cdots \qquad \lambda_n = A_{nn}^* \tag{9.7}$$

$$\mathbf{x}_1^* = \begin{bmatrix} 1 \\ 0 \\ \vdots \\ 0 \end{bmatrix} \qquad \mathbf{x}_2^* = \begin{bmatrix} 0 \\ 1 \\ \vdots \\ 0 \end{bmatrix} \qquad \cdots \qquad \mathbf{x}_n^* = \begin{bmatrix} 0 \\ 0 \\ \vdots \\ 1 \end{bmatrix}$$

or

$$\mathbf{X}^* = \begin{bmatrix} \mathbf{x}_1^* & \mathbf{x}_2^* & \cdots & \mathbf{x}_n^* \end{bmatrix} = \mathbf{I}$$

According to Eq. (9.5) the eigenvectors of $\mathbf{A}$ are

$$\mathbf{X} = \mathbf{P}\mathbf{X}^* = \mathbf{P}\mathbf{I} = \mathbf{P} \tag{9.8}$$

Hence the transformation matrix $\mathbf{P}$ contains the eigenvectors of $\mathbf{A}$, and the eigenvalues of $\mathbf{A}$ are the diagonal terms of $\mathbf{A}^*$.

Jacobi Rotation

A special similarity transformation is the plane rotation

$$\mathbf{x} = \mathbf{R}\mathbf{x}^* \tag{9.9}$$

where

$$
\mathbf{R} = \begin{array}{c} \\ \\ \\ \\ \\ \\ \\ \\ \end{array} \begin{bmatrix} 1 & 0 & 0 & 0 & 0 & 0 & 0 & 0 \\ 0 & 1 & 0 & 0 & 0 & 0 & 0 & 0 \\ 0 & 0 & c & 0 & 0 & s & 0 & 0 \\ 0 & 0 & 0 & 1 & 0 & 0 & 0 & 0 \\ 0 & 0 & 0 & 0 & 1 & 0 & 0 & 0 \\ 0 & 0 & -s & 0 & 0 & c & 0 & 0 \\ 0 & 0 & 0 & 0 & 0 & 0 & 1 & 0 \\ 0 & 0 & 0 & 0 & 0 & 0 & 0 & 1 \end{bmatrix} \begin{array}{c} \\ \\ k \\ \\ \\ \ell \\ \\ \\ \end{array} \tag{9.10}
$$

is called the *Jacobi rotation matrix*. Note that $\mathbf{R}$ is an identity matrix modified by the terms $c = \cos\theta$ and $s = \sin\theta$ appearing at the intersections of columns/rows k and ℓ, where θ is the rotation angle. The rotation matrix has the useful property of being *orthogonal*, meaning that

$$
\mathbf{R}^{-1} = \mathbf{R}^T \tag{9.11}
$$

One consequence of orthogonality is that the transformation in Eq. (9.5) has the essential characteristic of a rotation: It preserves the magnitude of the vector; that is, $|\mathbf{x}| = |\mathbf{x}^*|$.

The similarity transformation corresponding to the plane rotation in Eq. (9.9) is

$$
\mathbf{A}^* = \mathbf{R}^{-1}\mathbf{A}\mathbf{R} = \mathbf{R}^T\mathbf{A}\mathbf{R} \tag{9.12}
$$

The matrix $\mathbf{A}^*$ not only has the same eigenvalues as the original matrix $\mathbf{A}$ but thanks to orthogonality of $\mathbf{R}$, it is also symmetric. The transformation in Eq. (9.12) changes only the rows/columns k and ℓ of $\mathbf{A}$. The formulas for these changes are

$$
\begin{aligned}
A^*_{kk} &= c^2 A_{kk} + s^2 A_{\ell\ell} - 2cs A_{k\ell} \\
A^*_{\ell\ell} &= c^2 A_{\ell\ell} + s^2 A_{kk} + 2cs A_{k\ell} \\
A^*_{k\ell} &= A^*_{\ell k} = (c^2 - s^2)A_{k\ell} + cs(A_{kk} - A_{\ell\ell}) \\
A^*_{ki} &= A^*_{ik} = cA_{ki} - sA_{\ell i}, \quad i \neq k, \quad i \neq \ell \\
A^*_{\ell i} &= A^*_{i\ell} = cA_{\ell i} + sA_{ki}, \quad i \neq k, \quad i \neq \ell
\end{aligned} \tag{9.13}
$$

Jacobi Diagonalization

The angle θ in the Jacobi rotation matrix can be chosen so that $A^*_{k\ell} = A^*_{\ell k} = 0$. This suggests the following idea: Why not diagonalize $\mathbf{A}$ by looping through all the off-diagonal terms and zero them one by one? This is exactly what the Jacobi method does. However, there is a major snag—the transformation that annihilates an off-diagonal term also undoes some of the previously created zeroes. Fortunately, it turns out that the off-diagonal terms that reappear will be smaller than before. Thus the Jacobi method is an iterative procedure that repeatedly applies Jacobi rotations until the off-diagonal terms have virtually vanished. The final transformation matrix $\mathbf{P}$ is the accumulation of individual rotations $\mathbf{R}_i$:

$$\mathbf{P} = \mathbf{R}_1 \cdot \mathbf{R}_2 \cdot \mathbf{R}_3 \cdots \tag{9.14}$$

The columns of $\mathbf{P}$ finish up as the eigenvectors of $\mathbf{A}$, and the diagonal elements of $\mathbf{A}^* = \mathbf{P}^T \mathbf{A} \mathbf{P}$ become the eigenvectors.

Let us now look at details of a Jacobi rotation. From Eq. (9.13) we see that $A_{k\ell}^* = 0$ if

$$(c^2 - s^2)A_{k\ell} + cs(A_{kk} - A_{\ell\ell}) = 0 \tag{a}$$

Using the trigonometric identities $c^2 - s^2 = \cos^2\theta - \sin^2\theta = \cos 2\theta$ and $cs = \cos\theta\sin\theta = (1/2)\sin 2\theta$, Eq. (a) yields

$$\tan 2\theta = -\frac{2A_{k\ell}}{A_{kk} - A_{\ell\ell}} \tag{b}$$

which could be solved for θ, followed by computation of $c = \cos\theta$ and $s = \sin\theta$. However, the procedure described next leads to a better algorithm.[1]

Introducing the notation

$$\phi = \cot 2\theta = -\frac{A_{kk} - A_{\ell\ell}}{2A_{k\ell}} \tag{9.15}$$

and using the trigonometric identity

$$\tan 2\theta = \frac{2t}{(1 - t^2)}$$

where $t = \tan\theta$, Eq. (b) can be written as

$$t^2 + 2\phi t - 1 = 0$$

which has the roots

$$t = -\phi \pm \sqrt{\phi^2 + 1}$$

It has been found that the root $|t| \leq 1$, which corresponds to $|\theta| \leq 45°$, leads to the more stable transformation. Therefore, we choose the plus sign if $\phi > 0$ and the minus sign if $\phi \leq 0$, which is equivalent to using

$$t = \text{sgn}(\phi)\left(-|\phi| + \sqrt{\phi^2 + 1}\right)$$

To forestall excessive roundoff error if ϕ is large, we multiply both sides of the equation by $|\phi| + \sqrt{\phi^2 + 1}$, which yields

$$t = \frac{\text{sgn}(\phi)}{|\phi| + \sqrt{\phi^2 + 1}} \tag{9.16a}$$

In the case of very large ϕ, we should replace Eq. (9.16a) by the approximation

$$t = \frac{1}{2\phi} \tag{9.16b}$$

---

[1] The procedure is adapted from Press, W.H. et al., *Numerical Recipes in Fortran*, 2nd ed., Cambridge University Press, 1992.

to prevent overflow in the computation of ϕ^2. Having computed t, we can use the trigonometric relationship $\tan\theta = \sin\theta/\cos\theta = \sqrt{1 - \cos^2\theta}/\cos\theta$ to obtain

$$c = \frac{1}{\sqrt{1 + t^2}} \qquad s = tc \tag{9.17}$$

We now improve the transformation formulas in Eqs. (9.13). Solving Eq. (a) for $A_{\ell\ell}$, we obtain

$$A_{\ell\ell} = A_{kk} + A_{k\ell}\frac{c^2 - s^2}{cs} \tag{c}$$

Replacing all occurrences of $A_{\ell\ell}$ by Eq. (c) and simplifying, the transformation formulas in Eqs. (9.13) can be written as

$$
\begin{aligned}
A_{kk}^* &= A_{kk} - tA_{k\ell} \\
A_{\ell\ell}^* &= A_{\ell\ell} + tA_{k\ell} \\
A_{k\ell}^* &= A_{\ell k}^* = 0 \\
A_{ki}^* &= A_{ik}^* = A_{ki} - s(A_{\ell i} + \tau A_{ki}), \quad i \neq k, \quad i \neq \ell \\
A_{\ell i}^* &= A_{i\ell}^* = A_{\ell i} + s(A_{ki} - \tau A_{\ell i}), \quad i \neq k, \quad i \neq \ell
\end{aligned}
\tag{9.18}
$$

where

$$\tau = \frac{s}{1 + c} \tag{9.19}$$

The introduction of τ allowed us to express each formula in the form, (original value) + (change), which is helpful in reducing the roundoff error.

At the start of Jacobi's diagonalization process the transformation matrix $\mathbf{P}$ is initialized to the identity matrix. Each Jacobi rotation changes this matrix from $\mathbf{P}$ to $\mathbf{P}^* = \mathbf{PR}$. The corresponding changes in the elements of $\mathbf{P}$ can be shown to be (only the columns k and ℓ are affected)

$$
\begin{aligned}
P_{ik}^* &= P_{ik} - s(P_{i\ell} + \tau P_{ik}) \\
P_{i\ell}^* &= P_{i\ell} + s(P_{ik} - \tau P_{i\ell})
\end{aligned}
\tag{9.20}
$$

We still have to decide the order in which the off-diagonal elements of $\mathbf{A}$ are to be eliminated. Jacobi's original idea was to attack the largest element because doing so results in the fewest number of rotations. The problem here is that $\mathbf{A}$ has to be searched for the largest element before every rotation, which is a time-consuming process. If the matrix is large, it is faster to sweep through it by rows or columns and annihilate every element above some threshold value. In the next sweep the threshold is lowered and the process repeated.

There are several ways to choose the threshold. Our implementation starts by computing the sum S of the elements above the principal diagonal of $\mathbf{A}$:

$$S = \sum_{i=1}^{n-1}\sum_{j=i+1}^{n} |A_{ij}| \tag{a}$$

Because there are $n(n-1)/2$ such elements, the average magnitude of the off-diagonal elements is

$$\frac{2S}{n(n-1)}$$

The threshold we use is

$$\mu = \frac{0.5S}{n(n-1)} \tag{b}$$

which represents 0.25 times the average magnitude of the off-diagonal elements.

In summary, one sweep of Jacobi's diagonalization procedure (which uses only the upper half of the matrix), is as follows:

> Calculate the threshold μ using Eqs. (a) and (b).
> Sweep over off-diagonal terms of **A**:
> If $|A_{ij}| \geq \mu$:
> Compute ϕ, t, c, and s from Eqs.(9.15)–(9.17).
> Compute τ from Eq. (9.19).
> Modify the elements of **A** according to Eqs. (9.18).
> Update the transformation matrix **P** using Eqs. (9.20).

The sweeps are repeated until $\mu \leq \varepsilon$, where ε is the error tolerance. It takes usually 6 to 10 sweeps to achieve convergence.

■ jacobi

This function computes all eigenvalues λ_i and eigenvectors $\mathbf{x}_i$ of a symmetric, $n \times n$ matrix **A** by the Jacobi method. The algorithm works exclusively with the upper triangular part of **A**, which is destroyed in the process. The principal diagonal of **A** is replaced by the eigenvalues, and the columns of the transformation matrix **P** become the normalized eigenvectors.

```
## module jacobi
''' lam,x = jacobi(a,tol = 1.0e-8).
    Solution of std. eigenvalue problem [a]{x} = lam{x}
    by Jacobi's method. Returns eigenvalues in vector {lam}
    and the eigenvectors as columns of matrix [x].
'''
import numpy as np
import math

def jacobi(a,tol = 1.0e-8): # Jacobi method

    def threshold(a):
        sum = 0.0
        for i in range(n-1):
```

```
                for j in range (i+1,n):
                    sum = sum + abs(a[i,j])
            return 0.5*sum/n/(n-1)


    def rotate(a,p,k,l): # Rotate to make a[k,l] = 0
        aDiff = a[l,l] - a[k,k]
        if abs(a[k,l]) < abs(aDiff)*1.0e-36: t = a[k,l]/aDiff
        else:
            phi = aDiff/(2.0*a[k,l])
            t = 1.0/(abs(phi) + math.sqrt(phi**2 + 1.0))
            if phi < 0.0: t = -t
        c = 1.0/math.sqrt(t**2 + 1.0); s = t*c
        tau = s/(1.0 + c)
        temp = a[k,l]
        a[k,l] = 0.0
        a[k,k] = a[k,k] - t*temp
        a[l,l] = a[l,l] + t*temp
        for i in range(k):      # Case of i < k
            temp = a[i,k]
            a[i,k] = temp - s*(a[i,l] + tau*temp)
            a[i,l] = a[i,l] + s*(temp - tau*a[i,l])
        for i in range(k+1,l):  # Case of k < i < l
            temp = a[k,i]
            a[k,i] = temp - s*(a[i,l] + tau*a[k,i])
            a[i,l] = a[i,l] + s*(temp - tau*a[i,l])
        for i in range(l+1,n):  # Case of i > l
            temp = a[k,i]
            a[k,i] = temp - s*(a[l,i] + tau*temp)
            a[l,i] = a[l,i] + s*(temp - tau*a[l,i])
        for i in range(n):      # Update transformation matrix
            temp = p[i,k]
            p[i,k] = temp - s*(p[i,l] + tau*p[i,k])
            p[i,l] = p[i,l] + s*(temp - tau*p[i,l])

    n = len(a)
    p = np.identity(n,float)
    for k in range(20):
        mu = threshold(a)          # Compute new threshold
        for i in range(n-1):       # Sweep through matrix
            for j in range(i+1,n):
                if abs(a[i,j]) >= mu:
                    rotate(a,p,i,j)
        if mu <= tol: return np.diagonal(a),p
    print('Jacobi method did not converge')
```

■ sortJacobi

The eigenvalues/eigenvectors returned by `jacobi` are not ordered. The function listed next can be used to sort the eigenvalues and eigenvectors into ascending order of eigenvalues.

```
## module sortJacobi
''' sortJacobi(lam,x).
    Sorts the eigenvalues {lam} and eigenvectors [x]
    in order of ascending eigenvalues.
'''
import swap

def sortJacobi(lam,x):
    n = len(lam)
    for i in range(n-1):
        index = i
        val = lam[i]
        for j in range(i+1,n):
            if lam[j] < val:
                index = j
                val = lam[j]
        if index != i:
            swap.swapRows(lam,i,index)
            swap.swapCols(x,i,index)
```

Transformation to Standard Form

Physical problems often give rise to eigenvalue problems of the form

$$\mathbf{A}\mathbf{x} = \lambda \mathbf{B}\mathbf{x} \tag{9.21}$$

where $\mathbf{A}$ and $\mathbf{B}$ are symmetric $n \times n$ matrices. We assume that $\mathbf{B}$ is also positive definite. Such problems must be transformed into the standard form before they can be solved by Jacobi diagonalization.

Because $\mathbf{B}$ is symmetric and positive definite, we can apply Choleski decomposition $\mathbf{B} = \mathbf{L}\mathbf{L}^T$, where $\mathbf{L}$ is the lower triangular matrix (see Section 3.3). Then we introduce the transformation

$$\mathbf{x} = (\mathbf{L}^{-1})^T \mathbf{z} \tag{9.22}$$

Substituting into Eq. (9.21), we get

$$\mathbf{A}(\mathbf{L}^{-1})^T \mathbf{z} = \lambda \mathbf{L}\mathbf{L}^T (\mathbf{L}^{-1})^T \mathbf{z}$$

Pre-multiplying both sides by $\mathbf{L}^{-1}$ results in

$$\mathbf{L}^{-1}\mathbf{A}(\mathbf{L}^{-1})^T \mathbf{z} = \lambda \mathbf{L}^{-1}\mathbf{L}\mathbf{L}^T (\mathbf{L}^{-1})^T \mathbf{z}$$

Noting that $\mathbf{L}^{-1}\mathbf{L} = \mathbf{L}^T(\mathbf{L}^{-1})^T = \mathbf{I}$, the last equation reduces to the standard form

$$\mathbf{Hz} = \lambda\mathbf{z} \tag{9.23}$$

where

$$\mathbf{H} = \mathbf{L}^{-1}\mathbf{A}(\mathbf{L}^{-1})^T \tag{9.24}$$

An important property of this transformation is that it does not destroy the symmetry of the matrix (i.e., symmetric $\mathbf{A}$ results in symmetric $\mathbf{H}$).

The general procedure for solving eigenvalue problems of the form $\mathbf{Ax} = \lambda\mathbf{Bx}$ is shown in the following box:

> Use Choleski decomposition $\mathbf{B} = \mathbf{LL}^T$ to compute $\mathbf{L}$.
> Compute $\mathbf{L}^{-1}$ (a triangular matrix is easy to invert).
> Compute $\mathbf{H} = \mathbf{L}^{-1}\mathbf{A}(\mathbf{L}^{-1})^T$.
> Solve the standard eigenvalue problem $\mathbf{Hz} = \lambda\mathbf{z}$.
> Recover the eigenvectors of the original problem from $\mathbf{X} = (\mathbf{L}^{-1})^T\mathbf{Z}$.

Note that the eigenvalues were untouched by the transformation.

An important special case is where $\mathbf{B}$ is a diagonal matrix:

$$\mathbf{B} = \begin{bmatrix} \beta_1 & 0 & \cdots & 0 \\ 0 & \beta_2 & \cdots & 0 \\ \vdots & \vdots & \ddots & \vdots \\ 0 & 0 & \cdots & \beta_n \end{bmatrix} \tag{9.25}$$

Here

$$\mathbf{L} = \begin{bmatrix} \beta_1^{1/2} & 0 & \cdots & 0 \\ 0 & \beta_2^{1/2} & \cdots & 0 \\ \vdots & \vdots & \ddots & \vdots \\ 0 & 0 & \cdots & \beta_n^{1/2} \end{bmatrix} \qquad \mathbf{L}^{-1} = \begin{bmatrix} \beta_1^{-1/2} & 0 & \cdots & 0 \\ 0 & \beta_2^{-1/2} & \cdots & 0 \\ \vdots & \vdots & \ddots & \vdots \\ 0 & 0 & \cdots & \beta_n^{-1/2} \end{bmatrix} \tag{9.26a}$$

and

$$H_{ij} = \frac{A_{ij}}{\sqrt{\beta_i \beta_j}} \tag{9.26b}$$

■ stdForm

Given the matrices $\mathbf{A}$ and $\mathbf{B}$, the function stdForm returns $\mathbf{H}$ and the transformation matrix $\mathbf{T} = (\mathbf{L}^{-1})^T$. The inversion of $\mathbf{L}$ is carried out by invert (the triangular shape of $\mathbf{L}$ allows this to be done by back substitution). Note that original $\mathbf{A}$, $\mathbf{B}$, and $\mathbf{L}$ are destroyed.

```
## module stdForm
''' h,t = stdForm(a,b).
    Transforms the eigenvalue problem [a]{x} = lam*[b]{x}
```

to the standard form [h]{z} = lam*{z}. The eigenvectors
are related by {x} = [t]{z}.
'''

```
import numpy as np
from choleski import *

def stdForm(a,b):

    def invert(L): # Inverts lower triangular matrix L
        n = len(L)
        for j in range(n-1):
            L[j,j] = 1.0/L[j,j]
            for i in range(j+1,n):
                L[i,j] = -np.dot(L[i,j:i],L[j:i,j])/L[i,i]
        L[n-1,n-1] = 1.0/L[n-1,n-1]

    n = len(a)
    L = choleski(b)
    invert(L)
    h = np.dot(b,np.inner(a,L))
    return h,np.transpose(L)
```

EXAMPLE 9.2
The stress matrix (tensor) in Example 9.1 was

$$\mathbf{S} = \begin{bmatrix} 80 & 30 & 0 \\ 30 & 40 & 0 \\ 0 & 0 & 60 \end{bmatrix} \text{MPa}$$

(1) Determine the principal stresses by diagonalizing $\mathbf{S}$ with one Jacobi rotation; and (2) compute the eigenvectors.

Solution of Part(1). To eliminate S_{12} we must apply a rotation in the 1-2 plane. With $k = 1$ and $\ell = 2$, Eq. (9.15) is

$$\phi = -\frac{S_{11} - S_{22}}{2 S_{12}} = -\frac{80 - 40}{2(30)} = -\frac{2}{3}$$

Equation (9.16a) then yields

$$t = \frac{\text{sgn}(\phi)}{|\phi| + \sqrt{\phi^2 + 1}} = \frac{-1}{2/3 + \sqrt{(2/3)^2 + 1}} = -0.535\,18$$

According to Eqs. (9.18), the changes in $\mathbf{S}$ due to the rotation are

$$S_{11}^* = S_{11} - t S_{12} = 80 - (-0.535\,18)(30) = 96.055 \text{ MPa}$$

$$S_{22}^* = S_{22} + t S_{12} = 40 + (-0.535\,18)(30) = 23.945 \text{ MPa}$$

$$S_{12}^* = S_{21}^* = 0$$

Hence the diagonalized stress matrix is

$$\mathbf{S}^* = \begin{bmatrix} 96.055 & 0 & 0 \\ 0 & 23.945 & 0 \\ 0 & 0 & 60 \end{bmatrix}$$

where the diagonal terms are the principal stresses.

Solution of Part (2). To compute the eigenvectors, we start with Eqs. (9.17) and (9.19), which yield

$$c = \frac{1}{\sqrt{1 + t^2}} = \frac{1}{\sqrt{1 + (-0.535\,18)^2}} = 0.88168$$

$$s = tc = (-0.535\,18)\,(0.881\,68) = -0.471\,86$$

$$\tau = \frac{s}{1 + c} = \frac{-0.47186}{1 + 0.881\,68} = -0.250\,77$$

We obtain the changes in the transformation matrix $\mathbf{P}$ from Eqs. (9.20). Recalling that $\mathbf{P}$ is initialized to the identity matrix, the first equation gives us

$$P_{11}^* = P_{11} - s(P_{12} + \tau P_{11})$$

$$= 1 - (-0.471\,86)\,(0 + (-0.250\,77)\,(1)) = 0.881\,67$$

$$P_{21}^* = P_{21} - s(P_{22} + \tau P_{21})$$

$$= 0 - (-0.471\,86)\,[1 + (-0.250\,77)\,(0)] = 0.471\,86$$

Similarly, the second equation of Eqs. (9.20) yields

$$P_{12}^* = -0.471\,86 \qquad P_{22}^* = 0.881\,67$$

The third row and column of $\mathbf{P}$ are not affected by the transformation. Thus

$$\mathbf{P}^* = \begin{bmatrix} 0.88167 & -0.47186 & 0 \\ 0.47186 & 0.88167 & 0 \\ 0 & 0 & 1 \end{bmatrix}$$

The columns of $\mathbf{P}^*$ are the eigenvectors of $\mathbf{S}$.

EXAMPLE 9.3

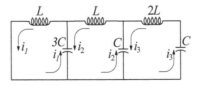

(1) Show that the analysis of the electric circuit shown leads to a matrix eigenvalue problem. (2) Determine the circular frequencies and the relative amplitudes of the currents.

Solution of Part(1). Kirchoff's equations for the three loops are

$$L\frac{di_1}{dt} + \frac{q_1 - q_2}{3C} = 0$$

$$L\frac{di_2}{dt} + \frac{q_2 - q_1}{3C} + \frac{q_2 - q_3}{C} = 0$$

$$2L\frac{di_3}{dt} + \frac{q_3 - q_2}{C} + \frac{q_3}{C} = 0$$

Differentiating and substituting $dq_k/dt = i_k$, we get

$$\frac{1}{3}i_1 - \frac{1}{3}i_2 = -LC\frac{d^2 i_1}{dt^2}$$

$$-\frac{1}{3}i_1 + \frac{4}{3}i_2 - i_3 = -LC\frac{d^2 i_2}{dt^2}$$

$$-i_2 + 2i_3 = -2LC\frac{d^2 i_3}{dt^2}$$

These equations admit the solution

$$i_k(t) = u_k \sin \omega t$$

where ω is the circular frequency of oscillation (measured in rad/s) and u_k are the relative amplitudes of the currents. Substitution into Kirchoff's equations yields $\mathbf{Au} = \lambda \mathbf{Bu}$ ($\sin \omega t$ cancels out), where

$$\mathbf{A} = \begin{bmatrix} 1/3 & -1/3 & 0 \\ -1/3 & 4/3 & -1 \\ 0 & -1 & 2 \end{bmatrix} \qquad \mathbf{B} = \begin{bmatrix} 1 & 0 & 0 \\ 0 & 1 & 0 \\ 0 & 0 & 2 \end{bmatrix} \qquad \lambda = LC\omega^2$$

which represents an eigenvalue problem of the nonstandard form.

Solution of Part (2). Because $\mathbf{B}$ is a diagonal matrix, we could readily transform the problem into the standard form $\mathbf{Hz} = \lambda \mathbf{z}$ using Eqs. (9.26). However, we chose to let the computer do all the work.

```
#!/usr/bin/python
## example9_3
import numpy
from jacobi import *
import math
from sortJacobi import *
from stdForm import *

A = np.array([[ 1/3, -1/3,  0.0], \
              [-1/3,  4/3, -1.0], \
              [ 0.0, -1.0,  2.0]])
B = np.array([[1.0, 0.0, 0.0],    \
              [0.0, 1.0, 0.0],    \
              [0.0, 0.0, 2.0]])
```

```
H,T = stdForm(A,B)     # Transform into std. form
lam,Z = jacobi(H)      # Z = eigenvecs. of H
X = np.dot(T,Z)        # Eigenvecs. of original problem
sortJacobi(lam,X)      # Arrange in ascending order of eigenvecs.
for i in range(3):     # Normalize eigenvecs.
    X[:,i] = X[:,i]/math.sqrt(np.dot(X[:,i],X[:,i]))

print('Eigenvalues:\n',lam)
print('Eigenvectors:\n',X)
input ("Press return to exit")


Eigenvalues:
 [ 0.1477883   0.58235144  1.93652692]
Eigenvectors:
 [[ 0.84021782 -0.65122529 -0.18040571]
 [ 0.46769473  0.48650067  0.86767582]
 [ 0.27440056  0.58242829 -0.46324126]]
```

The circular frequencies are given by $\omega_i = \sqrt{\lambda_i/LC}$:

$$\omega_1 = \frac{0.3844}{\sqrt{LC}} \qquad \omega_2 = \frac{0.7631}{\sqrt{LC}} \qquad \omega_3 = \frac{1.3916}{\sqrt{LC}}$$

9.3   Power and Inverse Power Methods

Inverse Power Method

The inverse power method is a simple and efficient algorithm that finds the smallest eigenvalue λ_1 and the corresponding eigenvector $\mathbf{x}_1$ of the standard eigenvalue problem

$$\mathbf{Ax} = \lambda\mathbf{x} \tag{9.27}$$

The method works like this:

Let $\mathbf{v}$ be an unit vector (a random vector will do).
Do the following until the change in $\mathbf{v}$ is negligible:
    Solve $\mathbf{Az} = \mathbf{v}$ for the vector $\mathbf{z}$.
    Compute $|\mathbf{z}|$.
    Compute $\mathbf{v} = \mathbf{z}/|\mathbf{z}|$.

At the conclusion of the procedure we have $|\mathbf{z}| = \pm 1/\lambda_1$ and $\mathbf{v} = \mathbf{x}_1$. The sign of λ_1 is determined as follows: If $\mathbf{z}$ changes sign between successive iterations, λ_1 is negative; otherwise, use the plus sign.

Let us now investigate why the method works. Since the eigenvectors $\mathbf{x}_i$ of $\mathbf{A}$ are orthonormal (linearly independent), they can be used as the basis for any n-dimensional vector. Thus $\mathbf{v}$ and $\mathbf{z}$ admit the unique representations

$$\mathbf{v} = \sum_{i=1}^{n} v_i \mathbf{x}_i \qquad \mathbf{z} = \sum_{i=1}^{n} z_i \mathbf{x}_i \tag{9.28}$$

where v_i and z_i are the components of $\mathbf{v}$ and $\mathbf{z}$ with respect to the eigenvectors $\mathbf{x}_i$. Substitution into $\mathbf{Az} = \mathbf{v}$ yields

$$\mathbf{A} \sum_{i=1}^{n} z_i \mathbf{x}_i - \sum_{i=1}^{n} v_i \mathbf{x}_i = \mathbf{0}$$

But $\mathbf{Ax}_i = \lambda_i \mathbf{x}_i$, so that

$$\sum_{i=1}^{n} (z_i \lambda_i - v_i) \mathbf{x}_i = \mathbf{0}$$

Hence

$$z_i = \frac{v_i}{\lambda_i}$$

It follows from Eq. (9.28) that

$$\mathbf{z} = \sum_{i=1}^{n} \frac{v_i}{\lambda_i} \mathbf{x}_i = \frac{1}{\lambda_1} \sum_{i=1}^{n} v_i \frac{\lambda_1}{\lambda_i} \mathbf{x}_i$$

$$= \frac{1}{\lambda_1} \left(v_1 \mathbf{x}_1 + v_2 \frac{\lambda_1}{\lambda_2} \mathbf{x}_2 + v_3 \frac{\lambda_1}{\lambda_3} \mathbf{x}_3 + \cdots \right) \tag{9.29}$$

Since $\lambda_1/\lambda_i < 1$ ($i \neq 1$), we observe that the coefficient of $\mathbf{x}_1$ has become more prominent in $\mathbf{z}$ than it was in $\mathbf{v}$; hence $\mathbf{z}$ is a better approximation to $\mathbf{x}_1$. This completes the first iterative cycle.

In subsequent cycles we set $\mathbf{v} = \mathbf{z}/|\mathbf{z}|$ and repeat the process. Each iteration increases the dominance of the first term in Eq. (9.29) so that the process converges to

$$\mathbf{z} = \frac{1}{\lambda_1} v_1 \mathbf{x}_1 = \frac{1}{\lambda_1} \mathbf{x}_1$$

(at this stage $v_1 = 1$ since $\mathbf{v} = \mathbf{x}_1$, so that $v_1 = 1$, $v_2 = v_3 = \cdots = 0$).

The inverse power method also works with the nonstandard eigenvalue problem

$$\mathbf{Ax} = \lambda \mathbf{Bx} \tag{9.30}$$

provided that $\mathbf{Az} = \mathbf{v}$ in the algorithm is replaced by

$$\mathbf{Az} = \mathbf{Bv} \tag{9.31}$$

The alternative is, of course, to transform the problem to standard form before applying the power method.

Eigenvalue Shifting

By inspection of Eq. (9.29) we see that the speed of convergence is determined by the strength of the inequality $|\lambda_1/\lambda_2| < 1$ (the second term in the equation). If $|\lambda_2|$ is well separated from $|\lambda_1|$, the inequality is strong and the convergence is rapid. In contrast, close proximity of these two eigenvalues results in very slow convergence.

The rate of convergence can be improved by a technique called *eigenvalue shifting*. Letting

$$\lambda = \lambda^* + s \tag{9.32}$$

where s is a predetermined "shift," the eigenvalue problem in Eq. (9.27) is transformed to

$$\mathbf{Ax} = (\lambda^* + s)\mathbf{x}$$

or

$$\mathbf{A}^*\mathbf{x} = \lambda^*\mathbf{x} \tag{9.33}$$

where

$$\mathbf{A}^* = \mathbf{A} - s\mathbf{I} \tag{9.34}$$

Solving the transformed problem in Eq. (9.33) by the inverse power method yields λ_1^* and $\mathbf{x}_1$, where λ_1^* is the smallest eigenvalue of $\mathbf{A}^*$. The corresponding eigenvalue of the original problem, $\lambda = \lambda_1^* + s$, is thus the *eigenvalue closest to s*.

Eigenvalue shifting has two applications. An obvious one is the determination of the eigenvalue closest to a certain value s. For example, if the working speed of a shaft is s rpm, it is imperative to assure that there are no natural frequencies (which are related to the eigenvalues) close to that speed.

Eigenvalue shifting is also used to speed up convergence. Suppose that we are computing the smallest eigenvalue λ_1 of the matrix $\mathbf{A}$. The idea is to introduce a shift s that makes λ_1^*/λ_2^* as small as possible. Since $\lambda_1^* = \lambda_1 - s$, we should choose $s \approx \lambda_1$ ($s = \lambda_1$ should be avoided to prevent division by zero). Of course, this method works only if we have a prior estimate of λ_1.

The inverse power method with eigenvalue shifting is a particularly powerful tool for finding eigenvectors if the eigenvalues are known. By shifting very close to an eigenvalue, the corresponding eigenvector can be computed in one or two iterations.

Power Method

The power method converges to the eigenvalue *farthest from zero* and the associated eigenvector. It is very similar to the inverse power method; the only difference between the two methods is the interchange of $\mathbf{v}$ and $\mathbf{z}$ in Eq. (9.28).

The outline of the procedure is

> Let **v** be an unit vector (a random vector will do).
> Do the following until the change in **v** is negligible:
>   Compute the vector $\mathbf{z} = \mathbf{Av}$.
>   Compute $|\mathbf{z}|$.
>   Compute $\mathbf{v} = \mathbf{z}/|\mathbf{z}|$.

At the conclusion of the procedure we have $|\mathbf{z}| = \pm\lambda_n$ and $\mathbf{v} = \mathbf{x}_n$ (the sign of λ_n is determined in the same way as in the inverse power method).

■ inversePower

Given the matrix **A** and the shift s, the function `inversePower` returns the eigenvalue of **A** closest to s and the corresponding eigenvector. The matrix $\mathbf{A}^* = \mathbf{A} - s\mathbf{I}$ is decomposed as soon as it is formed, so that only the solution phase (forward and back substitution) is needed in the iterative loop. If **A** is banded, the efficiency of the program could be improved by replacing `LUdecomp` and `LUsolve` by functions that specialize in banded matrices (e.g., `LUdecomp5` and `LUsolve5`)—see Example 9.6. The program line that forms $\mathbf{A}^*$ must also be modified to be compatible with the storage scheme used for **A**.

```
## module inversePower
''' lam,x = inversePower(a,s,tol=1.0e-6).
    Inverse power method for solving the eigenvalue problem
    [a]{x} = lam{x}. Returns 'lam' closest to 's' and the
    corresponding eigenvector {x}.
'''
import numpy as np
from LUdecomp import *
import math
from random import random

def inversePower(a,s,tol=1.0e-6):
    n = len(a)
    aStar = a - np.identity(n)*s   # Form [a*] = [a] - s[I]
    aStar = LUdecomp(aStar)        # Decompose [a*]
    x = np.zeros(n)
    for i in range(n):             # Seed [x] with random numbers
        x[i] = random()
    xMag = math.sqrt(np.dot(x,x))  # Normalize [x]
    x =x/xMag
    for i in range(50):            # Begin iterations
        xOld = x.copy()            # Save current [x]
```

```
    x = LUsolve(aStar,x)        # Solve [a*][x] = [xOld]
    xMag = math.sqrt(np.dot(x,x)) # Normalize [x]
    x = x/xMag
    if np.dot(xOld,x) < 0.0:    # Detect change in sign of [x]
        sign = -1.0
        x = -x
    else: sign = 1.0
    if math.sqrt(np.dot(xOld - x,xOld - x)) < tol:
        return s + sign/xMag,x
print('Inverse power method did not converge')
```

EXAMPLE 9.4

The stress matrix describing the state of stress at a point is

$$
\mathbf{S} = \begin{bmatrix} -30 & 10 & 20 \\ 10 & 40 & -50 \\ 20 & -50 & -10 \end{bmatrix} \text{MPa}
$$

Determine the largest principal stress (the eigenvalue of $\mathbf{S}$ farthest from zero) by the power method.

Solution. First Iteration:

Let $v = \begin{bmatrix} 1 & 0 & 0 \end{bmatrix}^T$ be the initial guess for the eigenvector. Then

$$
\mathbf{z} = \mathbf{Sv} = \begin{bmatrix} -30 & 10 & 20 \\ 10 & 40 & -50 \\ 20 & -50 & -10 \end{bmatrix} \begin{bmatrix} 1 \\ 0 \\ 0 \end{bmatrix} = \begin{bmatrix} -30.0 \\ 10.0 \\ 20.0 \end{bmatrix}
$$

$$
|\mathbf{z}| = \sqrt{30^2 + 10^2 + 20^2} = 37.417
$$

$$
\mathbf{v} = \frac{\mathbf{z}}{|\mathbf{z}|} = \begin{bmatrix} -30.0 \\ 10.0 \\ 20.0 \end{bmatrix} \frac{1}{37.417} = \begin{bmatrix} -0.801\,77 \\ 0.267\,26 \\ 0.534\,52 \end{bmatrix}
$$

Second Iteration:

$$
\mathbf{z} = \mathbf{Sv} = \begin{bmatrix} -30 & 10 & 20 \\ 10 & 40 & -50 \\ 20 & -50 & -10 \end{bmatrix} \begin{bmatrix} -0.801\,77 \\ 0.267\,26 \\ 0.534\,52 \end{bmatrix} = \begin{bmatrix} 37.416 \\ -24.053 \\ -34.744 \end{bmatrix}
$$

$$
|\mathbf{z}| = \sqrt{37.416^2 + 24.053^2 + 34.744^2} = 56.442
$$

$$
\mathbf{v} = \frac{\mathbf{z}}{|\mathbf{z}|} = \begin{bmatrix} 37.416 \\ -24.053 \\ -34.744 \end{bmatrix} \frac{1}{56.442} = \begin{bmatrix} 0.66291 \\ -0.42615 \\ -0.61557 \end{bmatrix}
$$

Third Iteration:

$$\mathbf{z} = \mathbf{Sv} = \begin{bmatrix} -30 & 10 & 20 \\ 10 & 40 & -50 \\ 20 & -50 & -10 \end{bmatrix} \begin{bmatrix} 0.66291 \\ -0.42615 \\ -0.61557 \end{bmatrix} = \begin{bmatrix} -36.460 \\ 20.362 \\ 40.721 \end{bmatrix}$$

$$|\mathbf{z}| = \sqrt{36.460^2 + 20.362^2 + 40.721^2} = 58.328$$

$$\mathbf{v} = \frac{\mathbf{z}}{|\mathbf{z}|} = \begin{bmatrix} -36.460 \\ 20.362 \\ 40.721 \end{bmatrix} \frac{1}{58.328} = \begin{bmatrix} -0.62509 \\ 0.34909 \\ 0.69814 \end{bmatrix}$$

At this point the approximation of the eigenvalue we seek is $\lambda = -58.328$ MPa (the negative sign is determined by the sign reversal of $\mathbf{z}$ between iterations). This is actually close to the second largest eigenvalue $\lambda_2 = -58.39$ MPa! By continuing the iterative process we would eventually end up with the largest eigenvalue $\lambda_3 = 70.94$ MPa. But because $|\lambda_2|$ and $|\lambda_3|$ are rather close, the convergence is too slow from this point on for manual labor. Here is a program that does the calculations for us:

```
#!/usr/bin/python
## example9_4
import numpy as np
import math

s = np.array([[-30.0,   10.0,   20.0], \
              [ 10.0,   40.0, -50.0], \
              [ 20.0, -50.0, -10.0]])
v = np.array([1.0, 0.0, 0.0])
for i in range(100):
    vOld = v.copy()
    z = np.dot(s,v)
    zMag = math.sqrt(np.dot(z,z))
    v = z/zMag
    if np.dot(vOld,v) < 0.0:
        sign = -1.0
        v = -v
    else: sign = 1.0
    if math.sqrt(np.dot(vOld - v,vOld - v)) < 1.0e-6: break
lam = sign*zMag
print("Number of iterations =",i)
print("Eigenvalue =",lam)
input("Press return to exit")
```

The output from the program is

```
Number of iterations = 92
Eigenvalue = 70.94348330679053
```

EXAMPLE 9.5

Determine the smallest eigenvalue λ_1 and the corresponding eigenvector of

$$\mathbf{A} = \begin{bmatrix} 11 & 2 & 3 & 1 & 4 \\ 2 & 9 & 3 & 5 & 2 \\ 3 & 3 & 15 & 4 & 3 \\ 1 & 5 & 4 & 12 & 4 \\ 4 & 2 & 3 & 4 & 17 \end{bmatrix}$$

Use the inverse power method with eigenvalue shifting knowing that $\lambda_1 \approx 5$.

Solution

```
#!/usr/bin/python
## example9_5
import numpy as np
from inversePower import *

s = 5.0
a = np.array([[ 11.0, 2.0,   3.0,   1.0,   4.0], \
              [  2.0, 9.0,   3.0,   5.0,   2.0], \
              [  3.0, 3.0, 15.0,    4.0,   3.0], \
              [  1.0, 5.0,   4.0, 12.0,    4.0], \
              [  4.0, 2.0,   3.0,   4.0, 17.0]])
lam,x = inversePower(a,s)
print("Eigenvalue =",lam)
print("\nEigenvector:\n",x)
input("\nPrint press return to exit")
```

Here is the output:

```
Eigenvalue = 4.873946378649195

Eigenvector:
 [ 0.26726604 -0.74142853 -0.05017272  0.59491453 -0.14970634]
```

Convergence was achieved with four iterations. Without the eigenvalue shift 26 iterations would be required.

EXAMPLE 9.6

The propped cantilever beam carries a compressive axial load P. The lateral displacement $u(x)$ of the beam can be shown to satisfy the differential equation

$$u^{(4)} + \frac{P}{EI}u'' = 0 \tag{a}$$

where EI is the bending rigidity. The boundary conditions are

$$u(0) = u''(0) = 0 \qquad u(L) = u'(L) = 0 \tag{b}$$

(1) Show that buckling analysis of the beam results in a matrix eigenvalue problem if the derivatives are approximated by finite differences. (2) Write a program that computes the smallest buckling load of the beam, making full use of banded matrices. Run the program with 100 interior nodes ($n = 100$).

Solution of Part (1). We divide the beam into $n + 1$ segments of length $L/(n+1)$ each as shown. Replacing the derivatives of u in Eq. (a) by central finite differences of $\mathcal{O}(h^2)$ at the interior nodes (nodes 1 to n), we obtain

$$\frac{u_{i-2} - 4u_{i-1} + 6u_i - 4u_{i+1} + u_{i+2}}{h^4}$$

$$= \frac{P}{EI} \frac{-u_{i-1} + 2u_i - u_{i-1}}{h^2}, i = 1, 2, \ldots, n$$

After multiplication by h^4, the equations become

$$u_{-1} - 4u_0 + 6u_1 - 4u_2 + u_3 = \lambda(-u_0 + 2u_1 - u_2)$$

$$u_0 - 4u_1 + 6u_2 - 4u_3 + u_4 = \lambda(-u_1 + 2u_2 - u_3)$$

$$\vdots \tag{c}$$

$$u_{n-3} - 4u_{n-2} + 6u_{n-1} - 4u_n + u_{n+1} = \lambda(-u_{n-2} + 2u_{n-1} - u_n)$$

$$u_{n-2} - 4u_{n-1} + 6u_n - 4u_{n+1} + u_{n+2} = \lambda(-u_{n-1} + 2u_n - u_{n+1})$$

where

$$\lambda = \frac{Ph^2}{EI} = \frac{PL^2}{(n+1)^2 EI}$$

The displacements u_{-1}, u_0, u_{n+1} and u_{n+2} can be eliminated by using the prescribed boundary conditions. Referring to Table 8.1, the finite difference approximations to the boundary conditions are

$$u_0 = 0 \qquad u_{-1} = -u_1 \qquad u_{n+1} = 0 \qquad u_{n+2} = u_n$$

Substitution into Eqs. (c) yields the matrix eigenvalue problem $\mathbf{A}x = \lambda \mathbf{B}x$, where

$$\mathbf{A} = \begin{bmatrix}
5 & -4 & 1 & 0 & 0 & \cdots & 0 \\
-4 & 6 & -4 & 1 & 0 & \cdots & 0 \\
1 & -4 & 6 & -4 & 1 & \cdots & 0 \\
\vdots & \ddots & \ddots & \ddots & \ddots & \ddots & \vdots \\
0 & \cdots & 1 & -4 & 6 & -4 & 1 \\
0 & \cdots & 0 & 1 & -4 & 6 & -4 \\
0 & \cdots & 0 & 0 & 1 & -4 & 7
\end{bmatrix}$$

$$
\mathbf{B} = \begin{bmatrix}
2 & -1 & 0 & 0 & 0 & \cdots & 0 \\
-1 & 2 & -1 & 0 & 0 & \cdots & 0 \\
0 & -1 & 2 & -1 & 0 & \cdots & 0 \\
\vdots & \ddots & \vdots & \ddots & \ddots & \ddots & \vdots \\
0 & \cdots & 0 & -1 & 2 & -1 & 0 \\
0 & \cdots & 0 & 0 & -1 & 2 & -1 \\
0 & \cdots & 0 & 0 & 0 & -1 & 2
\end{bmatrix}
$$

Solution of Part (2). The function inversePower5 listed next returns the smallest eigenvalue and the corresponding eigenvector of $\mathbf{Ax} = \lambda \mathbf{Bx}$, where $\mathbf{A}$ is a pentadiagonal matrix and $\mathbf{B}$ is a sparse matrix (in this problem it is tridiagonal). The matrix $\mathbf{A}$ is input by its diagonals $\mathbf{d}$, $\mathbf{e}$, and $\mathbf{f}$ as was done in Section 2.4 in conjunction with the LU decomposition. The algorithm does not use $\mathbf{B}$ directly, but calls the function Bv(v) that supplies the product $\mathbf{Bv}$. Eigenvalue shifting is not used.

```
## module inversePower5
''' lam,x = inversePower5(Bv,d,e,f,tol=1.0e-6).
    Inverse power method for solving the eigenvalue problem
    [A]{x} = lam[B]{x}, where [A] is pentadiagonal and [B]
    is sparse. User must supply the function Bv(v) that
    returns the vector [B]{v}.
'''
import numpy as np
from LUdecomp5 import *
import math
from numpy.random import rand

def inversePower5(Bv,d,e,f,tol=1.0e-6):
    n = len(d)
    d,e,f = LUdecomp5(d,e,f)
    x = rand(n)                      # Seed x with random numbers
    xMag = math.sqrt(np.dot(x,x))    # Normalize {x}
    x = x/xMag
    for i in range(30):              # Begin iterations
        xOld = x.copy()              # Save current {x}
        x = Bv(xOld)                 # Compute [B]{x}
        x = LUsolve5(d,e,f,x)        # Solve [A]{z} = [B]{x}
        xMag = math.sqrt(np.dot(x,x))  # Normalize {z}
        x = x/xMag
        if np.dot(xOld,x) < 0.0:     # Detect change in sign of {x}
            sign = -1.0
            x = -x
        else: sign = 1.0
```

```
        if math.sqrt(np.dot(xOld - x,xOld - x)) < tol:
            return sign/xMag,x
    print('Inverse power method did not converge')
```

The program that utilizes inversePower5 is

```
#!/usr/bin/python
## example9_6
import numpy as np
from inversePower5 import *

def Bv(v):                # Compute {z} = [B]{v}
    n = len(v)
    z = np.zeros(n)
    z[0] = 2.0*v[0] - v[1]
    for i in range(1,n-1):
        z[i] = -v[i-1] + 2.0*v[i] - v[i+1]
    z[n-1] = -v[n-2] + 2.0*v[n-1]
    return z

n = 100                   # Number of interior nodes
d = np.ones(n)*6.0        # Specify diagonals of [A] = [f\e\d\e\f]
d[0] = 5.0
d[n-1] = 7.0
e = np.ones(n-1)*(-4.0)
f = np.ones(n-2)*1.0
lam,x = inversePower5(Bv,d,e,f)
print("PL^2/EI =",lam*(n+1)**2)
input("\nPress return to exit")
```

The output, which agrees with the analytical value, is

```
PL^2/EI = 20.1867306935764
```

PROBLEM SET 9.1

1. Given

$$
\mathbf{A} = \begin{bmatrix} 7 & 3 & 1 \\ 3 & 9 & 6 \\ 1 & 6 & 8 \end{bmatrix} \qquad \mathbf{B} = \begin{bmatrix} 4 & 0 & 0 \\ 0 & 9 & 0 \\ 0 & 0 & 4 \end{bmatrix}
$$

convert the eigenvalue problem $\mathbf{Ax} = \lambda \mathbf{Bx}$ to the standard form $\mathbf{Hz} = \lambda \mathbf{z}$. What is the relationship between $\mathbf{x}$ and $\mathbf{z}$?

2. Convert the eigenvalue problem $\mathbf{Ax} = \lambda\mathbf{Bx}$, where

$$\mathbf{A} = \begin{bmatrix} 4 & -1 & 0 \\ -1 & 4 & -1 \\ 0 & -1 & 4 \end{bmatrix} \qquad \mathbf{B} = \begin{bmatrix} 2 & -1 & 0 \\ -1 & 2 & -1 \\ 0 & -1 & 1 \end{bmatrix}$$

to the standard form.

3. An eigenvalue of the problem in Prob. 2 is roughly 2.5. Use the inverse power method with eigenvalue shifting to compute this eigenvalue to four decimal places. Start with $\mathbf{x} = \begin{bmatrix} 1 & 0 & 0 \end{bmatrix}^T$. *Hint*: Two iterations should be sufficient.

4. The stress matrix at a point is

$$\mathbf{S} = \begin{bmatrix} 150 & -60 & 0 \\ -60 & 120 & 0 \\ 0 & 0 & 80 \end{bmatrix} \text{MPa}$$

Compute the principal stresses (eigenvalues of $\mathbf{S}$).

5.

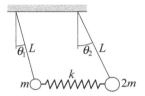

The two pendulums are connected by a spring that is undeformed when the pendulums are vertical. The equations of motion of the system can be shown to be

$$kL(\theta_2 - \theta_1) - mg\theta_1 = mL\ddot{\theta}_1$$
$$-kL(\theta_2 - \theta_1) - 2mg\theta_2 = 2mL\ddot{\theta}_2$$

where θ_1 and θ_2 are the angular displacements and k is the spring stiffness. Determine the circular frequencies of vibration and the relative amplitudes of the angular displacements. Use $m = 0.25$ kg, $k = 20$ N/m. $L = 0.75$ m, and $g = 9.806\,65$ m/s^2.

6.

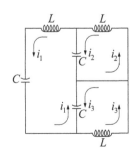

Kirchoff's laws for the electric circuit are

$$3i_1 - i_2 - i_3 = -LC\frac{d^2 i_1}{dt^2}$$

$$-i_1 + i_2 = -LC\frac{d^2 i_2}{dt^2}$$

$$-i_1 + i_3 = -LC\frac{d^2 i_3}{dt^2}$$

Compute the circular frequencies of the circuit and the relative amplitudes of the loop currents.

7. Compute the matrix $\mathbf{A}^*$ that results from annihilation A_{14} and A_{41} in the matrix

$$\mathbf{A} = \begin{bmatrix} 4 & -1 & 0 & 1 \\ -1 & 6 & -2 & 0 \\ 0 & -2 & 3 & 2 \\ 1 & 0 & 2 & 4 \end{bmatrix}$$

by a Jacobi rotation.

8. ■ Use the Jacobi method to determine the eigenvalues and eigenvectors of

$$\mathbf{A} = \begin{bmatrix} 4 & -1 & 2 \\ -1 & 3 & 3 \\ -2 & 3 & 1 \end{bmatrix}$$

9. ■ Find the eigenvalues and eigenvectors of

$$\mathbf{A} = \begin{bmatrix} 4 & -2 & 1 & -1 \\ -2 & 4 & -2 & 1 \\ 1 & -2 & 4 & -2 \\ -1 & 1 & -2 & 4 \end{bmatrix}$$

with the Jacobi method.

10. ■ Use the power method to compute the largest eigenvalue and the corresponding eigenvector of the matrix $\mathbf{A}$ given in Prob. 9.

11. ■ Find the smallest eigenvalue and the corresponding eigenvector of the matrix $\mathbf{A}$ in Prob. 9. Use the inverse power method.

12. ■ Let

$$\mathbf{A} = \begin{bmatrix} 1.4 & 0.8 & 0.4 \\ 0.8 & 6.6 & 0.8 \\ 0.4 & 0.8 & 5.0 \end{bmatrix} \qquad \mathbf{B} = \begin{bmatrix} 0.4 & -0.1 & 0.0 \\ -0.1 & 0.4 & -0.1 \\ 0.0 & -0.1 & 0.4 \end{bmatrix}$$

Find the eigenvalues and eigenvectors of $\mathbf{Ax} = \lambda \mathbf{Bx}$ by the Jacobi method.

13. ■ Use the inverse power method to compute the smallest eigenvalue in Prob. 12.

14. ■ Use the Jacobi method to compute the eigenvalues and eigenvectors of the following matrix:

$$
A = \begin{bmatrix}
11 & 2 & 3 & 1 & 4 & 2 \\
2 & 9 & 3 & 5 & 2 & 1 \\
3 & 3 & 15 & 4 & 3 & 2 \\
1 & 5 & 4 & 12 & 4 & 3 \\
4 & 2 & 3 & 4 & 17 & 5 \\
2 & 1 & 2 & 3 & 5 & 8
\end{bmatrix}
$$

15. ■ Find the eigenvalues of $\mathbf{Ax} = \lambda \mathbf{Bx}$ by the Jacobi method, where

$$
A = \begin{bmatrix}
6 & -4 & 1 & 0 \\
-4 & 6 & -4 & 1 \\
1 & -4 & 6 & -4 \\
0 & 1 & -4 & 7
\end{bmatrix}
\qquad
B = \begin{bmatrix}
1 & -2 & 3 & -1 \\
-2 & 6 & -2 & 3 \\
3 & -2 & 6 & -2 \\
-1 & 3 & -2 & 9
\end{bmatrix}
$$

Warning: **B** is not positive definite. *Hint*: Solve $\mathbf{Bx} = (1/\lambda)\mathbf{Ax}$.

16. ■

The figure shows a cantilever beam with a superimposed finite difference mesh. If $u(x, t)$ is the lateral displacement of the beam, the differential equation of motion governing bending vibrations is

$$
u^{(4)} = -\frac{\gamma}{EI}\ddot{u}
$$

where γ is the mass per unit length and EI is the bending rigidity. The boundary conditions are $u(0, t) = u'(0, t) = u''(L, t) = u'''(L, t) = 0$. With $u(x, t) = y(x)\sin \omega t$ the problem becomes

$$
y^{(4)} = \frac{\omega^2 \gamma}{EI} y \qquad y(0) = y'(0) = y''(L) = y'''(L) = 0
$$

The corresponding finite difference equations are

$$
A = \begin{bmatrix}
7 & -4 & 1 & 0 & 0 & \cdots & 0 \\
-4 & 6 & -4 & 1 & 0 & \cdots & 0 \\
1 & -4 & 6 & -4 & 1 & \cdots & 0 \\
\vdots & \ddots & \ddots & \ddots & \ddots & \ddots & \vdots \\
0 & \cdots & 1 & -4 & 6 & -4 & 1 \\
0 & \cdots & 0 & 1 & -4 & 5 & -2 \\
0 & \cdots & 0 & 0 & 1 & -2 & 1
\end{bmatrix}
\begin{bmatrix}
y_1 \\ y_2 \\ y_3 \\ \vdots \\ y_{n-2} \\ y_{n-1} \\ y_n
\end{bmatrix}
= \lambda
\begin{bmatrix}
y_1 \\ y_2 \\ y_3 \\ \vdots \\ y_{n-2} \\ y_{n-1} \\ y_n/2
\end{bmatrix}
$$

where

$$
\lambda = \frac{\omega^2 \gamma}{EI}\left(\frac{L}{n}\right)^4
$$

(a) Write down the matrix $\mathbf{H}$ of the standard form $\mathbf{Hz} = \lambda\mathbf{z}$ and the transformation matrix $\mathbf{P}$ as in $\mathbf{y} = \mathbf{Pz}$. (b) Write a program that computes the lowest two circular frequencies of the beam and the corresponding mode shapes (eigenvectors) using the Jacobi method. Run the program with $n = 10$. *Note*: The analytical solution for the lowest circular frequency is $\omega_1 = (3.515/L^2)\sqrt{EI/\gamma}$.

17. ■

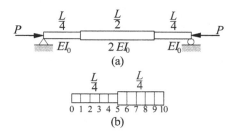

The simply supported column in Figure (a) consists of three segments with the bending rigidities shown. If only the first buckling mode is of interest, it is sufficient to model half of the beam as shown in Figure (b). The differential equation for the lateral displacement $u(x)$ is

$$u'' = -\frac{P}{EI}u$$

with the boundary conditions $u(0) = u'(0) = 0$. The corresponding finite difference equations are

$$
\begin{bmatrix}
2 & -1 & 0 & 0 & 0 & 0 & 0 & \cdots & 0 \\
-1 & 2 & -1 & 0 & 0 & 0 & 0 & \cdots & 0 \\
0 & -1 & 2 & -1 & 0 & 0 & 0 & \cdots & 0 \\
0 & 0 & -1 & 2 & -1 & 0 & 0 & \cdots & 0 \\
0 & 0 & 0 & -1 & 2 & -1 & 0 & \cdots & 0 \\
0 & 0 & 0 & 0 & -1 & 2 & -1 & \cdots & 0 \\
\vdots & \vdots & \vdots & \vdots & \vdots & \ddots & \ddots & \ddots & \vdots \\
0 & \cdots & 0 & 0 & 0 & 0 & -1 & 2 & -1 \\
0 & \cdots & 0 & 0 & 0 & 0 & 0 & -1 & 1
\end{bmatrix}
\begin{bmatrix}
u_1 \\ u_2 \\ u_3 \\ u_4 \\ u_5 \\ u_6 \\ \vdots \\ u_9 \\ u_{10}
\end{bmatrix}
= \lambda
\begin{bmatrix}
u_1 \\ u_2 \\ u_3 \\ u_4 \\ u_5/1.5 \\ u_6/2 \\ \vdots \\ u_9/2 \\ u_{10}/4
\end{bmatrix}
$$

where

$$\lambda = \frac{P}{EI_0}\left(\frac{L}{20}\right)^2$$

Write a program that computes the lowest buckling load P of the column with the inverse power method. Use the banded forms of the matrices.

18. ■

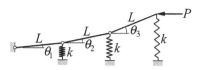

The springs supporting the three-bar linkage are undeformed when the linkage is horizontal. The equilibrium equations of the linkage in the presence of the horizontal force P can be shown to be

$$
\begin{bmatrix} 6 & 5 & 3 \\ 3 & 3 & 2 \\ 1 & 1 & 1 \end{bmatrix} \begin{bmatrix} \theta_1 \\ \theta_2 \\ \theta_3 \end{bmatrix} = \frac{P}{kL} \begin{bmatrix} 1 & 1 & 1 \\ 0 & 1 & 1 \\ 0 & 0 & 1 \end{bmatrix} \begin{bmatrix} \theta_1 \\ \theta_2 \\ \theta_3 \end{bmatrix}
$$

where k is the spring stiffness. Determine the smallest buckling load P and the corresponding mode shape. *Hint*: The equations can be easily rewritten in the standard form $\mathbf{A}\boldsymbol{\theta} = \lambda\boldsymbol{\theta}$, where $\mathbf{A}$ is symmetric.

19. ■

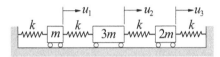

The differential equations of motion for the mass-spring system are

$$
k(-2u_1 + u_2) = m\ddot{u}_1
$$

$$
k(u_1 - 2u_2 + u_3) = 3m\ddot{u}_2
$$

$$
k(u_2 - 2u_3) = 2m\ddot{u}_3
$$

where $u_i(t)$ is the displacement of mass i from its equilibrium position and k is the spring stiffness. Determine the circular frequencies of vibration and the corresponding mode shapes.

20. ■

Kirchoff's equations for the circuit are

$$
L\frac{d^2i_1}{dt^2} + \frac{1}{C}i_1 + \frac{2}{C}(i_1 - i_2) = 0
$$

$$
L\frac{d^2i_2}{dt^2} + \frac{2}{C}(i_2 - i_1) + \frac{3}{C}(i_2 - i_3) = 0
$$

$$
L\frac{d^2i_3}{dt^2} + \frac{3}{C}(i_3 - i_2) + \frac{4}{C}(i_3 - i_4) = 0
$$

$$
L\frac{d^2i_4}{dt^2} + \frac{4}{C}(i_4 - i_3) + \frac{5}{C}i_4 = 0
$$

Find the circular frequencies of the current.

21. ■

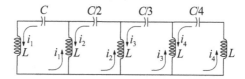

Determine the circular frequencies of oscillation for the circuit shown, given these Kirchoff equations:

$$L\frac{d^2 i_1}{dt^2} + L\left(\frac{d^2 i_1}{dt^2} - \frac{d^2 i_2}{dt^2}\right) + \frac{1}{C}i_1 = 0$$

$$L\left(\frac{d^2 i_2}{dt^2} - \frac{d^2 i_1}{dt^2}\right) + L\left(\frac{d^2 i_2}{dt^2} - \frac{d^2 i_3}{dt^2}\right) + \frac{2}{C} = 0$$

$$L\left(\frac{d^2 i_3}{dt^2} - \frac{d^2 i_2}{dt^2}\right) + L\left(\frac{d^2 i_3}{dt^2} - \frac{d^2 i_4}{dt^2}\right) + \frac{3}{C}i_3 = 0$$

$$L\left(\frac{d^2 i_4}{dt^2} - \frac{d^2 i_3}{dt^2}\right) + L\frac{d^2 i_4}{dt^2} + \frac{4}{C}i_4 = 0$$

22. ■ Several iterative methods exist for finding the eigenvalues of a matrix $\mathbf{A}$. One of these is the *LR method*, which requires the matrix to be symmetric and positive definite. Its algorithm is very simple:

> Do the following until the change in $\mathbf{A}$ is insignificant:
> Use Choleski's decomposition $\mathbf{A} = \mathbf{LL}^T$ to compute $\mathbf{L}$.
> Compute $\mathbf{A} = \mathbf{L}^T\mathbf{L}$.

It can be shown that the diagonal elements of $\mathbf{A}$ converge to the eigenvalues of $\mathbf{A}$. Write a program that implements the LR method and test it with

$$\mathbf{A} = \begin{bmatrix} 4 & 3 & 1 \\ 3 & 4 & 2 \\ 1 & 2 & 3 \end{bmatrix}$$

9.4 Householder Reduction to Tridiagonal Form

It was mentioned easier that similarity transformations can be used to transform an eigenvalue problem to a form that is easier to solve. The most desirable of the "easy" forms is, of course, the diagonal form that results from the Jacobi method. However, the Jacobi method requires about $10n^3$ to $20n^3$ multiplications, so that the amount of computation increases very rapidly with n. We are generally better off by reducing the matrix to the tridiagonal form, which can be done in precisely $n - 2$ transformations by the Householder method. Once the tridiagonal form is achieved, we still have to extract the eigenvalues and the eigenvectors, but there are effective means of dealing with that, as we seen the next section.

Householder Matrix

Each Householder transformation uses the *Householder matrix*

$$Q = I - \frac{\mathbf{u}\mathbf{u}^T}{H} \tag{9.36}$$

where $\mathbf{u}$ is a vector and

$$H = \frac{1}{2}\mathbf{u}^T\mathbf{u} = \frac{1}{2}|\mathbf{u}|^2 \tag{9.37}$$

Note that $\mathbf{u}\mathbf{u}^T$ in Eq. (9.36) is the outer product; that is, a matrix with the elements $\left(\mathbf{u}\mathbf{u}^T\right)_{ij} = u_i u_j$. Because Q is obviously symmetric ($Q^T = Q$), we can write

$$Q^T Q = QQ = \left(I - \frac{\mathbf{u}\mathbf{u}^T}{H}\right)\left(I - \frac{\mathbf{u}\mathbf{u}^T}{H}\right) = I - 2\frac{\mathbf{u}\mathbf{u}^T}{H} + \frac{\mathbf{u}\left(\mathbf{u}^T\mathbf{u}\right)\mathbf{u}^T}{H^2}$$

$$= I - 2\frac{\mathbf{u}\mathbf{u}^T}{H} + \frac{\mathbf{u}\left(2H\right)\mathbf{u}^T}{H^2} = I$$

which shows that Q is also orthogonal.

Now let $\mathbf{x}$ be an arbitrary vector and consider the transformation $Q\mathbf{x}$. Choosing

$$\mathbf{u} = \mathbf{x} + k\mathbf{e}_1 \tag{9.38}$$

where

$$k = \pm|\mathbf{x}| \qquad \mathbf{e}_1 = \begin{bmatrix} 1 & 0 & 0 & \cdots & 0 \end{bmatrix}^T$$

we get

$$Q\mathbf{x} = \left(I - \frac{\mathbf{u}\mathbf{u}^T}{H}\right)\mathbf{x} = \left[I - \frac{\mathbf{u}\left(\mathbf{x} + k\mathbf{e}_1\right)^T}{H}\right]\mathbf{x}$$

$$= \mathbf{x} - \frac{\mathbf{u}\left(\mathbf{x}^T\mathbf{x} + k\mathbf{e}_1^T\mathbf{x}\right)}{H} = \mathbf{x} - \frac{\mathbf{u}\left(k^2 + kx_1\right)}{H}$$

But

$$2H = \left(\mathbf{x} + k\mathbf{e}_1\right)^T\left(\mathbf{x} + k\mathbf{e}_1\right) = |\mathbf{x}|^2 + k\left(\mathbf{x}^T\mathbf{e}_1 + \mathbf{e}_1^T\mathbf{x}\right) + k^2\mathbf{e}_1^T\mathbf{e}_1$$

$$= k^2 + 2kx_1 + k^2 = 2\left(k^2 + kx_1\right)$$

so that

$$Q\mathbf{x} = \mathbf{x} - \mathbf{u} = -k\mathbf{e}_1 = \begin{bmatrix} -k & 0 & 0 & \cdots & 0 \end{bmatrix}^T \tag{9.39}$$

Hence the transformation eliminates all elements of $\mathbf{x}$ except the first one.

Householder Reduction of a Symmetric Matrix

Let us now apply the following transformation to a symmetric $n \times n$ matrix $\mathbf{A}$:

$$\mathbf{P}_1\mathbf{A} = \begin{bmatrix} 1 & \mathbf{0}^T \\ \mathbf{0} & \mathbf{Q} \end{bmatrix} \begin{bmatrix} A_{11} & \mathbf{x}^T \\ \mathbf{x} & \mathbf{A}' \end{bmatrix} = \begin{bmatrix} A_{11} & \mathbf{x}^T \\ \mathbf{Qx} & \mathbf{QA}' \end{bmatrix} \tag{9.40}$$

Here $\mathbf{x}$ represents the first column of $\mathbf{A}$ with the first element omitted, and $\mathbf{A}'$ is simply $\mathbf{A}$ with its first row and column removed. The matrix $\mathbf{Q}$ of dimensions $(n-1) \times (n-1)$ is constructed using Eqs. (9.36)–(9.38). Referring to Eq. (9.39), we see that the transformation reduces the first column of $\mathbf{A}$ to

$$\begin{bmatrix} A_{11} \\ \mathbf{Qx} \end{bmatrix} = \begin{bmatrix} A_{11} \\ -k \\ 0 \\ \vdots \\ 0 \end{bmatrix}$$

The transformation

$$\mathbf{A} \leftarrow \mathbf{P}_1\mathbf{AP}_1 = \begin{bmatrix} A_{11} & (\mathbf{Qx})^T \\ \mathbf{Qx} & \mathbf{QA}'\mathbf{Q} \end{bmatrix} \tag{9.41}$$

thus tridiagonalizes the first row as well as the first column of $\mathbf{A}$. Here is a diagram of the transformation for a 4×4 matrix:

$$
\begin{bmatrix} 1 & 0 & 0 & 0 \\ 0 & & & \\ 0 & & \mathbf{Q} & \\ 0 & & & \end{bmatrix}
\cdot
\begin{bmatrix} A_{11} & A_{12} & A_{13} & A_{14} \\ A_{21} & & & \\ A_{31} & & \mathbf{A}' & \\ A_{41} & & & \end{bmatrix}
\cdot
\begin{bmatrix} 1 & 0 & 0 & 0 \\ 0 & & & \\ 0 & & \mathbf{Q} & \\ 0 & & & \end{bmatrix}
$$

$$
= \begin{bmatrix} A_{11} & -k & 0 & 0 \\ -k & & & \\ 0 & & \mathbf{QA}'\mathbf{Q} & \\ 0 & & & \end{bmatrix}
$$

The second row and column of $\mathbf{A}$ are reduced next by applying the transformation to the 3×3 lower right portion of the matrix. This transformation can be expressed as $\mathbf{A} \leftarrow \mathbf{P}_2\mathbf{AP}_2$, where now

$$\mathbf{P}_2 = \begin{bmatrix} \mathbf{I}_2 & \mathbf{0}^T \\ \mathbf{0} & \mathbf{Q} \end{bmatrix} \tag{9.42}$$

In Eq. (9.42) $\mathbf{I}_2$ is a 2×2 identity matrix and $\mathbf{Q}$ is a $(n-2) \times (n-2)$ matrix constructed by choosing for $\mathbf{x}$ the bottom $n-2$ elements of the second column of $\mathbf{A}$. It takes a total of $n-2$ transformations with

$$\mathbf{P}_i = \begin{bmatrix} \mathbf{I}_i & \mathbf{0}^T \\ \mathbf{0} & \mathbf{Q} \end{bmatrix}, \quad i = 1, 2, \ldots, n-2$$

to attain the tridiagonal form.

It is wasteful to form $\mathbf{P}_i$ and then carry out the matrix multiplication $\mathbf{P}_i\mathbf{AP}_i$. We note that

$$\mathbf{A'Q} = \mathbf{A'}\left(\mathbf{I} - \frac{\mathbf{uu}^T}{H}\right) = \mathbf{A'} - \frac{\mathbf{A'u}}{H}\mathbf{u}^T = \mathbf{A'} - \mathbf{vu}^T$$

where

$$\mathbf{v} = \frac{\mathbf{A'u}}{H} \tag{9.43}$$

Therefore,

$$\mathbf{QA'Q} = \left(\mathbf{I} - \frac{\mathbf{uu}^T}{H}\right)\left(\mathbf{A'} - \mathbf{vu}^T\right) = \mathbf{A'} - \mathbf{vu}^T - \frac{\mathbf{uu}^T}{H}\left(\mathbf{A'} - \mathbf{vu}^T\right)$$

$$= \mathbf{A'} - \mathbf{vu}^T - \frac{\mathbf{u}\left(\mathbf{u}^T\mathbf{A'}\right)}{H} + \frac{\mathbf{u}\left(\mathbf{u}^T\mathbf{v}\right)\mathbf{u}^T}{H}$$

$$= \mathbf{A'} - \mathbf{vu}^T - \mathbf{uv}^T + 2g\mathbf{uu}^T$$

where

$$g = \frac{\mathbf{u}^T\mathbf{v}}{2H} \tag{9.44}$$

Letting

$$\mathbf{w} = \mathbf{v} - g\mathbf{u} \tag{9.45}$$

it can be easily verified that the transformation can be written as

$$\mathbf{QA'Q} = \mathbf{A'} - \mathbf{wu}^T - \mathbf{uw}^T \tag{9.46}$$

which gives us the following computational procedure that is to be carried out with $i = 1, 2, \ldots, n-2$:

Do with $i = 1, 2, \ldots, n-2$:

Let $\mathbf{A'}$ be the $(n-i) \times (n-i)$ lower right-hand portion of $\mathbf{A}$.

Let $\mathbf{x} = \begin{bmatrix} A_{i+1,i} & A_{i+2,i} & \cdots & A_{n,i} \end{bmatrix}^T$
(the column of length $n-i$ just left of $\mathbf{A'}$).

Compute $|\mathbf{x}|$. Let $k = |\mathbf{x}|$ if $x_1 > 0$ and $k = -|\mathbf{x}|$ if $x_1 < 0$
(this choice of sign minimizes the roundoff error).

Let $\mathbf{u} = \begin{bmatrix} k+x_1 & x_2 & x_3 & \cdots & x_{n-i} \end{bmatrix}^T$.

Compute $H = |\mathbf{u}|/2$.

Compute $\mathbf{v} = \mathbf{A'u}/H$.

Compute $g = \mathbf{u}^T\mathbf{v}/(2H)$.

Compute $\mathbf{w} = \mathbf{v} - g\mathbf{u}$.

Compute the transformation $\mathbf{A'} \leftarrow \mathbf{A'} - \mathbf{w}^T\mathbf{u} - \mathbf{u}^T\mathbf{w}$.

Set $A_{i,i+1} = A_{i+1,i} = -k$.

Accumulated Transformation Matrix

Because we used similarity transformations, the eigenvalues of the tridiagonal matrix are the same as those of the original matrix. However, to determine the eigenvectors $\mathbf{X}$ of original $\mathbf{A}$ we must use the transformation

$$\mathbf{X} = \mathbf{P}\mathbf{X}_{\text{tridiag}}$$

where $\mathbf{P}$ is the accumulation of the individual transformations:

$$\mathbf{P} = \mathbf{P}_1 \mathbf{P}_2 \cdots \mathbf{P}_{n-2}$$

We build up the accumulated transformation matrix by initializing $\mathbf{P}$ to a $n \times n$ identity matrix and then applying the transformation

$$\mathbf{P} \leftarrow \mathbf{P}\mathbf{P}_i = \begin{bmatrix} \mathbf{P}_{11} & \mathbf{P}_{12} \\ \mathbf{P}_{21} & \mathbf{P}_{22} \end{bmatrix} \begin{bmatrix} \mathbf{I}_i & \mathbf{0}^T \\ \mathbf{0} & \mathbf{Q} \end{bmatrix} = \begin{bmatrix} \mathbf{P}_{11} & \mathbf{P}_{21}\mathbf{Q} \\ \mathbf{P}_{12} & \mathbf{P}_{22}\mathbf{Q} \end{bmatrix} \tag{b}$$

with $i = 1, 2, \ldots, n - 2$. It can be seen that each multiplication affects only the rightmost $n - i$ columns of $\mathbf{P}$ (since the first row of $\mathbf{P}_{12}$ contains only zeroes, it can also be omitted in the multiplication). Using the notation

$$\mathbf{P}' = \begin{bmatrix} \mathbf{P}_{12} \\ \mathbf{P}_{22} \end{bmatrix}$$

we have

$$\begin{bmatrix} \mathbf{P}_{12}\mathbf{Q} \\ \mathbf{P}_{22}\mathbf{Q} \end{bmatrix} = \mathbf{P}'\mathbf{Q} = \mathbf{P}'\left(\mathbf{I} - \frac{\mathbf{u}\mathbf{u}^T}{H}\right) = \mathbf{P}' - \frac{\mathbf{P}'\mathbf{u}}{H}\mathbf{u}^T = \mathbf{P}' - \mathbf{y}\mathbf{u}^T \tag{9.47}$$

where

$$\mathbf{y} = \frac{\mathbf{P}'\mathbf{u}}{H} \tag{9.48}$$

The procedure for carrying out the matrix multiplication in Eq. (b) is as follows:

> Retrieve $\mathbf{u}$ ($\mathbf{u}$'s are stored by columns below the principal diagonal of $\mathbf{A}$).
> Compute $H = |\mathbf{u}|/2$.
> Compute $\mathbf{y} = \mathbf{P}'\mathbf{u}/H$.
> Compute the transformation $\mathbf{P}' \leftarrow \mathbf{P}' - \mathbf{y}\mathbf{u}^T$.

■ householder

The function `householder` in this module does reduction to tridiagonal form. It returns $(\mathbf{d}, \mathbf{c})$, where $\mathbf{d}$ and $\mathbf{c}$ are vectors that contain the elements of the principal diagonal and the subdiagonal, respectively. Only the upper triangular portion is reduced to the triangular form. The part below the principal diagonal is used to store the vectors **u**. This is done automatically by the statement u = a[k+1:n,k], which does not create a new object u, but simply sets up a reference to a[k+1:n,k] (makes a deep copy). Thus any changes made to u are reflected in a[k+1:n,k].

The function computeP returns the accumulated transformation matrix **P**. There is no need to call it if only the eigenvalues are to be computed.

```
## module householder
''' d,c = householder(a).
    Householder similarity transformation of matrix [a] to
    tridiagonal form.

    p = computeP(a).
    Computes the acccumulated transformation matrix [p]
    after calling householder(a).
'''
import numpy as np
import math

def householder(a):
    n = len(a)
    for k in range(n-2):
        u = a[k+1:n,k]
        uMag = math.sqrt(np.dot(u,u))
        if u[0] < 0.0: uMag = -uMag
        u[0] = u[0] + uMag
        h = np.dot(u,u)/2.0
        v = np.dot(a[k+1:n,k+1:n],u)/h
        g = np.dot(u,v)/(2.0*h)
        v = v - g*u
        a[k+1:n,k+1:n] = a[k+1:n,k+1:n] - np.outer(v,u)  \
                        -np.outer(u,v)
        a[k,k+1] = -uMag
    return np.diagonal(a),np.diagonal(a,1)

def computeP(a):
    n = len(a)
    p = np.identity(n)*1.0
    for k in range(n-2):
        u = a[k+1:n,k]
        h = np.dot(u,u)/2.0
        v = np.dot(p[1:n,k+1:n],u)/h
        p[1:n,k+1:n] = p[1:n,k+1:n] - np.outer(v,u)
    return p
```

EXAMPLE 9.7

Transform the matrix

$$A = \begin{bmatrix} 7 & 2 & 3 & -1 \\ 2 & 8 & 5 & 1 \\ 3 & 5 & 12 & 9 \\ -1 & 1 & 9 & 7 \end{bmatrix}$$

into tridiagonal form.

Solution. Reduce first row and column:

$$A' = \begin{bmatrix} 8 & 5 & 1 \\ 5 & 12 & 9 \\ 1 & 9 & 7 \end{bmatrix} \qquad x = \begin{bmatrix} 2 \\ 3 \\ -1 \end{bmatrix} \qquad k = |\mathbf{x}| = 3.7417$$

$$\mathbf{u} = \begin{bmatrix} k + x_1 \\ x_2 \\ x_3 \end{bmatrix} = \begin{bmatrix} 5.7417 \\ 3 \\ -1 \end{bmatrix} \qquad H = \frac{1}{2}|\mathbf{u}|^2 = 21.484$$

$$\mathbf{uu}^T = \begin{bmatrix} 32.967 & 17.225 & -5.7417 \\ 17.225 & 9 & -3 \\ -5.7417 & -3 & 1 \end{bmatrix}$$

$$Q = I - \frac{\mathbf{uu}^T}{H} = \begin{bmatrix} -0.53450 & -0.80176 & 0.26725 \\ -0.80176 & 0.58108 & 0.13964 \\ 0.26725 & 0.13964 & 0.95345 \end{bmatrix}$$

$$QA'Q = \begin{bmatrix} 10.642 & -0.1388 & -9.1294 \\ -0.1388 & 5.9087 & 4.8429 \\ -9.1294 & 4.8429 & 10.4480 \end{bmatrix}$$

$$A \leftarrow \begin{bmatrix} A_{11} & (\mathbf{Qx})^T \\ \mathbf{Qx} & QA'Q \end{bmatrix} = \begin{bmatrix} 7 & -3.7417 & 0 & 0 \\ -3.7417 & 10.642 & -0.1388 & -9.1294 \\ 0 & -0.1388 & 5.9087 & 4.8429 \\ 0 & -9.1294 & 4.8429 & 10.4480 \end{bmatrix}$$

In the last step we used the formula $\mathbf{Qx} = \begin{bmatrix} -k & 0 & \cdots & 0 \end{bmatrix}^T$.

Reduce the second row and column:

$$A' = \begin{bmatrix} 5.9087 & 4.8429 \\ 4.8429 & 10.4480 \end{bmatrix} \qquad x = \begin{bmatrix} -0.1388 \\ -9.1294 \end{bmatrix} \qquad k = -|\mathbf{x}| = -9.1305$$

where the negative sign on k was determined by the sign of x_1.

$$\mathbf{u} = \begin{bmatrix} k + x_1 \\ -9.1294 \end{bmatrix} = \begin{bmatrix} -9.2693 \\ -9.1294 \end{bmatrix} \qquad H = \frac{1}{2}|\mathbf{u}|^2 = 84.633$$

$$\mathbf{u}\mathbf{u}^T = \begin{bmatrix} 85.920 & 84.623 \\ 84.623 & 83.346 \end{bmatrix}$$

$$\mathbf{Q} = \mathbf{I} - \frac{\mathbf{u}\mathbf{u}^T}{H} = \begin{bmatrix} 0.01521 & -0.99988 \\ -0.99988 & 0.01521 \end{bmatrix}$$

$$\mathbf{Q}\mathbf{A}'\mathbf{Q} = \begin{bmatrix} 10.594 & 4.772 \\ 4.772 & 5.762 \end{bmatrix}$$

$$\mathbf{A} \leftarrow \begin{bmatrix} A_{11} & A_{12} & \mathbf{0}^T \\ A_{21} & A_{22} & (\mathbf{Q}\mathbf{x})^T \\ \mathbf{0} & \mathbf{Q}\mathbf{x} & \mathbf{Q}\mathbf{A}'\mathbf{Q} \end{bmatrix} \begin{bmatrix} 7 & -3.742 & 0 & 0 \\ -3.742 & 10.642 & 9.131 & 0 \\ 0 & 9.131 & 10.594 & 4.772 \\ 0 & 0 & 4.772 & 5.762 \end{bmatrix}$$

EXAMPLE 9.8

Use the function `householder` to tridiagonalize the matrix in Example 9.7; also determine the transformation matrix **P**.

Solution

```
#!/usr/bin/python
## example9_8
import numpy as np
from householder import *

a = np.array([[ 7.0, 2.0,  3.0, -1.0], \
              [ 2.0, 8.0,  5.0,  1.0], \
              [ 3.0, 5.0, 12.0,  9.0], \
              [-1.0, 1.0,  9.0,  7.0]])
d,c = householder(a)
print("Principal diagonal {d}:\n", d)
print("\nSubdiagonal {c}:\n",c)
print("\nTransformation matrix [P]:")
print(computeP(a))
input("\nPress return to exit")
```

The results of running the above program are as follows:

```
Principal diagonal {d}:
[  7.          10.64285714  10.59421525   5.76292761]

Subdiagonal {c}:
[-3.74165739  9.13085149  4.77158058]
```

```
Transformation matrix [P]:
[[ 1.            0.            0.            0.          ]
 [ 0.           -0.53452248  -0.25506831   0.80574554]
 [ 0.           -0.80178373  -0.14844139  -0.57888514]
 [ 0.            0.26726124  -0.95546079  -0.12516436]]
```

9.5 Eigenvalues of Symmetric Tridiagonal Matrices

Sturm Sequence

In principle, the eigenvalues of a matrix $\mathbf{A}$ can be determined by finding the roots of the characteristic equation $|\mathbf{A} - \lambda\mathbf{I}| = 0$. This method is impractical for large matrices, because the evaluation of the determinant involves $n^3/3$ multiplications. However, if the matrix is tridiagonal (we also assume it to be symmetric), its characteristic polynomial

$$P_n(\lambda) = |\mathbf{A} - \lambda\mathbf{I}| = \begin{vmatrix} d_1 - \lambda & c_1 & 0 & 0 & \cdots & 0 \\ c_1 & d_2 - \lambda & c_2 & 0 & \cdots & 0 \\ 0 & c_2 & d_3 - \lambda & c_3 & \cdots & 0 \\ 0 & 0 & c_3 & d_4 - \lambda & \cdots & 0 \\ \vdots & \vdots & \vdots & \vdots & \ddots & \vdots \\ 0 & 0 & \cdots & 0 & c_{n-1} & d_n - \lambda \end{vmatrix}$$

can be computed with only $3(n-1)$ multiplications using the following sequence of operations:

$$P_0(\lambda) = 1$$
$$P_1(\lambda) = d_1 - \lambda \qquad\qquad (9.49)$$
$$P_i(\lambda) = (d_i - \lambda)P_{i-1}(\lambda) - c_{i-1}^2 P_{i-2}(\lambda), \quad i = 2, 3, \ldots, n$$

The polynomials $P_0(\lambda)$, $P_1(\lambda)$, ..., $P_n(\lambda)$ form a *Sturm sequence* that has the following property: The number of sign changes in the sequence $P_0(a)$, $P_1(a)$, ..., $P_n(a)$ is equal to the number of roots of $P_n(\lambda)$ that are smaller than a. If a member $P_i(a)$ of the sequence is zero, its sign is to be taken opposite to that of $P_{i-1}(a)$.

As we see later, Sturm sequence property makes it possible to bracket the eigenvalues of a tridiagonal matrix.

■ sturmSeq

Given $\mathbf{d}$, $\mathbf{c}$, and λ, the function sturmSeq returns the Sturm sequence

$$P_0(\lambda), \ P_1(\lambda), \ldots P_n(\lambda)$$

The function numLambdas returns the number of sign changes in the sequence (as noted earlier, this equals the number of eigenvalues that are smaller than λ).

```
## module sturmSeq
''' p = sturmSeq(c,d,lam).
    Returns the Sturm sequence {p[0],p[1],...,p[n]}
    associated with the characteristic polynomial
    |[A] - lam[I]| = 0, where [A] is a n x n
    tridiagonal matrix.

    numLam = numLambdas(p).
    Returns the number of eigenvalues of a tridiagonal
    matrix that are smaller than 'lam'.
    Uses the Sturm sequence {p} obtained from 'sturmSeq'.
'''
import numpy as np

def sturmSeq(d,c,lam):
    n = len(d) + 1
    p = np.ones(n)
    p[1] = d[0] - lam
    for i in range(2,n):
        p[i] = (d[i-1] - lam)*p[i-1] - (c[i-2]**2)*p[i-2]
    return p

def numLambdas(p):
    n = len(p)
    signOld = 1
    numLam = 0
    for i in range(1,n):
        if p[i] > 0.0: sign = 1
        elif p[i] < 0.0: sign = -1
        else: sign = -signOld
        if sign*signOld < 0: numLam = numLam + 1
        signOld = sign
    return numLam
```

EXAMPLE 9.9

Use the Sturm sequence property to show that the smallest eigenvalue of **A** is in the interval (0.25, 0.5), where

$$
\mathbf{A} = \begin{bmatrix} 2 & -1 & 0 & 0 \\ -1 & 2 & -1 & 0 \\ 0 & -1 & 2 & -1 \\ 0 & 0 & -1 & 2 \end{bmatrix}
$$

Solution. Taking $\lambda = 0.5$, we have $d_i - \lambda = 1.5$ and $c_{i-1}^2 = 1$, and the Sturm sequence in Eqs. (9.49) becomes

$$P_0(0.5) = 1$$

$$P_1(0.5) = 1.5$$

$$P_2(0.5) = 1.5(1.5) - 1 = 1.25$$

$$P_3(0.5) = 1.5(1.25) - 1.5 = 0.375$$

$$P_4(0.5) = 1.5(0.375) - 1.25 = -0.6875$$

Since the sequence contains one sign change, there exists one eigenvalue smaller than 0.5.

Repeating the process with $\lambda = 0.25$, we get $d_i - \lambda = 1.75$ and $c_i^2 = 1$, which results in the Sturm sequence:

$$P_0(0.25) = 1$$

$$P_1(0.25) = 1.75$$

$$P_2(0.25) = 1.75(1.75) - 1 = 2.0625$$

$$P_3(0.25) = 1.75(2.0625) - 1.75 = 1.8594$$

$$P_4(0.25) = 1.75(1.8594) - 2.0625 = 1.1915$$

There are no sign changes in the sequence, so that all the eigenvalues are greater than 0.25. We thus conclude that $0.25 < \lambda_1 < 0.5$.

Gerschgorin's Theorem

Gerschgorin's theorem is useful in determining the *global bounds* on the eigenvalues of a $n \times n$ matrix $\mathbf{A}$. The term "global" means the bounds that enclose all the eigenvalues. We give here a simplified version for a symmetric matrix.

If λ is an eigenvalue of $\mathbf{A}$, then

$$a_i - r_i \le \lambda \le a_i + r_i, \quad i = 1, 2, \ldots, n$$

where

$$a_i = A_{ii} \qquad r_i = \sum_{\substack{j=1 \\ j \ne i}}^{n} |A_{ij}| \qquad (9.50)$$

It follows that the limits on the smallest and the largest eigenvalues are given by

$$\lambda_{\min} \ge \min_i(a_i - r_i) \qquad \lambda_{\max} \le \max_i(a_i + r_i) \qquad (9.51)$$

■ gerschgorin

The function gerschgorin returns the lower and upper global bounds on the eigenvalues of a symmetric tridiagonal matrix $\mathbf{A} = [\mathbf{c}\backslash\mathbf{d}\backslash\mathbf{c}]$.

```
## module gerschgorin
''' lamMin,lamMax = gerschgorin(d,c).
    Applies Gerschgorin's theorem to find the global bounds on
    the eigenvalues of a symmetric tridiagonal matrix.
'''
def gerschgorin(d,c):
    n = len(d)
    lamMin = d[0] - abs(c[0])
    lamMax = d[0] + abs(c[0])
    for i in range(1,n-1):
        lam = d[i] - abs(c[i]) - abs(c[i-1])
        if lam < lamMin: lamMin = lam
        lam = d[i] + abs(c[i]) + abs(c[i-1])
        if lam > lamMax: lamMax = lam
    lam = d[n-1] - abs(c[n-2])
    if lam < lamMin: lamMin = lam
    lam = d[n-1] + abs(c[n-2])
    if lam > lamMax: lamMax = lam
    return lamMin,lamMax
```

EXAMPLE 9.10
Use Gerschgorin's theorem to determine the bounds on the eigenvalues of the matrix

$$\mathbf{A} = \begin{bmatrix} 4 & -2 & 0 \\ -2 & 4 & -2 \\ 0 & -2 & 5 \end{bmatrix}$$

Solution. Referring to Eqs. (9.50), we get

$$a_1 = 4 \qquad a_2 = 4 \qquad a_3 = 5$$

$$r_1 = 2 \qquad r_2 = 4 \qquad r_3 = 2$$

Hence

$$\lambda_{\min} \geq \min(a_i - r_i) = 4 - 4 = 0$$

$$\lambda_{\max} \leq \max(a_i + r_i) = 4 + 4 = 8$$

Bracketing Eigenvalues

The Sturm sequence property used together with Gerschgorin's theorem provides us a convenient tool for bracketing each eigenvalue of a symmetric tridiagonal matrix.

■ lamRange

The function `lamRange` brackets the N smallest eigenvalues of a symmetric tridi-
agonal matrix $\mathbf{A} = [\mathbf{c} \backslash \mathbf{d} \backslash \mathbf{c}]$. It returns the sequence $r_0, r_1, \ldots, r_N$, where each interval
(r_{i-1}, r_i) contains exactly one eigenvalue. The algorithm first finds the bounds on all
the eigenvalues by Gerschgorin's theorem. Then the method of bisection in conjunction with the Sturm sequence property is used to determine $r_N, r_{N-1}, \ldots, r_0$ in that
order.

```
## module lamRange
''' r = lamRange(d,c,N).
    Returns the sequence {r[0],r[1],...,r[N]} that
    separates the N lowest eigenvalues of the tridiagonal
    matrix; that is, r[i] < lam[i] < r[i+1].
'''
import numpy as np
from sturmSeq import *
from gerschgorin import *

def lamRange(d,c,N):
    lamMin,lamMax = gerschgorin(d,c)
    r = np.ones(N+1)
    r[0] = lamMin
  # Search for eigenvalues in descending order
    for k in range(N,0,-1):
      # First bisection of interval(lamMin,lamMax)
        lam = (lamMax + lamMin)/2.0
        h = (lamMax - lamMin)/2.0
        for i in range(1000):
          # Find number of eigenvalues less than lam
            p = sturmSeq(d,c,lam)
            numLam = numLambdas(p)
          # Bisect again & find the half containing lam
            h = h/2.0
            if numLam < k: lam = lam + h
            elif numLam > k: lam = lam - h
            else: break
      # If eigenvalue located, change the upper limit
      # of search and record it in [r]
        lamMax = lam
        r[k] = lam
    return r
```

EXAMPLE 9.11
Bracket each eigenvalue of the matrix **A** in Example 9.10.

Solution. In Example 9.10 we found that all the eigenvalues lie in $(0, 8)$. We now bisect this interval and use the Sturm sequence to determine the number of eigenvalues in $(0, 4)$. With $\lambda = 4$, the sequence is—see Eqs. (9.49)—

$$P_0(4) = 1$$
$$P_1(4) = 4 - 4 = 0$$
$$P_2(4) = (4 - 4)(0) - 2^2(1) = -4$$
$$P_3(4) = (5 - 4)(-4) - 2^2(0) = -4$$

Because a zero value is assigned to the sign opposite to that of the preceding member, the signs in this sequence are $(+, -, -, -)$. The one sign change shows the presence of one eigenvalue in $(0, 4)$.

Next we bisect the interval $(4, 8)$ and compute the Sturm sequence with $\lambda = 6$:

$$P_0(6) = 1$$
$$P_1(6) = 4 - 6 = -2$$
$$P_2(6) = (4 - 6)(-2) - 2^2(1) = 0$$
$$P_3(6) = (5 - 6)(0) - 2^2(-2) = 8$$

In this sequence the signs are $(+, -, +, +)$, indicating two eigenvalues in $(0, 6)$.

Therefore

$$0 \leq \lambda_1 \leq 4 \qquad 4 \leq \lambda_2 \leq 6 \qquad 6 \leq \lambda_3 \leq 8$$

Computation of Eigenvalues

Once the desired eigenvalues are bracketed, they can be found by determining the roots of $P_n(\lambda) = 0$ with bisection or Ridder's method.

■ eigenvals3

The function `eigenvals3` computes N smallest eigenvalues of a symmetric tridiagonal matrix with the method of Ridder.

```
## module eigenvals3
''' lam = eigenvals3(d,c,N).
    Returns the N smallest eigenvalues of a symmetric
    tridiagonal matrix defined by its diagonals d and c.
'''
from lamRange import *
from ridder import *
from sturmSeq import sturmSeq
from numpy import zeros

def eigenvals3(d,c,N):
```

```
def f(x):                # f(x) = |[A] - x[I]|
    p = sturmSeq(d,c,x)
    return p[len(p)-1]

lam = zeros(N)
r = lamRange(d,c,N)    # Bracket eigenvalues
for i in range(N):     # Solve by Ridder's method
    lam[i] = ridder(f,r[i],r[i+1])
return lam
```

EXAMPLE 9.12

Use `eigenvals3` to determine the three smallest eigenvalues of the 100×100 matrix:

$$\mathbf{A} = \begin{bmatrix} 2 & -1 & 0 & \cdots & 0 \\ -1 & 2 & -1 & \cdots & 0 \\ 0 & -1 & 2 & \cdots & 0 \\ \vdots & \vdots & \ddots & \ddots & \vdots \\ 0 & 0 & \cdots & -1 & 2 \end{bmatrix}$$

Solution

```
#!/usr/bin/python
## example9_12
import numpy as np
from eigenvals3 import *

N = 3
n = 100
d = np.ones(n)*2.0
c = np.ones(n-1)*(-1.0)
lambdas = eigenvals3(d,c,N)
print lambdas
raw_input("\nPress return to exit")
```

Here are the eigenvalues:

```
[ 0.00096744  0.00386881  0.0087013 ]
```

Computation of Eigenvectors

If the eigenvalues are known (approximate values will be good enough), the best means of computing the corresponding eigenvectors is the inverse power method with eigenvalue shifting. This method was discussed earlier, but the algorithm listed did not take advantage of banding. Here we present a version of the method written for symmetric tridiagonal matrices.

■ inversePower3

This function is very similar to inversePower listed in Section 9.3, but it executes much faster because it exploits the tridiagonal structure of the matrix.

```
## module inversePower3
''' lam,x = inversePower3(d,c,s,tol=1.0e-6).
    Inverse power method applied to a symmetric tridiagonal
    matrix. Returns the eigenvalue closest to 's'
    and the corresponding eigenvector.
'''
from LUdecomp3 import *
import math
import numpy as np
from numpy.random import rand

def inversePower3(d,c,s,tol=1.0e-6):
    n = len(d)
    e = c.copy()
    dStar = d - s                      # Form [A*] – [A]   s[I]
    LUdecomp3(c,dStar,e)               # Decompose [A*]
    x = rand(n)                        # Seed x with random numbers
    xMag = math.sqrt(np.dot(x,x))      # Normalize [x]
    x =x/xMag
    flag = 0
    for i in range(30):                   # Begin iterations
        xOld = x.copy()                   # Save current [x]
        LUsolve3(c,dStar,e,x)             # Solve [A*][x] = [xOld]
        xMag = math.sqrt(np.dot(x,x))  # Normalize [x]
        x = x/xMag
        if np.dot(xOld,x) < 0.0: # Detect change in sign of [x]
            sign = -1.0
            x = -x
        else: sign = 1.0
        if math.sqrt(np.dot(xOld - x,xOld - x)) < tol:
            return s + sign/xMag,x
    print('Inverse power method did not converge')
```

EXAMPLE 9.13
Compute the 10th smallest eigenvalue of the matrix **A** given in Example 9.12.

Solution. The following program extracts the *N*th eigenvalue of **A** by the inverse power method with eigenvalue shifting:

```
#!/usr/bin/python
## example9_13
```

```
import numpy as np
from lamRange import *
from inversePower3 import *


N = 10
n = 100
d = np.ones(n)*2.0
c = np.ones(n-1)*(-1.0)
r = lamRange(d,c,N)              # Bracket N smallest eigenvalues
s = (r[N-1] + r[N])/2.0          # Shift to midpoint of Nth bracket
lam,x = inversePower3(d,c,s)   # Inverse power method
print("Eigenvalue No.",N," =",lam)
input("\nPress return to exit")
```

The result is

```
Eigenvalue No. 10  = 0.0959737849345
```

EXAMPLE 9.14
Compute the three smallest eigenvalues and the corresponding eigenvectors of the matrix **A** in Example 9.5.

Solution

```
#!/usr/bin/python
## example9_14
from householder import *
from eigenvals3 import *
from inversePower3 import *
import numpy as np


N = 3    # Number of eigenvalues requested
a = np.array([[ 11.0, 2.0,   3.0,   1.0,   4.0], \
              [  2.0, 9.0,   3.0,   5.0,   2.0], \
              [  3.0, 3.0, 15.0,   4.0,   3.0], \
              [  1.0, 5.0,   4.0, 12.0,   4.0], \
              [  4.0, 2.0,   3.0,   4.0, 17.0]])
xx = np.zeros((len(a),N))
d,c = householder(a)                    # Tridiagonalize [A]
p = computeP(a)                         # Compute transformation matrix
lambdas = eigenvals3(d,c,N)             # Compute eigenvalues
for i in range(N):
    s = lambdas[i]*1.0000001            # Shift very close to eigenvalue
    lam,x = inversePower3(d,c,s) # Compute eigenvector [x]
    xx[:,i] = x                         # Place [x] in array [xx]
xx = np.dot(p,xx)                       # Recover eigenvectors of [A]
```

```
print("Eigenvalues:\n",lambdas)
print("\nEigenvectors:\n",xx)
input("Press return to exit")
```

```
Eigenvalues:
[   4.87394638    8.66356791   10.93677451]
```

```
Eigenvectors:
[[ 0.26726603   0.72910002   0.50579164]
 [-0.74142854   0.41391448  -0.31882387]
 [-0.05017271  -0.4298639    0.52077788]
 [ 0.59491453   0.06955611  -0.60290543]
 [-0.14970633  -0.32782151  -0.08843985]]
```

PROBLEM SET 9.2

1. Use Gerschgorin's theorem to determine bounds on the eigenvalues of

$$
\text{(a)} \quad \mathbf{A} = \begin{bmatrix} 10 & 4 & -1 \\ 4 & 2 & 3 \\ -1 & 3 & 6 \end{bmatrix}
\qquad
\text{(b)} \quad \mathbf{B} = \begin{bmatrix} 4 & 2 & -2 \\ 2 & 5 & 3 \\ -2 & 3 & 4 \end{bmatrix}
$$

2. Use the Sturm sequence to show that

$$
\mathbf{A} = \begin{bmatrix} 5 & -2 & 0 & 0 \\ -2 & 4 & -1 & 0 \\ 0 & -1 & 4 & -2 \\ 0 & 0 & -2 & 5 \end{bmatrix}
$$

has one eigenvalue in the interval (2, 4).

3. Bracket each eigenvalue of

$$
\mathbf{A} = \begin{bmatrix} 4 & -1 & 0 \\ -1 & 4 & -1 \\ 0 & -1 & 4 \end{bmatrix}
$$

4. Bracket each eigenvalue of

$$
\mathbf{A} = \begin{bmatrix} 6 & 1 & 0 \\ 1 & 8 & 2 \\ 0 & 2 & 9 \end{bmatrix}
$$

5. Bracket every eigenvalue of

$$
\mathbf{A} = \begin{bmatrix} 2 & -1 & 0 & 0 \\ -1 & 2 & -1 & 0 \\ 0 & -1 & 2 & -1 \\ 0 & 0 & -1 & 1 \end{bmatrix}
$$

6. Tridiagonalize the matrix

$$A = \begin{bmatrix} 12 & 4 & 3 \\ 4 & 9 & 3 \\ 3 & 3 & 15 \end{bmatrix}$$

with the Householder's reduction.

7. Use the Householder's reduction to transform the matrix

$$A = \begin{bmatrix} 4 & -2 & 1 & -1 \\ -2 & 4 & -2 & 1 \\ 1 & -2 & 4 & -2 \\ -1 & 1 & -2 & 4 \end{bmatrix}$$

to tridiagonal form.

8. ■ Compute all the eigenvalues of

$$A = \begin{bmatrix} 6 & 2 & 0 & 0 & 0 \\ 2 & 5 & 2 & 0 & 0 \\ 0 & 2 & 7 & 4 & 0 \\ 0 & 0 & 4 & 6 & 1 \\ 0 & 0 & 0 & 1 & 3 \end{bmatrix}$$

9. ■ Find the smallest two eigenvalues of

$$A = \begin{bmatrix} 4 & -1 & 0 & 1 \\ -1 & 6 & -2 & 0 \\ 0 & -2 & 3 & 2 \\ 1 & 0 & 2 & 4 \end{bmatrix}$$

10. ■ Compute the three smallest eigenvalues of

$$A = \begin{bmatrix} 7 & -4 & 3 & -2 & 1 & 0 \\ -4 & 8 & -4 & 3 & -2 & 1 \\ 3 & -4 & 9 & -4 & 3 & -2 \\ -2 & 3 & -4 & 10 & -4 & 3 \\ 1 & -2 & 3 & -4 & 11 & -4 \\ 0 & 1 & -2 & 3 & -4 & 12 \end{bmatrix}$$

and the corresponding eigenvectors.

11. ■ Find the two smallest eigenvalues of the 6×6 Hilbert matrix

$$A = \begin{bmatrix} 1 & 1/2 & 1/3 & \cdots & 1/6 \\ 1/2 & 1/3 & 1/4 & \cdots & 1/7 \\ 1/3 & 1/4 & 1/5 & \cdots & 1/7 \\ \vdots & \vdots & \vdots & \ddots & \vdots \\ 1/8 & 1/9 & 1/10 & \cdots & 1/11 \end{bmatrix}$$

Recall that this matrix is ill conditioned.

12. ■ Rewrite the function `lamRange(d,c,N)` so that it will bracket the N *largest* eigenvalues of a tridiagonal matrix. Use this function to compute the two largest eigenvalues of the Hilbert matrix in Example 9.11.

13. ■

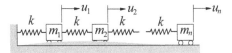

The differential equations governing free vibration of the mass-spring system are

$$k(-2u_1 + u_2) = m_1\ddot{u}_1$$

$$k(u_{i-1} - 2u_i + u_{i+1}) = m_i\ddot{u}_i \quad (i = 2, 3, \ldots, n-1)$$

$$k(u_{n-1} - u_n) = m_n\ddot{u}_n$$

where $u_i(t)$ is the displacement of mass i from its equilibrium position and k is the spring stiffness. Given k and the masses $m = \begin{bmatrix} m_1 & m_2 & \cdots & m_n \end{bmatrix}^T$, write a program that computes N lowest circular frequencies of the system and the corresponding relative displacements of the masses. Run the program with $N = 2$, $k = 500$ kN/m, and

$$\mathbf{m} = \begin{bmatrix} 1.0 & 1.0 & 1.0 & 8.0 & 1.0 & 1.0 & 8.0 \end{bmatrix}^T \text{ kg}$$

14. ■

The figure shows n identical masses connected by springs of different stiffnesses. The equations governing free vibration of the system are

$$-(k_1 + k_2)u_1 + k_2u_2 = m\ddot{u}_1$$

$$k_iu_{i-1} - (k_i + k_{i+1})u_i + k_{i+1}u_{i+1} = m\ddot{u}_i \quad (i = 2, 3, \ldots, n-1)$$

$$k_nu_{n-1} - k_nu_n = m\ddot{u}_n$$

where u_i is the displacement of mass i from its equilibrium position. Given m and the spring stiffnesses $\mathbf{k} = \begin{bmatrix} k_1 & k_2 & \cdots & k_n \end{bmatrix}^T$, write a program that computes N lowest circular frequencies and the corresponding relative displacements of the masses. Run the program with $N = 2$, $m = 2$ kg, and

$$\mathbf{k} = \begin{bmatrix} 400 & 400 & 400 & 0.2 & 400 & 400 & 200 \end{bmatrix}^T \text{ kN/m}$$

Note that the system is weakly coupled, k_4 being small. Do the results make sense?

15. ■

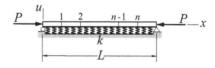

The differential equation of motion of the axially vibrating bar is

$$u'' = \frac{\rho}{E}\ddot{u}$$

where $u(x, t)$ is the axial displacement, ρ represents the mass density, and E is the modulus of elasticity. The boundary conditions are $u(0, t) = u'(L, t) = 0$. Letting $u(x, t) = y(x)\sin\omega t$, we obtain

$$y'' = -\omega^2\frac{\rho}{E}y \qquad y(0) = y'(L) = 0$$

The corresponding finite difference equations are

$$
\begin{bmatrix}
2 & -1 & 0 & 0 & \cdots & 0 \\
-1 & 2 & -1 & 0 & \cdots & 0 \\
0 & -1 & 2 & -1 & \cdots & 0 \\
\vdots & \vdots & \ddots & \ddots & \ddots & \vdots \\
0 & 0 & \cdots & -1 & 2 & -1 \\
0 & 0 & \cdots & 0 & -1 & 1
\end{bmatrix}
\begin{bmatrix}
y_1 \\ y_2 \\ y_3 \\ \vdots \\ y_{n-1} \\ y_n
\end{bmatrix}
= \left(\frac{\omega L}{n}\right)^2\frac{\rho}{E}
\begin{bmatrix}
y_1 \\ y_2 \\ y_3 \\ \vdots \\ y_{n-1} \\ y_n/2
\end{bmatrix}
$$

(a) If the standard form of these equations is $\mathbf{Hz} = \lambda\mathbf{z}$, write down $\mathbf{H}$ and the transformation matrix $\mathbf{P}$ in $\mathbf{y} = \mathbf{Pz}$. (b) Compute the lowest circular frequency of the bar with $n = 10$, 100, and 1,000 using the module `inversePower3`. *Note*: The analytical solution is $\omega_1 = \pi\sqrt{E/\rho}/(2L)$. (This beam also appeared in Prob. 16, Problem Set 9.1.)

16. ■

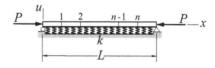

The simply supported column is resting on an elastic foundation of stiffness k (N/m per meter length). An axial force P acts on the column. The differential equation and the boundary conditions for the lateral displacement u are

$$u^{(4)} + \frac{P}{EI}u'' + \frac{k}{EI}u = 0$$

$$u(0) = u''(0) = u(L) = u''(L) = 0$$

Using the mesh shown, the finite difference approximation of these equations is

$$(5 + \alpha)u_1 - 4u_2 + u_3 = \lambda(2u_1 - u_2)$$

$$-4u_1 + (6 + \alpha)u_2 - 4u_3 + u_4 = \lambda(-u_1 + 2u_2 + u_3)$$

$$u_1 - 4u_2 + (6 + \alpha)u_3 - 4u_4 + u_5 = \lambda(-u_2 + 2u_3 - u_4)$$

$$\vdots$$

$$u_{n-3} - 4u_{n-2} + (6 + \alpha)u_{n-1} - 4u_n = \lambda(-u_{n-2} + 4u_{n-1} - u_n)$$

$$u_{n-2} - 4u_{n-1} + (5 + \alpha)u_n = \lambda(-u_{n-1} + 2u_n)$$

where

$$\alpha = \frac{kh^4}{EI} = \frac{1}{(n+1)^4}\frac{kL^4}{EI} \qquad \lambda = \frac{Ph^2}{EI} = \frac{1}{(n+1)^2}\frac{PL^2}{EI}$$

Write a program that computes the lowest three buckling loads P and the corresponding mode shapes. Run the program with $kL^4/(EI) = 1{,}000$ and $n = 25$.

17. ■ Find smallest five eigenvalues of the 20×20 matrix

$$\mathbf{A} = \begin{bmatrix} 2 & 1 & 0 & 0 & \cdots & 0 & 1 \\ 1 & 2 & 1 & 0 & \cdots & 0 & 0 \\ 0 & 1 & 2 & 1 & \cdots & 0 & 0 \\ \vdots & \vdots & \ddots & \ddots & \ddots & \vdots & \vdots \\ 0 & 0 & \cdots & 1 & 2 & 1 & 0 \\ 0 & 0 & \cdots & 0 & 1 & 2 & 1 \\ 1 & 0 & \cdots & 0 & 0 & 1 & 2 \end{bmatrix}$$

Note: This is a difficult matrix that has many pairs of double eigenvalues.

18. ■

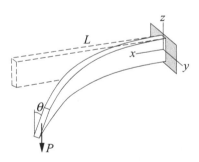

When the depth/width ratio of a beam is large, lateral buckling may occur. The differential equation that governs lateral buckling of the cantilever beam shown is

$$\frac{d^2\theta}{dx^2} + \gamma^2 \left(1 - \frac{x}{L}\right)^2 \theta = 0$$

where θ is the angle of rotation of the cross section and

$$\gamma^2 = \frac{P^2 L^2}{(GJ)(EI_z)}$$

$$GJ = \text{torsional rigidity}$$

$$EI_z = \text{bending rigidity about the } z\text{-axis}$$

The boundary conditions are $\theta|_{x=0} = 0$ and $d\theta/dx|_{x=L} = 0$. Using the finite difference approximation of the differential equation, determine the buckling load P_{cr}. The analytical solution is

$$P_{cr} = 4.013 \frac{\sqrt{(GJ)(EI_z)}}{L^2}$$

19. ■ Determine the value of z so that the smallest eigenvalue of the matrix

$$\begin{bmatrix} z & 4 & 3 & 5 & 2 & 1 \\ 4 & z & 2 & 4 & 3 & 4 \\ 3 & 2 & z & 4 & 1 & 8 \\ 5 & 4 & 4 & z & 2 & 5 \\ 2 & 3 & 1 & 2 & z & 3 \\ 1 & 4 & 8 & 5 & 3 & z \end{bmatrix}$$

is equal to 1.0. *Hint*: This is a root-finding problem.

9.6   Other Methods

On occasions when all the eigenvalues and eigenvectors of a matrix are required, the *OR algorithm* is a worthy contender. It is based on the decomposition $\mathbf{A} = \mathbf{QR}$ where $\mathbf{Q}$ and $\mathbf{R}$ are orthogonal and upper triangular matrices, respectively. The decomposition is carried out in conjunction with the Householder transformation. There is also a *QL algorithm*, $\mathbf{A} = \mathbf{QL}$, that works in the same manner, but here $\mathbf{L}$ is a lower triangular matrix.

Schur's factorization is another solid technique for determining the eigenvalues of $\mathbf{A}$. Here the decomposition is $\mathbf{A} = \mathbf{Q}^T \mathbf{UQ}$, where $\mathbf{Q}$ is orthogonal and $\mathbf{U}$ is an upper triangular matrix. The diagonal terms of $\mathbf{U}$ are the eigenvalues of $\mathbf{A}$.

The *LR algorithm* is probably the fastest means of computing the eigenvalues; it is also very simple to implement—see Prob. 22 of Problem Set 9.1. But its stability is inferior to the other methods.

10 Introduction to Optimization

> Find $\mathbf{x}$ that minimizes $F(\mathbf{x})$ subject to $g(\mathbf{x}) = 0$, $h(\mathbf{x}) \geq 0$.

10.1 Introduction

Optimization is the term often used for minimizing or maximizing a function. It is sufficient to consider the problem of minimization only; maximization of $F(\mathbf{x})$ is achieved by simply minimizing $-F(\mathbf{x})$. In engineering, optimization is closely related to design. The function $F(\mathbf{x})$, called the *merit function* or *objective function*, is the quantity that we wish to keep as small as possible, such as the cost or weight. The components of $\mathbf{x}$, known as the *design variables*, are the quantities that we are free to adjust. Physical dimensions (lengths, areas, angles, and so on) are common examples of design variables.

Optimization is a large topic with many books dedicated to it. The best we can do in limited space is to introduce a few basic methods that are good enough for problems that are reasonably well behaved and do not involve too many design variables. By omitting the more sophisticated methods, we may actually not miss all that much. All optimization algorithms are unreliable to a degree—any one may work on one problem and fail on another. As a rule of thumb, by going up in sophistication we gain computational efficiency, but not necessarily reliability.

The algorithms for minimization are iterative procedures that require starting values of the design variables $\mathbf{x}$. If $F(\mathbf{x})$ has several local minima, the initial choice of $\mathbf{x}$ determines which of these will be computed. There is no guaranteed way of finding the global optimal point. One suggested procedure is to make several computer runs using different starting points and pick the best result.

More often than not, the design variables are also subjected to restrictions, or *constraints*, which may have the form of equalities or inequalities. As an example, take the minimum weight design of a roof truss that has to carry a certain loading. Assume that the layout of the members is given, so that the design variables are the cross-sectional areas of the members. Here the design is dominated by inequality constraints that consist of prescribed upper limits on the stresses and possibly the displacements.

The majority of available methods are designed for *unconstrained optimization*, where no restrictions are placed on the design variables. In these problems the minima, if they exit, are stationary points (points where the gradient vector of $F(\mathbf{x})$ vanishes). In the more difficult problem of *constrained optimization* the minima are usually located where the $F(\mathbf{x})$ surface meets the constraints. There are special algorithms for constrained optimization, but they are not easily accessible because of their complexity and specialization. One way to tackle a problem with constraints is to use an unconstrained optimization algorithm, but to modify the merit function so that any violation of constrains is heavily penalized.

Consider the problem of minimizing $F(\mathbf{x})$ where the design variables are subject to the constraints

$$g_i(\mathbf{x}) = 0, \quad i = 1, 2, \ldots, M$$

$$h_j(\mathbf{x}) \leq 0, \quad j = 1, 2, \ldots, N$$

We choose the new merit function be

$$F^*(\mathbf{x}) = F(\mathbf{x}) + \lambda P(\mathbf{x}) \tag{10.1a}$$

where

$$P(\mathbf{x}) = \sum_{i=1}^{M} [g_i(x)]^2 + \sum_{j=1}^{N} \left\{ \max\left[0, h_j(\mathbf{x}) \right] \right\}^2 \tag{10.1b}$$

is the *penalty function* and λ is a multiplier. The function max(a, b) returns the larger of a and b. It is evident that $P(\mathbf{x}) = 0$ if no constraints are violated. Violation of a constraint imposes a penalty proportional to the square of the violation. Hence the minimization algorithm tends to avoid the violations, the degree of avoidance being dependent on the magnitude of λ. If λ is small, optimization will proceed faster because there is more "space" in which the procedure can operate, but there may be significant violation of constraints. In contrast, large λ can result in a poorly conditioned procedure, but the constraints will be tightly enforced. It is advisable to run the optimization program with λ that is on the small side. If the results show unacceptable constraint violation, increase λ and run the program again, starting with the results of the previous run.

An optimization procedure may also become ill conditioned when the constraints have widely different magnitudes. This problem can be alleviated by *scaling* the offending constraints; that is, multiplying the constraint equations by suitable constants.

It is not always necessary (or even advisable) to employ an iterative minimization algorithm. In problems where the derivatives of $F(\mathbf{x})$ can be readily computed and inequality constraints are absent, the optimal point can always be found directly by calculus. For example, if there are no constraints, the coordinates of the point where $F(\mathbf{x})$ is minimized are given by the solution of the simultaneous (usually nonlinear) equations $\nabla F(\mathbf{x}) = \mathbf{0}$. The direct method for finding the minimum of $F(\mathbf{x})$ subject to equality constraints $g_i(\mathbf{x}) = 0$, $i = 1, 2, \ldots m$ is to form the function

$$F^*(\mathbf{x}, \lambda) = F(\mathbf{x}) + \sum_{i=1}^{m} \lambda_i g_i(\mathbf{x}) \tag{10.2a}$$

and solve the equations

$$\nabla F^*(\mathbf{x}) = \mathbf{0} \qquad g_i(\mathbf{x}) = 0, \quad i = 1, 2, \ldots, m \tag{10.2b}$$

for $\mathbf{x}$ and λ_i. The parameters λ_i are known as the *Lagrangian multipliers*. The direct method can also be extended to inequality constraints, but the solution of the resulting equations is not straightforward due to a lack of uniqueness.

This discussion exempts problems where the merit function and the constraints are linear functions of $\mathbf{x}$. These problems, classified as *linear programming* problems, can be solved without difficulty by specialized methods, such as the *simplex method*. Linear programming is used mainly for operations research and cost analysis; there are very few engineering applications. This is not to say that linear programming has no place in nonlinear optimization. There are several effective methods that rely in part on the simplex method. For example, problems with nonlinear constraints can often be solved by piecewise application of linear programming. The simplex method is also used to compute search directions in the method of feasible directions.

10.2 Minimization Along a Line

Consider the problem of minimizing a function $f(x)$ of a single variable x with the constraints $c \leq x \leq d$. A hypothetical plot of the function is shown in Figure 10.1. There are two minimum points: a stationary point characterized by $f'(x) = 0$ that represents a local minimum, and a global minimum at the constraint boundary. Finding the global minimum is simple. All the stationary points could be located by finding the roots of $df/dx = 0$, and each constraint boundary may be checked for a global minimum by evaluating $f(c)$ and $f(d)$. Then why do we need an optimization algorithm? We need it if $f(\mathbf{x})$ is difficult or impossible to differentiate; for example, if f represents a complex computer algorithm.

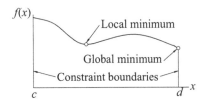

Figure 10.1. Example of local and global minima.

Bracketing

Before a minimization algorithm can be entered, the minimum point must be brack-eted. The procedure of bracketing is simple: Start with an initial value of x_0 and move *downhill* computing the function at $x_1, x_2, x_3, \ldots$ until we reach the point x_n where $f(x)$ increases for the first time. The minimum point is now bracketed in the inter-val (x_{n-2}, x_n). What should the step size $h_i = x_{i+1} - x_i$ be? It is not a good idea have a constant h_i because it often results in too many steps. A more efficient scheme is to increase the size with every step, the goal being to reach the minimum quickly, even if the resulting bracket is wide. In our algorithm we chose to increase the step size by a constant factor; that is, we use $h_{i+1} = ch_i, c > 1$.

Golden Section Search

The golden section search is the counterpart of bisection used in finding roots of equations. Suppose that the minimum of $f(x)$ has been bracketed in the interval (a, b) of length h. To telescope the interval, we evaluate the function at $x_1 = b - Rh$ and $x_2 = a + Rh$, as shown in Figure 10.2(a). The constant R is to be determined shortly. If $f_1 > f_2$ as indicated in the figure, the minimum lies in (x_1, b); otherwise it is located in (a, x_2).

Assuming that $f_1 > f_2$, we set $a \leftarrow x_1$ and $x_1 \leftarrow x_2$, which yields a new interval (a, b) of length $h' = Rh$, as illustrated in Figure 10.2(b). To carry out the next tele-scoping operation we evaluate the function at $x_2 = a + Rh'$ and repeat the process.

The procedure works only if Figures 10.1(a) and (b) are similar (i.e., if the same constant R locates x_1 and x_2 in both figures). Referring to Figure 10.2(a), we note that $x_2 - x_1 = 2Rh - h$. The same distance in Figure 10.2(b) is $x_1 - a = h' - Rh'$. Equating the two, we get

$$2Rh - h = h' - Rh'$$

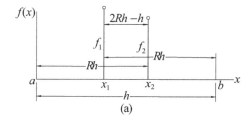

(a)

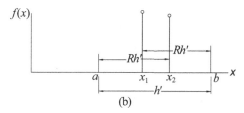

(b)

Figure 10.2. Golden section telescoping.

Substituting $h' = Rh$ and canceling h yields

$$2R - 1 = R(1 - R)$$

the solution of which is the *golden ratio.*[1]

$$R = \frac{-1 + \sqrt{5}}{2} = 0.618\,033\,989\ldots \tag{10.3}$$

Note that each telescoping decreases the interval containing the minimum by the factor R, which is not as good as the factor is 0.5 in bisection. However, the golden search method achieves this reduction with *one function evaluation*, whereas two evaluations would be needed in bisection.

The number of telescoping operations required to reduce h from $|b - a|$ to an error tolerance ε is given by

$$|b - a|\, R^n = \varepsilon$$

which yields

$$n = \frac{\ln(\varepsilon / |b - a|)}{\ln R} = -2.078\,087 \ln \frac{\varepsilon}{|b - a|} \tag{10.4}$$

■ goldSearch

This module contains the bracketing and the golden section search algorithms. For the factor that multiplies successive search intervals in `bracket` we chose $c = 1 + R$.

```
## module goldSearch
''' a,b = bracket(f,xStart,h)
    Finds the brackets (a,b) of a minimum point of the
    user-supplied scalar function f(x).
    The search starts downhill from xStart with a step
    length h.

    x,fMin = search(f,a,b,tol=1.0e-6)
    Golden section method for determining x that minimizes
    the user-supplied scalar function f(x).
    The minimum must be bracketed in (a,b).
'''
import math
def bracket(f,x1,h):
    c = 1.618033989
    f1 = f(x1)
    x2 = x1 + h; f2 = f(x2)
```

---

[1] R is the ratio of the sides of a "golden rectangle," considered by ancient Greeks to have the perfect proportions.

```
    # Determine downhill direction and change sign of h if needed
    if f2 > f1:
        h = -h
        x2 = x1 + h; f2 = f(x2)
      # Check if minimum between x1 - h and x1 + h
        if f2 > f1: return x2,x1 - h
  # Search loop
    for i in range (100):
        h = c*h
        x3 = x2 + h; f3 = f(x3)
        if f3 > f2: return x1,x3
        x1 = x2; x2 = x3
        f1 = f2; f2 = f3
    print("Bracket did not find a minimum")

def search(f,a,b,tol=1.0e-9):
    nIter = int(math.ceil(-2.078087*math.log(tol/abs(b-a))))
    R = 0.618033989
    C = 1.0 - R
  # First telescoping
    x1 = R*a + C*b; x2 = C*a + R*b
    f1 = f(x1); f2 = f(x2)
  # Main loop
    for i in range(nIter):
        if f1 > f2:
            a = x1
            x1 = x2; f1 = f2
            x2 = C*a + R*b; f2 = f(x2)
        else:
            b = x2
            x2 = x1; f2 = f1
            x1 = R*a + C*b; f1 = f(x1)
    if f1 < f2: return x1,f1
    else: return x2,f2
```

EXAMPLE 10.1

Use goldSearch to find x that minimizes

$$f(x) = 1.6x^3 + 3x^2 - 2x$$

subject to the constraint $x \geq 0$. Compare the result with the analytical solution.

Solution. This is a constrained minimization problem. The minimum of $f(x)$ is either a stationary point in $x \geq 0$, or it is located at the constraint boundary $x = 0$.

We handle the constraint with the penalty function method by minimizing $f(x) + \lambda \left[\min(0, x) \right]^2$.

Starting at $x = 1$ and choosing $h = 0.01$ for the first step size in bracket (both choices being rather arbitrary), we arrive at the following program:

```
#!/usr/bin/python
## example10_1
from goldSearch import *

def f(x):
    lam = 1.0          # Constraint multiplier
    c = min(0.0, x)   # Constraint function
    return 1.6*x**3 + 3.0*x**2 - 2.0*x + lam*c**2

xStart = 1.0
h = 0.01
x1,x2 = bracket(f,xStart,h)
x,fMin = search(f,x1,x2)
print("x =",x)
print("f(x) =",fMin)
input ("\nPress return to exit")
```

The result is

```
x = 0.2734941131714084
f(x) = -0.28985978554959224
```

Because the minimum was found to be a stationary point, the constraint was not active. Therefore, the penalty function was superfluous, but we did not know that at the beginning.

The locations of stationary points are obtained analytically by solving

$$f'(x) = 4.8x^2 + 6x - 2 = 0$$

The positive root of this equation is $x = 0.273\,49\,4$. Because this is the only positive root, there are no other stationary points in $x \geq 0$ that we must check out. The only other possible location of a minimum is the constraint boundary $x = 0$. But here $f(0) = 0$ is larger than the function at the stationary point, leading to the conclusion that the global minimum occurs at $x = 0.273\,49\,4$.

EXAMPLE 10.2

The trapezoid shown is the cross section of a beam. It is formed by removing the top from a triangle of base $B = 48$ mm and height $H = 60$ mm. The problem is to find the height y of the trapezoid that maximizes the section modulus

$$S = I_{\bar{x}}/c$$

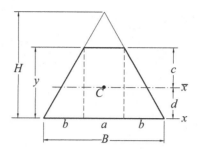

where $I_{\bar{x}}$ is the second moment of the cross-sectional area about the axis that passes through the centroid C of the cross section. By optimizing the section modulus, we minimize the maximum bending stress $\sigma_{\max} = M/S$ in the beam, M being the bending moment.

Solution. Considering the area of the trapezoid as a composite of a rectangle and two triangles, the section modulus is found through the following sequence of computations:

Base of rectangle	$a = B(H - y)/H$
Base of triangle	$b = (B - a)/2$
Area	$A = (B + a)y/2$
First moment of area about x-axis	$Q_x = (ay)y/2 + 2(by/2)y/3$
Location of centroid	$d = Q_x/A$
Distance involved in S	$c = y - d$
Second moment of area about x-axis	$I_x = ay^3/3 + 2(by^3/12)$
Parallel axis theorem	$I_{\bar{x}} = I_x - Ad^2$
Section modulus	$S = I_{\bar{x}}/c$

We could use the formulas in the table to derive S as an explicit function of y, but that would involve a lot of error-prone algebra and result in an overly complicated expression. It makes more sense to let the computer do the work.

　　The program we used and its output are listed next. Because we wish to maximize S with a minimization algorithm, the merit function is $-S$. There are no constraints in this problem.

```python
#!/usr/bin/python
## example10_2
from goldSearch import *

def f(y):
    B = 48.0
    H = 60.0
    a = B*(H - y)/H
    b = (B - a)/2.0
```

```
        A = (B + a)*y/2.0
        Q = (a*y**2)/2.0 + (b*y**2)/3.0
        d = Q/A
        c = y - d
        I = (a*y**3)/3.0 + (b*y**3)/6.0
        Ibar = I - A*d**2
        return -Ibar/c

yStart = 60.0  # Starting value of y
h = 1.0        # Size of first step used in bracketing
a,b = bracket(f,yStart,h)
yOpt,fOpt = search(f,a,b)
print("Optimal y =",yOpt)
print("Optimal S =",-fOpt)
print("S of triangle =",-f(60.0))
input("Press return to exit")

Optimal y = 52.17627387056691
Optimal S = 7864.430941364856
S of triangle = 7200.0
```

The printout includes the section modulus of the original triangle. The optimal section shows a 9.2% improvement over the triangle.

10.3 Powell's Method

Introduction

We now look at optimization in n-dimensional design space. The objective is to minimize $F(\mathbf{x})$, where the components of $\mathbf{x}$ are the n independent design variables. One way to tackle the problem is to use a succession of one-dimensional minimizations to close in on the optimal point. The basic strategy is as follows:

Choose a point $\mathbf{x}_0$ in the design space.
Loop over $i = 1, 2, 3, \ldots$:
    Choose a vector $\mathbf{v}_i$.
    Minimize $F(\mathbf{x})$ along the line through $\mathbf{x}_{i-1}$ in the direction of $\mathbf{v}_i$.
        Let the minimum point be $\mathbf{x}_i$.
    If $|\mathbf{x}_i - \mathbf{x}_{i-1}| < \varepsilon$ exit loop.

The minimization along a line can be accomplished with any one-dimensional optimization algorithm (such as the golden section search). The only question left open is how to choose the vectors $\mathbf{v}_i$.

Conjugate Directions

Consider the quadratic function

$$F(\mathbf{x}) = c - \sum_i b_i x_i + \frac{1}{2} \sum_i \sum_j A_{ij} x_i x_j$$

$$= c - \mathbf{b}^T \mathbf{x} + \frac{1}{2} \mathbf{x}^T \mathbf{A} \mathbf{x} \tag{10.5}$$

Differentiation with respect to x_i yields

$$\frac{\partial F}{\partial x_i} = -b_i + \sum_j A_{ij} x_j$$

which can be written in vector notation as

$$\nabla F = -\mathbf{b} + \mathbf{A} \mathbf{x} \tag{10.6}$$

where ∇F is called the *gradient* of F.

Now consider the change in the gradient as we move from point $\mathbf{x}_0$ in the direction of a vector $\mathbf{u}$. The motion takes place along the line

$$\mathbf{x} = \mathbf{x}_0 + s\mathbf{u}$$

where s is the distance moved. Substitution into Eq. (10.6) yields the expression for the gradient at $\mathbf{x}$:

$$\nabla F|_{\mathbf{x}_0 + s\mathbf{u}} = -\mathbf{b} + \mathbf{A} (\mathbf{x}_0 + s\mathbf{u}) = \nabla F|_{\mathbf{x}_0} + s\mathbf{A}\mathbf{u}$$

Note that the change in the gradient is $s\mathbf{A}\mathbf{u}$. If this change is perpendicular to a vector $\mathbf{v}$; that is, if

$$\mathbf{v}^T \mathbf{A} \mathbf{u} = 0, \tag{10.7}$$

the directions of $\mathbf{u}$ and $\mathbf{v}$ are said to be mutually *conjugate* (non-interfering). The implication is that once we have minimized $F(\mathbf{x})$ in the direction of $\mathbf{v}$, we can move along $\mathbf{u}$ without ruining the previous minimization.

For a quadratic function of n independent variables it is possible to construct n mutually conjugate directions. Therefore, it would take precisely n line minimizations along these directions to reach the minimum point. If $F(\mathbf{x})$ is not a quadratic function, Eq. (10.5) can be treated as a local approximation of the merit function, obtained by truncating the Taylor series expansion of $F(\mathbf{x})$ about $\mathbf{x}_0$ (see Appendix A1):

$$F(\mathbf{x}) \approx F(\mathbf{x}_0) + \nabla F(\mathbf{x}_0)(\mathbf{x} - \mathbf{x}_0) + \frac{1}{2}(\mathbf{x} - \mathbf{x}_0)^T \mathbf{H}(\mathbf{x}_0)(\mathbf{x} - \mathbf{x}_0)$$

Now the conjugate directions based on the quadratic form are only approximations, valid in the close vicinity of $\mathbf{x}_0$. Consequently, it would take several cycles of n line minimizations to reach the optimal point.

The various conjugate gradient methods use different techniques for constructing conjugate directions. The so-called *zero-order methods* work with $F(\mathbf{x})$ only, whereas the *first-order methods* use both $F(\mathbf{x})$ and ∇F. The first-order methods are computationally more efficient, of course, but the input of ∇F, if it is available at all, can be very tedious.

Powell's Algorithm

Powell's method is a zero-order method, requiring the evaluation of $F(\mathbf{x})$ only. The basic algorithm is as follows:

Choose a point $\mathbf{x}_0$ in the design space.
Choose the starting vectors $\mathbf{v}_i$, $1, 2, \ldots, n$ (the usual choice is $\mathbf{v}_i = \mathbf{e}_i$,
      where $\mathbf{e}_i$ is the unit vector in the x_i-coordinate direction).
Cycle:
      Loop over $i = 1, 2, \ldots, n$:
            Minimize $F(\mathbf{x})$ along the line through $\mathbf{x}_{i-1}$ in the direction of $\mathbf{v}_i$.
            Let the minimum point be $\mathbf{x}_i$.
      End loop.
      $\mathbf{v}_{n+1} \leftarrow \mathbf{x}_0 - \mathbf{x}_n$.
      Minimize $F(\mathbf{x})$ along the line through $\mathbf{x}_0$ in the direction of $\mathbf{v}_{n+1}$.
      Let the minimum point be $\mathbf{x}_{n+1}$.
      if $|\mathbf{x}_{n+1} - \mathbf{x}_0| < \varepsilon$ exit loop.
      Loop over $i = 1, 2, \ldots, n$:
            $\mathbf{v}_i \leftarrow \mathbf{v}_{i+1}$ ($\mathbf{v}_1$ is discarded; the other vectors are reused).
      End loop.
      $\mathbf{x}_0 \leftarrow \mathbf{x}_{n+1}$
End cycle.

Powell demonstrated that the vectors $\mathbf{v}_{n+1}$ produced in successive cycles are mutually conjugate, so that the minimum point of a quadratic surface is reached in precisely n cycles. In practice, the merit function is seldom quadratic, but as long as it can be approximated locally by Eq. (10.5), Powell's method will work. Of course, it usually takes more than n cycles to arrive at the minimum of a nonquadratic function. Note that it takes n line minimizations to construct each conjugate direction.

Figure 10.3(a) illustrates one typical cycle of the method in a two dimensional design space ($n = 2$). We start with point $\mathbf{x}_0$ and vectors $\mathbf{v}_1$ and $\mathbf{v}_2$. Then we find the distance s_1 that minimizes $F(\mathbf{x}_0 + s\mathbf{v}_1)$, finishing up at point $\mathbf{x}_1 = \mathbf{x}_0 + s_1\mathbf{v}_1$. Next, we determine s_2 that minimizes $F(\mathbf{x}_1 + s\mathbf{v}_2)$ which takes us to $\mathbf{x}_2 = \mathbf{x}_1 + s_2\mathbf{v}_2$. The last search direction is $\mathbf{v}_3 = \mathbf{x}_2 - \mathbf{x}_0$. After finding s_3 by minimizing $F(\mathbf{x}_0 + s\mathbf{v}_3)$ we get to $\mathbf{x}_3 = \mathbf{x}_0 + s_3\mathbf{v}_3$, completing the cycle.

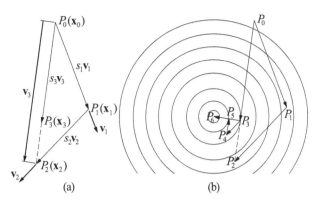

Figure 10.3. The method of Powell.

Figure 10.3(b) shows the moves carried out in two cycles superimposed on the contour map of a quadratic surface. As explained earlier, the first cycle starts at point P_0 and ends up at P_3. The second cycle takes us to P_6, which is the optimal point. The directions $P_0 P_3$ and $P_3 P_6$ are mutually conjugate.

Powell's method does have a major flaw that has to be remedied—if $F(\mathbf{x})$ is not a quadratic, the algorithm tends to produce search directions that gradually become linearly dependent, thereby ruining the progress toward the minimum. The source of the problem is the automatic discarding of $\mathbf{v}_1$ at the end of each cycle. It has been suggested that it is better to throw out the direction that resulted in the *largest decrease* of $F(\mathbf{x})$, a policy that we adopt. It seems counterintuitive to discard the best direction, but it is likely to be close to the direction added in the next cycle, thereby contributing to linear dependence. As a result of the change, the search directions cease to be mutually conjugate, so that a quadratic form is no longer minimized in n cycles. This is not a significant loss because in practice $F(\mathbf{x})$ is seldom a quadratic.

Powell suggested a few other refinements to speed up convergence. Because they complicate the bookkeeping considerably, we did not implement them.

■ powell

The algorithm for Powell's method is listed next. It uses two arrays: df contains the decreases of the merit function in the first n moves of a cycle, and the matrix u stores the corresponding direction vectors $\mathbf{v}_i$ (one vector per row).

```
## module powell
'''  xMin,nCyc = powell(F,x,h=0.1,tol=1.0e-6)
     Powell's method of minimizing user-supplied function F(x).
     x    = starting point
     h    = initial search increment used in 'bracket'
     xMin = mimimum point
```

```
    nCyc = number of cycles
'''

import numpy as np
from goldSearch import *
import math


def powell(F,x,h=0.1,tol=1.0e-6):

    def f(s): return F(x + s*v)    # F in direction of v

    n = len(x)                     # Number of design variables
    df = np.zeros(n)               # Decreases of F stored here
    u = np. identity(n)            # Vectors v stored here by rows
    for j in range(30):            # Allow for 30 cycles:
        xOld = x.copy()            # Save starting point
        fOld = F(xOld)
      # First n line searches record decreases of F
        for i in range(n):
            v = u[i]
            a,b = bracket(f,0.0,h)
            s,fMin = search(f,a,b)
            df[i] = fOld - fMin
            fOld = fMin
            x = x + s*v
      # Last line search in the cycle
        v = x - xOld
        a,b = bracket(f,0.0,h)
        s,fLast = search(f,a,b)
        x = x + s*v
      # Check for convergence
        if math.sqrt(np.dot(x-xOld,x-xOld)/n) < tol: return x,j+1
      # Identify biggest decrease & update search directions
        iMax = np.argmax(df)
        for i in range(iMax,n-1):
            u[i] = u[i+1]
        u[n-1] = v
    print("Powell did not converge")
```

EXAMPLE 10.3
Find the minimum of the function[2]

$$F = 100(y - x^2)^2 + (1 - x)^2$$

---

[2] From Shoup, T.E. and Mistree, F., *Optimization Methods with Applications for Personal Computers*, Prentice-Hall, 1987.

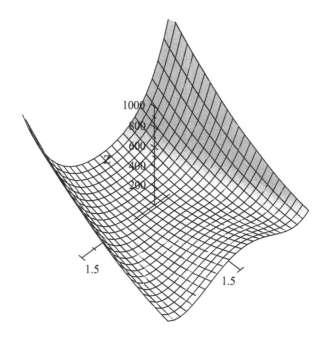

with Powell's method starting at the point $(-1, 1)$. This function has an interesting topology. The minimum value of F occurs at the point $(1, 1)$, But there is a considerable hump between the starting and minimum points that the algorithm must negotiate.

Solution. The program that solves this unconstrained optimization problem is as follows:

```
#!/usr/bin/python
## example10_3
from powell import *
from numpy import array

def F(x): return 100.0*(x[1] - x[0]**2)**2 + (1 - x[0])**2

xStart = array([-1.0, 1.0])
xMin,nIter = powell(F,xStart)
print("x =",xMin)
print("F(x) =",F(xMin))
print("Number of cycles =",nIter)
input ("Press return to exit")
```

As seen in the printout, the minimum point was obtained after 12 cycles.

```
x = [ 1.  1.]
F(x) = 3.71750701585e-029
Number of cycles = 12
```

EXAMPLE 10.4

Use `powell` to determine the smallest distance from the point (5, 8) to the curve $xy = 5$.

Solution. This is a constrained optimization problem: Minimize $F(x, y) = (x - 5)^2 + (y - 8)^2$ (the square of the distance) subject to the equality constraint $xy - 5 = 0$. The following program uses Powell's method with penalty function:

```
## example10_4
from powell import *
from numpy import array
from math import sqrt

def F(x):
    lam = 1.0                      # Penalty multiplier
    c = x[0]*x[1] - 5.0            # Constraint equation
    return  distSq(x) + lam*c**2   # Penalized merit function

def distSq(x): return (x[0] - 5)**2 + (x[1] - 8)**2

xStart = array([ 1.0,5.0])
x,numIter = powell(F,xStart,0.1)
print("Intersection point =",x)
print("Minimum distance =", sqrt(distSq(x)))
print("xy =", x[0]*x[1])
print("Number of cycles =",numIter)
input ("Press return to exit")
```

As mentioned earlier, the value of the penalty function multiplier λ (called `lam` in the program) can have a profound effect on the result. We chose $\lambda = 1$ (as in the program listing) with the following result:

```
Intersection point = [ 0.73306761   7.58776385]
Minimum distance = 4.28679958767
xy = 5.56234387462
Number of cycles = 5
```

The small value of λ favored speed of convergence over accuracy. Because the violation of the constraint $xy = 5$ is clearly unacceptable, we ran the program again with $\lambda = 10\,000$ and changed the starting point to (0.73306761, 7.58776385), the end point of the first run. The results shown next are now acceptable:

```
Intersection point = [ 0.65561311   7.62653592]
Minimum distance = 4.36040970945
xy = 5.00005696357
Number of cycles = 5
```

Could we have used $\lambda = 10\,000$ in the first run? In this case we would be lucky and obtain the minimum in 17 cycles. Hence we save seven cycles by using two runs. However, a large λ often causes the algorithm to hang up, so that it is generally wise to start with a small λ.

Check. Since we have an equality constraint, the optimal point can readily be found by calculus. The function in Eq. (10.2a) is (here λ is the Lagrangian multiplier)

$$F^*(x, y, \lambda) = (x - 5)^2 + (y - 8)^2 + \lambda(xy - 5)$$

so that Eqs. (10.2b) become

$$\frac{\partial F^*}{\partial x} = 2(x - 5) + \lambda y = 0$$

$$\frac{\partial F^*}{\partial y} = 2(y - 8) + \lambda x = 0$$

$$g(x) = xy - 5 = 0$$

which can be solved with the Newton-Raphson method (the function `newton-Raphson2` in Section 4.6). In the following program we used the notation $\mathbf{x} = [x \quad y \quad \lambda]^T$.

```python
#!/usr/bin/python
## example10_4_check
import numpy as np
from newtonRaphson2 import *

def F(x):
    return np.array([2.0*(x[0] - 5.0) + x[2]*x[1], \
                     2.0*(x[1] - 8.0) + x[2]*x[0], \
                     x[0]*x[1] - 5.0])

xStart = np.array([1.0, 5.0, 1.0])
print("x = ", newtonRaphson2(F,xStart))
input("Press return to exit")
```

The result is

```
x =  [ 0.6556053   7.62653992  1.13928328]
```

EXAMPLE 10.5

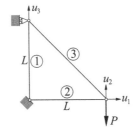

The displacement formulation of the truss shown results in the following simultaneous equations for the joint displacements $\mathbf{u}$:

$$\frac{E}{2\sqrt{2}L}\begin{bmatrix} 2\sqrt{2}A_2 + A_3 & -A_3 & A_3 \\ -A_3 & A_3 & -A_3 \\ A_3 & -A_3 & 2\sqrt{2}A_1 + A_3 \end{bmatrix}\begin{bmatrix} u_1 \\ u_2 \\ u_3 \end{bmatrix} = \begin{bmatrix} 0 \\ -P \\ 0 \end{bmatrix}$$

where E represents the modulus of elasticity of the material and A_i is the cross-sectional area of member i. Use Powell's method to minimize the structural volume (i.e., the weight) of the truss while keeping the displacement u_2 below a given value δ.

Solution. Introducing the dimensionless variables

$$v_i = \frac{u_i}{\delta} \qquad x_i = \frac{E\delta}{PL}A_i,$$

the equations become

$$\frac{1}{2\sqrt{2}}\begin{bmatrix} 2\sqrt{2}x_2 + x_3 & -x_3 & x_3 \\ -x_3 & x_3 & -x_3 \\ x_3 & -x_3 & 2\sqrt{2}x_1 + x_3 \end{bmatrix}\begin{bmatrix} v_1 \\ v_2 \\ v_3 \end{bmatrix} = \begin{bmatrix} 0 \\ -1 \\ 0 \end{bmatrix} \qquad \text{(a)}$$

The structural volume to be minimized is

$$V = L(A_1 + A_2 + \sqrt{2}A_3) = \frac{PL^2}{E\delta}(x_1 + x_2 + \sqrt{2}x_3)$$

In addition to the displacement constraint $|u_2| \leq \delta$, we should also prevent the cross-sectional areas from becoming negative by applying the constraints $A_i \geq 0$. Thus the optimization problem becomes: Minimize

$$F = x_1 + x_2 + \sqrt{2}x_3$$

with the inequality constraints

$$|v_2| \leq 1 \qquad x_i \geq 0 \ (i = 1, 2, 3)$$

Note that to obtain v_2 we must solve Eqs. (a).

Here is the program:

```
#!/usr/bin/python
## example10_5
from powell import *
from numpy import array
from math import sqrt
from gaussElimin import *

def F(x):
    global v, weight
    lam = 100.0
    c = 2.0*sqrt(2.0)
    A = array([[c*x[1] + x[2], -x[2],   x[2]],          \
              [-x[2],               x[2], -x[2]],          \
              [ x[2],              -x[2],   c*x[0] + x[2]]])/c
    b = array([0.0, -1.0, 0.0])
    v = gaussElimin(A,b)
    weight = x[0] + x[1] + sqrt(2.0)*x[2]
    penalty = max(0.0,abs(v[1]) - 1.0)**2   \
            + max(0.0,-x[0])**2             \
            + max(0.0,-x[1])**2             \
            + max(0.0,-x[2])**2
    return weight + penalty*lam

xStart = array([1.0, 1.0, 1.0])
x,numIter = powell(F,xStart)
print("x = ",x)
print("v = ",v)
print("Relative weight F = ",weight)
print("Number of cycles = ",numIter)
input("Press return to exit")
```

The subfunction F returns the penalized merit function. It includes the code that sets up and solves Eqs. (a). The displacement vector $\mathbf{v}$ is called u in the program.

The first run of the program started with $\mathbf{x} = \begin{bmatrix} 1 & 1 & 1 \end{bmatrix}^T$ and used $\lambda = 100$ for the penalty multiplier. The results were

```
x =  [ 3.73870376  3.73870366  5.28732564]
v =  [-0.26747239 -1.06988953 -0.26747238]
Relative weight F =  14.9548150471
Number of cycles =  10
```

Because the magnitude of v_2 is excessive, the penalty multiplier was increased to 10 000 and the program run again using the output **x** from the last run as the input. As seen, v_2 is now much closer to the constraint value.

```
x =  [ 3.99680758   3.9968077    5.65233961]
v =  [-0.25019968 -1.00079872 -0.25019969]
Relative weight F =   15.9872306185
Number of cycles =   11
```

In this problem the use of $\lambda = 10\,000$ at the outset would not work. You are invited to try it.

10.4 Downhill Simplex Method

The downhill simplex method is also known as the *Nelder-Mead method*. The idea is to employ a moving simplex in the design space to surround the optimal point and then shrink the simplex until its dimensions reach a specified error tolerance. In n-dimensional space a simplex is a figure of $n+1$ vertices connected by straight lines and bounded by polygonal faces. If $n = 2$, a simplex is a triangle; if $n = 3$, it is a tetrahedron.

The allowed moves of a two-dimensional simplex ($n = 2$) are illustrated in Figure 10.4. By applying these moves in a suitable sequence, the simplex can always hunt down the minimum point, enclose it, and then shrink around it. The direction of a move is determined by the values of $F(\mathbf{x})$ (the function to be minimized) at the vertices. The vertex with the highest value of F is labeled *Hi*, and *Lo* denotes the vertex with the lowest value. The magnitude of a move is controlled by the distance d measured from the *Hi* vertex to the centroid of the opposing face (in the case of the triangle, the middle of the opposing side).

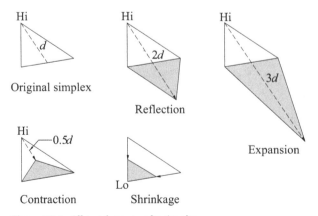

Figure 10.4. Allowed moves of a simplex.

The outline of the algorithm is as follows:

> Choose a starting simplex.
> Cycle until $d \leq \varepsilon$ (ε being the error tolerance):
>   Try reflection.
>   if new vertex $\leq$ old *Lo*: accept reflection
>   Try expansion.
>   if new vertex $\leq$ old *Lo*: accept expansion.
>   else:
>   if new vertex $>$ old *Hi*:
>   Try contraction.
>   if new vertex $\leq$ old *Hi*: accept contraction.
>   else: use shrinkage.
> End cycle.

The downhill simplex algorithm is much slower than Powell's method in most cases, but makes up for it in robustness. It often works in problems where Powell's method hangs up.

■ downhill

The implementation of the downhill simplex method is given next. The starting simplex has one of its vertices at $\mathbf{x}_0$ and the others at $\mathbf{x}_0 + \mathbf{e}_i b$ ($i = 1, 2, \ldots, n$), where $\mathbf{e}_i$ is the unit vector in the direction of the x_i-coordinate. The user inputs the vector $\mathbf{x}_0$ (called xStart in the program) and the edge length b of the simplex.

```
## module downhill
''' x = downhill(F,xStart,side,tol=1.0e-6)
    Downhill simplex method for minimizing the user-supplied
    scalar function F(x) with respect to the vector x.
    xStart = starting vector x.
    side   = side length of the starting simplex (default is 0.1)
'''
import numpy as np
import math

def downhill(F,xStart,side=0.1,tol=1.0e-6):
    n = len(xStart)                     # Number of variables
    x = np.zeros((n+1,n))
    f = np.zeros(n+1)

  # Generate starting simplex
    x[0] = xStart
```

```
    for i in range(1,n+1):
        x[i] = xStart
        x[i,i-1] = xStart[i-1] + side
# Compute values of F at the vertices of the simplex
    for i in range(n+1): f[i] = F(x[i])


# Main loop
    for k in range(500):
      # Find highest and lowest vertices
        iLo = np.argmin(f)
        iHi = np.argmax(f)
      # Compute the move vector d
        d = (-(n+1)*x[iHi] + np.sum(x,axis=0))/n
      # Check for convergence
        if math.sqrt(np.dot(d,d)/n) < tol: return x[iLo]


      # Try reflection
        xNew = x[iHi] + 2.0*d
        fNew = F(xNew)
        if fNew <= f[iLo]:        # Accept reflection
            x[iHi] = xNew
            f[iHi] = fNew
          # Try expanding the reflection
            xNew = x[iHi] + d
            fNew = F(xNew)
            if fNew <= f[iLo]:    # Accept expansion
                x[iHi] = xNew
                f[iHi] = fNew
        else:
          # Try reflection again
            if fNew <= f[iHi]:    # Accept reflection
                x[iHi] = xNew
                f[iHi] = fNew
            else:
              # Try contraction
                xNew = x[iHi] + 0.5*d
                fNew = F(xNew)
                if fNew <= f[iHi]: # Accept contraction
                    x[iHi] = xNew
                    f[iHi] = fNew
                else:
                  # Use shrinkage
                    for i in range(len(x)):
                        if i != iLo:
```

```
                              x[i] = (x[i] - x[iLo])*0.5
                              f[i] = F(x[i])
       print("Too many iterations in downhill")
       return x[iLo]
```

EXAMPLE 10.6

Use the downhill simplex method to minimize

$$F = 10x_1^2 + 3x_2^2 - 10x_1x_2 + 2x_1$$

The coordinates of the vertices of the starting simplex are $(0, 0)$, $(0, -0.2)$ and $(0.2, 0)$. Show graphically the first four moves of the simplex.

Solution. The figure shows the design space (the x_1-x_2 plane). The numbers in the figure are the values of F at the vertices. The move numbers are enclosed in circles. The starting move (move 1) is a reflection, followed by an expansion (move 2). The next two moves are reflections. At this stage, the simplex is still moving downhill. Contraction will not start until move 8, when the simplex has surrounded the optimal point at $(-0.6, -1.0)$.

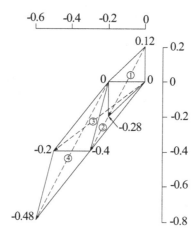

EXAMPLE 10.7

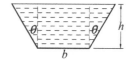

The figure shows the cross section of a channel carrying water. Use the downhill simplex to determine h, b, and θ that minimize the length of the wetted perimeter while maintaining a cross-sectional area of 8 m^2. (Minimizing the wetted perimeter results in the least resistance to the flow.) Check the answer with calculus.

Solution. The cross-sectional area of the channel is

$$A = \frac{1}{2}\left[b + (b + 2h\tan\theta)\right]h = (b + h\tan\theta)h$$

and the length of the wetted perimeter is

$$S = b + 2(h \sec \theta)$$

The optimization problem is to minimize S subject to the constraint $A - 8 = 0$. Using the penalty function to take care of the equality constraint, the function to be minimized is

$$S^* = b + 2h \sec \theta + \lambda \left[(b + h \tan \theta) h - 8 \right]^2$$

Letting $\mathbf{x} = \begin{bmatrix} b & h & \theta \end{bmatrix}^T$ and starting with $\mathbf{x}_0 = \begin{bmatrix} 4 & 2 & 0 \end{bmatrix}^T$, we arrive at the following program:

```
#!/usr/bin/python
## example10_7
import numpy as np
import math
from downhill import *

def S(x):
    global perimeter,area
    lam = 10000.0
    perimeter = x[0] + 2.0*x[1]/math.cos(x[2])
    area = (x[0] + x[1]*math.tan(x[2]))*x[1]
    return perimeter + lam*(area - 8.0)**2

xStart = np.array([4.0, 2.0, 0.0])
x = downhill(S,xStart)
area = (x[0] + x[1]*math.tan(x[2]))*x[1]
print("b = ",x[0])
print("h = ",x[1])
print("theta (deg) = ",x[2]*180.0/math.pi)
print("area = ",area)
print("perimeter = ",perimeter)
input("Finished. Press return to exit")
```

The results are

```
b =   2.4816069148
h =   2.14913738694
theta (deg) =   30.0000185796
area =   7.99997671775
perimeter =   7.44482803952
```

Check. Because we have an equality constraint, the problem can be solved by calculus with help from a Lagrangian multiplier. Referring to Eqs. (10.2a), we have $F = S$ and $g = A - 8$, so that

$$F^* = S + \lambda(A - 8)$$

$$= b + 2(h\sec\theta) + \lambda\left[(b + h\tan\theta)h - 8\right]$$

Therefore, Eqs. (10.2b) become

$$\frac{\partial F^*}{\partial b} = 1 + \lambda h = 0$$

$$\frac{\partial F^*}{\partial h} = 2\sec\theta + \lambda(b + 2h\tan\theta) = 0$$

$$\frac{\partial F^*}{\partial \theta} = 2h\sec\theta\tan\theta + \lambda h^2\sec^2\theta = 0$$

$$g = (b + h\tan\theta)h - 8 = 0$$

which can be solved with newtonRaphson2 as shown next.

```
#!/usr/bin/python
## example10_7_check
import numpy as np
import math
from newtonRaphson2 import *

def f(x):
    f = np.zeros(4)
    f[0] = 1.0 + x[3]*x[1]
    f[1] = 2.0/math.cos(x[2]) + x[3]*(x[0]          \
         + 2.0*x[1]*math.tan(x[2]))
    f[2] = 2.0*x[1]*math.tan(x[2])/math.cos(x[2]) \
         + x[3]*(x[1]/math.cos(x[2]))**2
    f[3] = (x[0] + x[1]*math.tan(x[2]))*x[1] - 8.0
    return f

xStart = np.array([3.0, 2.0, 0.0, 1.0])
print("x =",newtonRaphson2(f,xStart))
input("Press return to exit")
```

The solution $\mathbf{x} = \begin{bmatrix} b & h & \theta & \lambda \end{bmatrix}^T$ is

```
x = [ 2.48161296  2.14913986  0.52359878 -0.46530243]
```

EXAMPLE 10.8

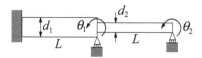

The fundamental circular frequency of the stepped shaft is required to be higher than ω_0 (a given value). Use the downhill simplex to determine the diameters d_1 and d_2 that minimize the volume of the material without violating the frequency constraint. The approximate value of the fundamental frequency can be computed by solving the eigenvalue problem (obtainable from the finite element approximation)

$$\begin{bmatrix} 4(d_1^4 + d_2^4) & 2d_2^4 \\ 2d_2^4 & 4d_2^4 \end{bmatrix} \begin{bmatrix} \theta_1 \\ \theta_2 \end{bmatrix} = \frac{4\gamma L^4 \omega^2}{105\,E} \begin{bmatrix} 4(d_1^2 + d_2^2) & -3d_2^2 \\ -3d_2^2 & 4d_2^2 \end{bmatrix} \begin{bmatrix} \theta_1 \\ \theta_2 \end{bmatrix}$$

where

$$\gamma = \text{mass density of the material}$$
$$\omega = \text{circular frequency}$$
$$E = \text{modulus of elasticity}$$
$$\theta_1, \theta_2 = \text{rotations at the simple supports}$$

Solution. We start by introducing the dimensionless variables $x_i = d_i/d_0$, where d_0 is an arbitrary "base" diameter. As a result, the eigenvalue problem becomes

$$\begin{bmatrix} 4(x_1^4 + x_2^4) & 2x_2^4 \\ 2x_2^4 & 4x_2^4 \end{bmatrix} \begin{bmatrix} \theta_1 \\ \theta_2 \end{bmatrix} = \lambda \begin{bmatrix} 4(x_1^2 + x_2^2) & -3x_2^2 \\ -3x_2^2 & 4x_2^2 \end{bmatrix} \begin{bmatrix} \theta_1 \\ \theta_2 \end{bmatrix} \qquad \text{(a)}$$

where

$$\lambda = \frac{4\gamma L^4 \omega^2}{105\,E d_0^2}$$

In the following program we assume that the constraint on the frequency ω is equivalent to $\lambda \geq 0.4$.

```python
#!/usr/bin/python
## example10_8
import numpy as np
from stdForm import *
from inversePower import *
from downhill import *

def F(x):
    global eVal
    lam = 1.0e6
    eVal_min = 0.4
    A = np.array([[4.0*(x[0]**4 + x[1]**4), 2.0*x[1]**4], \
                  [2.0*x[1]**4, 4.0*x[1]**4]])
    B = np.array([[4.0*(x[0]**2 + x[1]**2), -3.0*x[1]**2], \
                  [-3*x[1]**2, 4.0*x[1]**2]])
    H,t = stdForm(A,B)
    eVal,eVec = inversePower(H,0.0)
    return x[0]**2 + x[1]**2 + lam*(max(0.0,eVal_min - eVal))**2
```

```
xStart = np.array([1.0,1.0])
x = downhill(F,xStart,0.1)
print("x = ", x)
print("eigenvalue = ",eVal)
input ("Press return to exit")
```

Although a 2×2 eigenvalue problem can be solved easily, we avoid the work involved by employing functions that have been already prepared— using `stdForm` to turn the eigenvalue problem into standard form, and `inversePower` to compute the eigenvalue closest to zero.

The results shown next were obtained with $x_1 = x_2 = 1$ as the starting values and 10^6 for the penalty multiplier. The downhill simplex method is robust enough to alleviate the need for multiple runs with an increasing penalty multiplier.

```
x =   [ 1.07512696  0.79924677]
eigenvalue =  0.399997757238
```

PROBLEM SET 10.1

1. ■ The Lennard-Jones potential between two molecules is

$$V = 4\varepsilon \left[\left(\frac{\sigma}{r}\right)^{12} - \left(\frac{\sigma}{r}\right)^{6} \right]$$

where ε and σ are constants, and r is the distance between the molecules. Use the module `goldSearch` to find σ/r that minimizes the potential and verify the result analytically.

2. ■ One wave-function of a hydrogen atom is

$$\psi = C(27 - 18\sigma + 2\sigma^2)e^{-\sigma/3}$$

where

$$\sigma = zr/a_0$$

$$C = \frac{1}{81\sqrt{3\pi}} \left(\frac{z}{a_0}\right)^{2/3}$$

$$z = \text{nuclear charge}$$

$$a_0 = \text{Bohr radius}$$

$$r = \text{radial distance}$$

Find σ where ψ is at a minimum. Verify the result analytically,

3. ■ Determine the parameter p that minimizes the integral

$$\int_0^\pi \sin x \cos px \, dx$$

Hint: Use quadrature to evaluate the integral.

4. ■

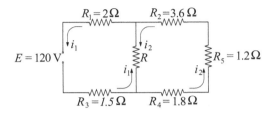

Kirchoff's equations for the two loops of the electrical circuit are

$$R_1 i_1 + R_3 i_1 + R(i_1 - i_2) = E$$

$$R_2 i_2 + R_4 i_2 + R_5 i_2 + R(i_2 - i_1) = 0$$

Find the resistance R that maximizes the power dissipated by R. *Hint*: Solve Kirchoff's equations numerically with one of the functions in Chapter 2.

5. ■

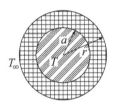

A wire carrying an electric current is surrounded by rubber insulation of outer radius r. The resistance of the wire generates heat, which is conducted through the insulation and convected into the surrounding air. The temperature of the wire can be shown to be

$$T = \frac{q}{2\pi} \left(\frac{\ln(r/a)}{k} + \frac{1}{hr} \right) + T_\infty$$

where

q = rate of heat generation in wire = 50 W/m

a = radius of wire = 5 mm

k = thermal conductivity of rubber = 0.16 W/m · K

h = convective heat-transfer coefficient = 20 W/m^2 · K

T_∞ = ambient temperature = 280 K

Find r that minimizes T.

6. ■ Minimize the function

$$F(x, y) = (x - 1)^2 + (y - 1)^2$$

subject to the constraints $x + y \geq 1$ and $x \geq 0.6$.

7. ■ Find the minimum of the function

$$F(x, y) = 6x^2 + y^3 + xy$$

in $y \geq 0$. Verify the result analytically.

8. ■ Solve Prob. 7 if the constraint is changed to $y \geq -2$.

9. ■ Determine the smallest distance from the point $(1, 2)$ to the parabola $y = x^2$.

10. ■

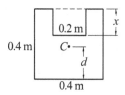

Determine x that minimizes the distance d between the base of the area shown and its centroid C.

11. ■

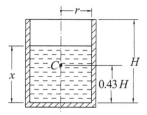

The cylindrical vessel of mass M has its center of gravity at C. The water in the vessel has a depth x. Determine x so that the center of gravity of the vessel-water combination is as low as possible. Use $M = 115$ kg, $H = 0.8$ m, and $r = 0.25$ m.

12. ■

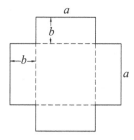

The sheet of cardboard is folded along the dashed lines to form a box with an open top. If the volume of the box is to be 1.0 m³, determine the dimensions a and b that would use the least amount of cardboard. Verify the result analytically.

13. ■

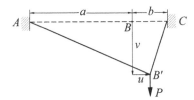

The elastic cord $A\,BC$ has an extensional stiffness k. When the vertical force P is applied at B, the cord deforms to the shape $A\,B'C$. The potential energy of the system in the deformed position is

$$V = -Pv + \frac{k(a+b)}{2a}\delta_{AB} + \frac{k(a+b)}{2b}\delta_{BC}$$

where

$$\delta_{AB} = \sqrt{(a+u)^2 + v^2} - a$$
$$\delta_{BC} = \sqrt{(b-u)^2 + v^2} - b$$

are the elongations of $A\,B$ and BC. Determine the displacements u and v by minimizing V (this is an application of the principle of minimum potential energy: A system is in stable equilibrium if its potential energy is at minimum). Use $a = 150$ mm, $b = 50$ mm, $k = 0.6$ N/mm, and $P = 5$ N.

14. ■

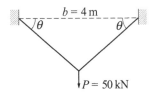

Each member of the truss has a cross-sectional area A. Find A and the angle θ that minimize the volume

$$V = \frac{bA}{\cos\theta}$$

of the material in the truss without violating the constraints

$$\sigma \le 150 \text{ MPa} \qquad \delta \le 5 \text{ mm}$$

where

$$\sigma = \frac{P}{2A\sin\theta} = \text{stress in each member}$$

$$\delta = \frac{Pb}{2\,EA\sin 2\theta\,\sin\theta} = \text{displacement at the load } P$$

and $E = 200 \times 10^9$.

15. ■ Solve Prob. 14 if the allowable displacement is changed to 2.5 mm.
16. ■

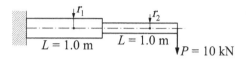

The cantilever beam of the circular cross section is to have the smallest volume possible subject to constraints

$$\sigma_1 \le 180 \text{ MPa} \qquad \sigma_2 \le 180 \text{ MPa} \qquad \delta \le 25 \text{ mm}$$

where

$$\sigma_1 = \frac{8PL}{\pi r_1^3} = \text{maximum stress in left half}$$

$$\sigma_2 = \frac{4PL}{\pi r_2^3} = \text{maximum stress in right half}$$

$$\delta = \frac{PL^3}{3\pi E}\left(\frac{7}{r_1^4} + \frac{1}{r_2^4}\right) = \text{displacement at free end}$$

and $E = 200$ GPa. Determine r_1 and r_2.

17. ■ Find the minimum of the function

$$F(x, y, z) = 2x^2 + 3y^2 + z^2 + xy + xz - 2y$$

and confirm the result analytically.

18. ■

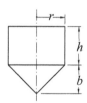

The cylindrical container has a conical bottom and an open top. If the volume V of the container is to be 1.0 m³, find the dimensions r, h, and b that minimize the surface area S. Note that

$$V = \pi r^2 \left(\frac{b}{3} + h\right)$$

$$S = \pi r \left(2h + \sqrt{b^2 + r^2}\right)$$

19. ■

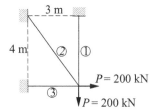

The equilibrium equations of the truss shown are

$$\sigma_1 A_1 + \frac{4}{5}\sigma_2 A_2 = P \qquad \frac{3}{5}\sigma_2 A_2 + \sigma_3 A_3 = P$$

where σ_i is the axial stress in member i and A_i is the cross-sectional area. The third equation is supplied by compatibility (geometric constraints on the elongations of the members):

$$\frac{16}{5}\sigma_1 - 5\sigma_2 + \frac{9}{5}\sigma_3 = 0$$

Find the cross-sectional areas of the members that minimize the weight of the truss without the stresses exceeding 150 MPa.

20. ■

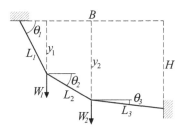

A cable supported at the ends carries the weights W_1 and W_2. The potential energy of the system is

$$V = -W_1 y_1 - W_2 y_2$$

$$= -W_1 L_1 \sin\theta_1 - W_2(L_1 \sin\theta_1 + L_2 \sin\theta_2)$$

and the geometric constraints are

$$L_1 \cos\theta_1 + L_2 \cos\theta_2 + L_3 \cos\theta_3 = B$$

$$L_1 \sin\theta_1 + L_2 \sin\theta_2 + L_3 \sin\theta_3 = H$$

The principle of minimum potential energy states that the equilibrium configuration of the system is the one that satisfies geometric constraints and minimizes the potential energy. Determine the equilibrium values of θ_1, θ_2, and θ_3 given that $L_1 = 1.2$ m, $L_2 = 1.5$ m, $L_3 = 1.0$ m, $B = 3.5$ m, $H = 0$, $W_1 = 20$ kN, and $W_2 = 30$ kN.

21. ■

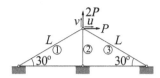

The displacement formulation of the truss results in the equations

$$\frac{E}{4L}\begin{bmatrix} 3A_1 + 3A_3 & \sqrt{3}A_1 + \sqrt{3}A_3 \\ \sqrt{3}A_1 + \sqrt{3}A_3 & A_1 + 8A_2 + A_3 \end{bmatrix}\begin{bmatrix} u \\ v \end{bmatrix} = \begin{bmatrix} P \\ 2P \end{bmatrix}$$

where E is the modulus of elasticity, A_i is the cross-sectional area of member i, and u, v are the displacement components of the loaded joint. Letting $A_1 = A_3$ (a symmetric truss), determine the cross-sectional areas that minimize the structural volume without violating the constraints $u \leq \delta$ and $v \leq \delta$. *Hint*: Nondimensionalize the problem as in Example 10.5.

22. ■ Solve Prob. 21 if the three cross-sectional areas are independent.

23. ■ A beam of rectangular cross section is cut from a cylindrical log of diameter d. Calculate the height h and the width b of the cross section that maximizes the cross-sectional moment of inertia $I = bh^3/12$. Check the result with calculus.

Appendices

A1 **Taylor Series**

Function of a Single Variable

Taylor series expansion of a function $f(x)$ about the point $x = a$ is the infinite series

$$f(x) = f(a) + f'(a)(x - a) + f''(a)\frac{(x - a)^2}{2!} + f'''(a)\frac{(x - a)^3}{3!} + \cdots \quad \text{(A1)}$$

In the special case $a = 0$ the series is also known as the *MacLaurin series*. It can be shown that Taylor series expansion is unique in the sense that no two functions have identical Taylor series.

A Taylor series is meaningful only if all the derivatives of $f(x)$ exist at $x = a$ and the series converges. In general, convergence occurs only if x is sufficiently close to a; that is, if $|x - a| \leq \varepsilon$, where ε is called the *radius of convergence*. In many cases ε is infinite.

Another useful form of a Taylor series is the expansion about an arbitrary value of x:

$$f(x + h) = f(x) + f'(x)h + f''(x)\frac{h^2}{2!} + f'''(x)\frac{h^3}{3!} + \cdots \quad \text{(A2)}$$

Because it is not possible to evaluate all the terms of an infinite series, the effect of truncating the series in Eq. (A2) is of great practical importance. Keeping the first $n + 1$ terms, we have

$$f(x + h) = f(x) + f'(x)h + f''(x)\frac{h^2}{2!} + \cdots + f^{(n)}(x)\frac{h^n}{n!} + E_n \quad \text{(A3)}$$

where E_n is the *truncation error* (sum of the truncated terms). The bounds on the truncation error are given by *Taylor's theorem*,

$$E_n = f^{(n+1)}(\xi)\frac{h^{n+1}}{(n + 1)!} \quad \text{(A4)}$$

where ξ is some point in the interval $(x, x + h)$. Note that the expression for E_n is identical to the first discarded term of the series, but with x replaced by ξ. Since the value of ξ is undetermined (only its limits are known), the most we can get out of Eq. (A4) are the upper and lower bounds on the truncation error.

If the expression for $f^{(n+1)}(\xi)$ is not available, the information conveyed by Eq. (A4) is reduced to

$$E_n = \mathcal{O}(h^{n+1}) \tag{A5}$$

which is a concise way of saying that the truncation error is *of the order of* h^{n+1} or behaves as h^{n+1}. If h is within the radius of convergence, then

$$\mathcal{O}(h^n) > \mathcal{O}(h^{n+1})$$

that is, the error is always reduced if a term is added to the truncated series (this may not be true for the first few terms).

In the special case $n = 1$, Taylor's theorem is known as the *mean value theorem*:

$$f(x + h) = f(x) + f'(\xi)h, \quad x \le \xi \le x + h \tag{A6}$$

Function of Several Variables

If f is a function of the m variables $x_1, x_2, \ldots, x_m$, then its Taylor series expansion about the point $\mathbf{x} = [x_1, x_2, \ldots, x_m]^T$ is

$$f(\mathbf{x} + \mathbf{h}) = f(\mathbf{x}) + \sum_{i=1}^{m} \frac{\partial f}{\partial x_i}\bigg|_{\mathbf{x}} h_i + \frac{1}{2!}\sum_{i=1}^{m}\sum_{j=1}^{m} \frac{\partial^2 f}{\partial x_i \partial x_j}\bigg|_{\mathbf{x}} h_i h_j + \cdots \tag{A7}$$

This is sometimes written as

$$f(\mathbf{x} + \mathbf{h}) = f(\mathbf{x}) + \nabla f(\mathbf{x}) \cdot \mathbf{h} + \frac{1}{2}\mathbf{h}^T \mathbf{H}(\mathbf{x})\mathbf{h} + \cdots \tag{A8}$$

The vector ∇f is known as the *gradient* of f, and the matrix $\mathbf{H}$ is called the *Hessian matrix* of f.

EXAMPLE A1
Derive the Taylor series expansion of $f(x) = \ln(x)$ about $x = 1$.

Solution. The derivatives of f are

$$f'(x) = \frac{1}{x} \qquad f''(x) = -\frac{1}{x^2} \qquad f'''(x) = \frac{2!}{x^3} \qquad f^{(4)} = -\frac{3!}{x^4} \text{ etc.}$$

Evaluating the derivatives at $x = 1$, we get

$$f'(1) = 1 \qquad f''(1) = -1 \qquad f'''(1) = 2! \qquad f^{(4)}(1) = -3! \text{ etc.}$$

which upon substitution into Eq. (A1), together with $a = 1$, yields

$$\ln(x) = 0 + (x - 1) - \frac{(x-1)^2}{2!} + 2!\frac{(x-1)^3}{3!} - 3!\frac{(x-1)^4}{4!} + \cdots$$

$$= (x - 1) - \frac{1}{2}(x-1)^2 + \frac{1}{3}(x-1)^3 - \frac{1}{4}(x-1)^4 + \cdots$$

EXAMPLE A2

Use the first five terms of the Taylor series expansion of e^x about $x = 0$:

$$e^x = 1 + x + \frac{x^2}{2!} + \frac{x^3}{3!} + \frac{x^4}{4!} + \cdots$$

together with the error estimate to find the bounds of e.

Solution

$$e = 1 + 1 + \frac{1}{2} + \frac{1}{6} + \frac{1}{24} + E_4 = \frac{65}{24} + E_4$$

$$E_4 = f^{(4)}(\xi)\frac{h^5}{5!} = \frac{e^\xi}{5!}, \quad 0 \le \xi \le 1$$

The bounds on the truncation error are

$$(E_4)_{\min} = \frac{e^0}{5!} = \frac{1}{120} \qquad (E_4)_{\max} = \frac{e^1}{5!} = \frac{e}{120}$$

Thus the lower bound on e is

$$e_{\min} = \frac{65}{24} + \frac{1}{120} = \frac{163}{60}$$

and the upper bound is given by

$$e_{\max} = \frac{65}{24} + \frac{e_{\max}}{120}$$

which yields

$$\frac{119}{120}e_{\max} = \frac{65}{24} \qquad e_{\max} = \frac{325}{119}$$

Therefore,

$$\frac{163}{60} \le e \le \frac{325}{119}$$

EXAMPLE A3

Compute the gradient and the Hessian matrix of

$$f(x, y) = \ln\sqrt{x^2 + y^2}$$

at the point $x = -2, y = 1$.

Solution

$$\frac{\partial f}{\partial x} = \frac{1}{\sqrt{x^2 + y^2}}\left(\frac{1}{2}\frac{2x}{\sqrt{x^2 + y^2}}\right) = \frac{x}{x^2 + y^2} \qquad \frac{\partial f}{\partial y} = \frac{y}{x^2 + y^2}$$

$$\nabla f(x, y) = \left[\, x/(x^2 + y^2) \quad y/(x^2 + y^2) \,\right]^T$$

$$\nabla f(-2, 1) = \left[\, -0.4 \quad 0.2 \,\right]^T$$

$$\frac{\partial^2 f}{\partial x^2} = \frac{(x^2 + y^2) - x(2x)}{(x^2 + y^2)^2} = \frac{-x^2 + y^2}{(x^2 + y^2)^2}$$

$$\frac{\partial^2 f}{\partial y^2} = \frac{x^2 - y^2}{(x^2 + y^2)^2}$$

$$\frac{\partial^2 f}{\partial x \partial y} = \frac{\partial^2 f}{\partial y \partial x} = \frac{-2xy}{(x^2 + y^2)^2}$$

$$\mathbf{H}(x, y) = \begin{bmatrix} -x^2 + y^2 & -2xy \\ -2xy & x^2 - y^2 \end{bmatrix} \frac{1}{(x^2 + y^2)^2}$$

$$\mathbf{H}(-2, 1) = \begin{bmatrix} -0.12 & 0.16 \\ 0.16 & 0.12 \end{bmatrix}$$

A2   Matrix Algebra

A matrix is a rectangular array of numbers. The *size* of a matrix is determined by the number of rows and columns, also called the *dimensions* of the matrix. Thus a matrix of m rows and n columns is said to have the size $m \times n$ (the number of rows is always listed first). A particularly important matrix is the square matrix, which has the same number of rows and columns.

An array of numbers arranged in a single column is called a *column vector*, or simply a vector. If the numbers are set out in a row, the term *row vector* is used. Thus a column vector is a matrix of dimensions $n \times 1$, and a row vector can be viewed as a matrix of dimensions $1 \times n$.

We denote matrices by boldfaced uppercase letters. For vectors we use boldface lowercase letters. Here are examples of the notation:

$$\mathbf{A} = \begin{bmatrix} A_{11} & A_{12} & A_{13} \\ A_{21} & A_{22} & A_{23} \\ A_{31} & A_{32} & A_{33} \end{bmatrix} \qquad \mathbf{b} = \begin{bmatrix} b_1 \\ b_2 \\ b_3 \end{bmatrix} \tag{A9}$$

Indices of the elements of a matrix are displayed in the same order as its dimensions: The row number comes first, followed by the column number. Only one index is needed for the elements of a vector.

Transpose

The transpose of a matrix $\mathbf{A}$ is denoted by $\mathbf{A}^T$ and defined as

$$A_{ij}^T = A_{ji}$$

The transpose operation thus interchanges the rows and columns of the matrix. If applied to vectors, it turns a column vector into a row vector and vice versa.

For example, transposing $\mathbf{A}$ and $\mathbf{b}$ in Eq. (A9), we get

$$\mathbf{A}^T = \begin{bmatrix} A_{11} & A_{21} & A_{31} \\ A_{12} & A_{22} & A_{32} \\ A_{13} & A_{23} & A_{33} \end{bmatrix} \qquad \mathbf{b}^T = \begin{bmatrix} b_1 & b_2 & b_3 \end{bmatrix}$$

An $n \times n$ matrix is said to be *symmetric* if $\mathbf{A}^T = \mathbf{A}$. This means that the elements in the upper triangular portion (above the diagonal connecting A_{11} and A_{nn}) of a symmetric matrix are mirrored in the lower triangular portion.

Addition

The sum $\mathbf{C} = \mathbf{A} + \mathbf{B}$ of two $m \times n$ matrices $\mathbf{A}$ and $\mathbf{B}$ is defined as

$$C_{ij} = A_{ij} + B_{ij}, \quad i = 1, 2, \ldots, m; \quad j = 1, 2, \ldots, n \tag{A10}$$

Thus the elements of $\mathbf{C}$ are obtained by adding elements of $\mathbf{A}$ to the elements of $\mathbf{B}$. Note that addition is defined only for matrices that have the same dimensions.

Vector Products

The *dot* or *inner product* $c = \mathbf{a} \cdot \mathbf{b}$ of the vectors $\mathbf{a}$ and $\mathbf{b}$, each of size m, is defined as the scalar

$$c = \sum_{k=1}^{m} a_k b_k \tag{A11}$$

It can also be written in the form $c = \mathbf{a}^T \mathbf{b}$. In NumPy the function for the dot product is dot(a,b) or inner(a,b).

The *outer product* $\mathbf{C} = \mathbf{a} \otimes \mathbf{b}$ is defined as the matrix

$$C_{ij} = a_i b_j$$

An alternative notation is $\mathbf{C} = \mathbf{a}\mathbf{b}^T$. The NumPy function for the outer product is outer(a,b).

Array Products

The *matrix product* $\mathbf{C} = \mathbf{A}\mathbf{B}$ of an $l \times m$ matrix $\mathbf{A}$ and an $m \times n$ matrix $\mathbf{B}$ is defined by

$$C_{ij} = \sum_{k=1}^{m} A_{ik} B_{kj}, \quad i = 1, 2, \ldots, l; \quad j = 1, 2, \ldots, n \tag{A12}$$

The definition requires the number of columns in $\mathbf{A}$ (the dimension m) to be equal to the number of rows in $\mathbf{B}$. The matrix product can also be defined in terms of the dot product. Representing the ith row of $\mathbf{A}$ as the vector $\mathbf{a}_i$ and the jth column of $\mathbf{B}$ as the vector $\mathbf{b}_j$, we have

$$\mathbf{AB} = \begin{bmatrix} \mathbf{a}_1 \cdot \mathbf{b}_1 & \mathbf{a}_1 \cdot \mathbf{b}_2 & \cdots & \mathbf{a}_1 \cdot \mathbf{b}_n \\ \mathbf{a}_2 \cdot \mathbf{b}_1 & \mathbf{a}_2 \cdot \mathbf{b}_2 & \cdots & \mathbf{a}_2 \cdot \mathbf{b}_n \\ \vdots & \vdots & \ddots & \vdots \\ \mathbf{a}_\ell \cdot \mathbf{b}_1 & \mathbf{a}_\ell \cdot \mathbf{b}_2 & \cdots & \mathbf{a}_\ell \cdot \mathbf{b}_n \end{bmatrix} \tag{A13}$$

NumPy treats the matrix product as the dot product for arrays, so that the function dot(A,B) returns the matrix product of $\mathbf{A}$ and $\mathbf{B}$.

NumPy defines the *inner product* of matrices $\mathbf{A}$ and $\mathbf{B}$ to be $\mathbf{C} = \mathbf{AB}^T$. Equation (A13) still applies, but now $\mathbf{b}$ represents the jth *row* of B.

NumPy's definition of the *outer product* of matrices $\mathbf{A}$ (size $k \times \ell$) and $\mathbf{B}$ (size $m \times n$) is as follows. Let $\mathbf{a}_i$ be the ith row of $\mathbf{A}$, and let $\mathbf{b}_j$ represent the jth row of $\mathbf{B}$. Then the outer product is of $\mathbf{A}$ and $\mathbf{B}$ is

$$\mathbf{A} \otimes \mathbf{B} = \begin{bmatrix} \mathbf{a}_1 \otimes \mathbf{b}_1 & \mathbf{a}_1 \otimes \mathbf{b}_2 & \cdots & \mathbf{a}_1 \otimes \mathbf{b}_m \\ \mathbf{a}_2 \otimes \mathbf{b}_1 & \mathbf{a}_2 \otimes \mathbf{b}_2 & \cdots & \mathbf{a}_2 \otimes \mathbf{b}_m \\ \vdots & \vdots & \ddots & \vdots \\ \mathbf{a}_k \otimes \mathbf{b}_1 & \mathbf{a}_k \otimes \mathbf{b}_2 & \cdots & \mathbf{a}_k \otimes \mathbf{b}_m \end{bmatrix} \tag{A14}$$

The submatrices $\mathbf{a}_i \otimes \mathbf{b}_j$ are of dimensions $\ell \times n$. As you can see, the size of the outer product is much larger than either $\mathbf{A}$ or $\mathbf{B}$.

Identity Matrix

A square matrix of special importance is the identity or *unit matrix*:

$$\mathbf{I} = \begin{bmatrix} 1 & 0 & 0 & \cdots & 0 \\ 0 & 1 & 0 & \cdots & 0 \\ 0 & 0 & 1 & \cdots & 0 \\ \vdots & \vdots & \vdots & \ddots & \vdots \\ 0 & 0 & 0 & 0 & 1 \end{bmatrix} \tag{A15}$$

It has the property $\mathbf{AI} = \mathbf{IA} = \mathbf{A}$.

Inverse

The inverse of an $n \times n$ matrix $\mathbf{A}$, denoted by $\mathbf{A}^{-1}$, is defined to be an $n \times n$ matrix that has the property

$$\mathbf{A}^{-1}\mathbf{A} = \mathbf{A}\mathbf{A}^{-1} = \mathbf{I} \tag{A16}$$

Determinant

The determinant of a square matrix $\mathbf{A}$ is a scalar denoted by $|\mathbf{A}|$ or $\det(\mathbf{A})$. There is no concise definition of the determinant for a matrix of arbitrary size. We start with the determinant of a 2×2 matrix, which is defined as

$$\begin{vmatrix} A_{11} & A_{12} \\ A_{21} & A_{22} \end{vmatrix} = A_{11}A_{22} - A_{12}A_{21} \tag{A17}$$

The determinant of a 3×3 matrix is then defined as

$$\begin{vmatrix} A_{11} & A_{12} & A_{13} \\ A_{21} & A_{22} & A_{23} \\ A_{31} & A_{32} & A_{33} \end{vmatrix} = A_{11} \begin{vmatrix} A_{22} & A_{23} \\ A_{32} & A_{33} \end{vmatrix} - A_{12} \begin{vmatrix} A_{21} & A_{23} \\ A_{31} & A_{33} \end{vmatrix} + A_{13} \begin{vmatrix} A_{21} & A_{22} \\ A_{31} & A_{32} \end{vmatrix}$$

Having established the pattern, we can now define the determinant of an $n \times n$ matrix in terms of the determinant of an $(n-1) \times (n-1)$ matrix:

$$|\mathbf{A}| = \sum_{k=1}^{n} (-1)^{k+1} A_{1k} M_{1k} \tag{A18}$$

where M_{ik} is the determinant of the $(n-1) \times (n-1)$ matrix obtained by deleting the ith row and kth column of $\mathbf{A}$. The term $(-1)^{k+i} M_{ik}$ is called a *cofactor* of A_{ik}.

Equation (A18) is known as *Laplace's development* of the determinant on the first row of $\mathbf{A}$. Actually Laplace's development can take place on any convenient row. Choosing the ith row, we have

$$|\mathbf{A}| = \sum_{k=1}^{n} (-1)^{k+i} A_{ik} M_{ik} \tag{A19}$$

The matrix $\mathbf{A}$ is said to be *singular* if $|\mathbf{A}| = 0$.

Positive Definiteness

An $n \times n$ matrix A is said to be positive definite if

$$\mathbf{x}^T \mathbf{A} \mathbf{x} > 0 \tag{A20}$$

for all nonvanishing vectors $\mathbf{x}$. It can be shown that a matrix is positive definite if the determinants of all its leading minors are positive. The leading minors of $\mathbf{A}$ are the n square matrices

$$\begin{bmatrix} A_{11} & A_{12} & \cdots & A_{1k} \\ A_{12} & A_{22} & \cdots & A_{2k} \\ \vdots & \vdots & \ddots & \vdots \\ A_{k1} & A_{k2} & \cdots & A_{kk} \end{bmatrix}, \quad k = 1, 2, \ldots, n$$

Therefore, positive definiteness requires that

$$A_{11} > 0, \quad \begin{vmatrix} A_{11} & A_{12} \\ A_{21} & A_{22} \end{vmatrix} > 0, \quad \begin{vmatrix} A_{11} & A_{12} & A_{13} \\ A_{21} & A_{22} & A_{23} \\ A_{31} & A_{32} & A_{33} \end{vmatrix} > 0, \dots, |A| > 0 \qquad \text{(A21)}$$

Useful Theorems

We list without proof a few theorems that are used in the main body of the text. Most proofs are easy and could be attempted as exercises in matrix algebra.

$$(\mathbf{AB})^T = \mathbf{B}^T \mathbf{A}^T \qquad \text{(A22a)}$$

$$(\mathbf{AB})^{-1} = \mathbf{B}^{-1} \mathbf{A}^{-1} \qquad \text{(A22b)}$$

$$\left| \mathbf{A}^T \right| = |\mathbf{A}| \qquad \text{(A22c)}$$

$$|\mathbf{AB}| = |\mathbf{A}| \, |\mathbf{B}| \qquad \text{(A22d)}$$

$$\text{if } \mathbf{C} = \mathbf{A}^T \mathbf{B} \mathbf{A} \text{ where } \mathbf{B} = \mathbf{B}^T, \text{ then } \mathbf{C} = \mathbf{C}^T \qquad \text{(A22e)}$$

EXAMPLE A4
Letting

$$\mathbf{A} = \begin{bmatrix} 1 & 2 & 3 \\ 1 & 2 & 1 \\ 0 & 1 & 2 \end{bmatrix} \quad \mathbf{u} = \begin{bmatrix} 1 \\ 6 \\ -2 \end{bmatrix} \quad \mathbf{v} = \begin{bmatrix} 8 \\ 0 \\ -3 \end{bmatrix}$$

compute $\mathbf{u} + \mathbf{v}$, $\mathbf{u} \cdot \mathbf{v}$, $\mathbf{Av}$, and $\mathbf{u}^T\mathbf{Av}$.

Solution

$$\mathbf{u} + \mathbf{v} = \begin{bmatrix} 1+8 \\ 6+0 \\ -2-3 \end{bmatrix} = \begin{bmatrix} 9 \\ 6 \\ -5 \end{bmatrix}$$

$$\mathbf{u} \cdot \mathbf{v} = 1(8)) + 6(0) + (-2)(-3) = 14$$

$$\mathbf{Av} = \begin{bmatrix} \mathbf{a}_1 \cdot \mathbf{v} \\ \mathbf{a}_2 \cdot \mathbf{v} \\ \mathbf{a}_3 \cdot \mathbf{v} \end{bmatrix} = \begin{bmatrix} 1(8) + 2(0) + 3(-3) \\ 1(8) + 2(0) + 1(-3) \\ 0(8) + 1(0) + 2(-3) \end{bmatrix} = \begin{bmatrix} -1 \\ 5 \\ -6 \end{bmatrix}$$

$$\mathbf{u}^T\mathbf{Av} = \mathbf{u} \cdot (\mathbf{Av}) = 1(-1) + 6(5) + (-2)(-6) = 41$$

EXAMPLE A5
Compute $|\mathbf{A}|$, where $\mathbf{A}$ is given in Example A4. Is $\mathbf{A}$ positive definite?

Solution. Laplace's development of the determinant on the first row yields

$$|\mathbf{A}| = 1 \begin{vmatrix} 2 & 1 \\ 1 & 2 \end{vmatrix} - 2 \begin{vmatrix} 1 & 1 \\ 0 & 2 \end{vmatrix} + 3 \begin{vmatrix} 1 & 2 \\ 0 & 1 \end{vmatrix}$$

$$= 1(3) - 2(2) + 3(1) = 2$$

Development of the third row is somewhat easier because of the presence of the zero element:

$$|\mathbf{A}| = 0 \begin{vmatrix} 2 & 3 \\ 2 & 1 \end{vmatrix} - 1 \begin{vmatrix} 1 & 3 \\ 1 & 1 \end{vmatrix} + 2 \begin{vmatrix} 1 & 2 \\ 1 & 2 \end{vmatrix}$$

$$= 0(-4) - 1(-2) + 2(0) = 2$$

To verify positive definiteness, we evaluate the determinants of the leading minors:

$$A_{11} = 1 > 0 \quad \text{O.K.}$$

$$\begin{vmatrix} A_{11} & A_{12} \\ A_{21} & A_{22} \end{vmatrix} = \begin{vmatrix} 1 & 2 \\ 1 & 2 \end{vmatrix} = 0 \quad \text{Not O.K.}$$

A is not positive definite.

EXAMPLE A6

Evaluate the matrix product **AB**, where **A** is given in Example A4 and

$$\mathbf{B} = \begin{bmatrix} -4 & 1 \\ 1 & -4 \\ 2 & -2 \end{bmatrix}$$

Solution

$$\mathbf{AB} = \begin{bmatrix} \mathbf{a}_1 \cdot \mathbf{b}_1 & \mathbf{a}_1 \cdot \mathbf{b}_2 \\ \mathbf{a}_2 \cdot \mathbf{b}_1 & \mathbf{a}_2 \cdot \mathbf{b}_2 \\ \mathbf{a}_3 \cdot \mathbf{b}_1 & \mathbf{a}_3 \cdot \mathbf{b}_2 \end{bmatrix}$$

$$= \begin{bmatrix} 1(-4) + 2(1) + 3(2) & 1(1) + 2(-4) + 3(-2) \\ 1(-4) + 2(1) + 1(2) & 1(1) + 2(-4) + 1(-2) \\ 0(-4) + 1(1) + 2(2) & 0(1) + 1(-4) + 2(-2) \end{bmatrix} = \begin{bmatrix} 4 & -13 \\ 0 & -9 \\ 5 & -8 \end{bmatrix}$$

EXAMPLE A7

Compute $\mathbf{A} \otimes \mathbf{b}$, where

$$\mathbf{A} = \begin{bmatrix} 5 & -2 \\ -3 & 4 \end{bmatrix} \qquad \mathbf{b} = \begin{bmatrix} 1 \\ 3 \end{bmatrix}$$

Solution

$$\mathbf{A} \otimes \mathbf{b} = \begin{bmatrix} \mathbf{a}_1 \otimes \mathbf{b} \\ \mathbf{a}_2 \otimes \mathbf{b} \end{bmatrix}$$

$$\mathbf{a}_1 \otimes \mathbf{b} = \begin{bmatrix} 5 \\ -2 \end{bmatrix} \begin{bmatrix} 1 & 3 \end{bmatrix} = \begin{bmatrix} 5 & 15 \\ -2 & -6 \end{bmatrix}$$

$$\mathbf{a}_2 \otimes \mathbf{b} = \begin{bmatrix} -3 \\ 4 \end{bmatrix} \begin{bmatrix} 1 & 3 \end{bmatrix} = \begin{bmatrix} -3 & -9 \\ 4 & 12 \end{bmatrix}$$

$$\therefore \mathbf{A} \otimes \mathbf{b} = \begin{bmatrix} 5 & 15 \\ -2 & -6 \\ -3 & -9 \\ 4 & 12 \end{bmatrix}$$

List of Program Modules (by Chapter)

Chapter 1

1.7   `error`   Error-handling routine

Chapter 2

2.2	`gaussElimin`	Gauss elimination
2.3	`LUdecomp`	LU decomposition
2.3	`choleski`	Choleski decomposition
2.4	`LUdecomp3`	LU decomposition of tridiagonal matrices
2.4	`LUdecomp5`	LU decomposition of pentadiagonal matrices
2.5	`swap`	Interchanges rows or columns of a matrix
2.5	`gaussPivot`	Gauss elimination with row pivoting
2.5	`LUpivot`	LU decomposition with row pivoting
2.7	`gaussSeidel`	Gauss-Seidel method with relaxation
2.7	`conjGrad`	Conjugate gradient method

Chapter 3

3.2	`newtonPoly`	Newton's method of polynomial interpolation
3.2	`neville`	Neville's method of polynomial interpolation
3.2	`rational`	Rational function interpolation
3.3	`cubicSpline`	Cubic spline interpolation
3.4	`polyFit`	Polynomial curve fitting
3.4	`plotPoly`	Plots data points and the fitting polynomial

Chapter 4

4.2	`rootsearch`	Brackets a root of an equation
4.3	`bisection`	Method of bisection

Chapter 9

9.2	`jacobi`	Jacobi's method
9.2	`sortJacobi`	Sorts eigenvectors in ascending order of eigenvalues
9.2	`stdForm`	Transforms eigenvalue problem into standard form
9.3	`inversePower`	Inverse power method with eigenvalue shifting
9.3	`inversePower5`	Inverse power method for pentadiagonal matrices
9.4	`householder`	Householder reduction to tridiagonal form
9.5	`sturmSeq`	Sturm sequence for tridiagonal matrices
9.5	`gerschgorin`	Computes global bounds on eigenvalues
9.5	`lamRange`	Brackets m smallest eigenvalues of a 3-diag. matrix
9.5	`eigenvals3`	Finds m smallest eigenvalues of a tridiagonal matrix
9.5	`inversePower3`	Inverse power method for tridiagonal matrices

Chapter 10

10.2	`goldSearch`	Golden section search for the minimum of a function
10.3	`powell`	Powell's method of minimization
10.4	`downhill`	Downhill simplex method of minimization

Index

Printed in the United States
By Bookmasters